S. Peter Volpe and Peter J. Volpe
CONSTRUCTION BUSINESS MANAGEMENT

Julian R. Panek and John Philip Cook
CONSTRUCTION SEALANTS AND ADHESIVES, Third Edition

J. Patrick Powers
CONSTRUCTION DEWATERING: NEW METHODS AND
APPLICATIONS, Second Edition

William R. Park and Wayne B. Chapin
CONSTRUCTION BIDDING: STRATEGIC PRICING
FOR PROFIT, Second Edition

Ellis J. Krinitzky, James P. Gould and Peter D. Edinger
FUNDAMENTALS OF EARTHQUAKE RESISTANT
CONSTRUCTION

Ben C. Gerwick, Jr.
CONSTRUCTION OF PRESTRESSED CONCRETE STRUCTURES,
Second Edition

CONSTRUCTION OF PRESTRESSED CONCRETE STRUCTURES

CONSTRUCTION OF PRESTRESSED CONCRETE STRUCTURES

SECOND EDITION

BEN C. GERWICK, JR.
Professor Emeritus
University of California at Berkeley
and
Chairman
Ben C. Gerwick, Inc.
Consulting Engineers
San Francisco, California

A WILEY INTERSCIENCE PUBLICATION
JOHN WILEY & SONS, INC.
New York • Chichester • Brisbane • Toronto • Singapore

Copyright © 1993 by John Wiley & Sons, Inc.

Library of Congress Cataloging in Publication Data.
Gerwick, Ben C.
    Construction of prestressed concrete structures / Ben C. Gerwick,
Jr. — 2nd ed.
      p. cm — (Wiley series of practical construction guides)
    Includes index.
    ISBN 0-471-53915-5 (cloth)
    1. Prestressed concrete construction. I Title II. Series.
TA683.9 G47   1992
624.1'83412—dc20   92-20200

SERIES PREFACE

Congratulations! You've just bought a profit-making tool that is inexpensive and requires no maintenance, no overhead, and no amortization. Actually, it will increase in value for you each time you use this volume in the Wiley Series of Practical Construction Guides. This book should contribute toward getting your project done under budget, ahead of schedule, and out of court.

For nearly a quarter of a century, over 50 books on various aspects of construction and contracting have appeared in this series. If one is still valid, it is "updated" to stay on the cutting edge. If it ceases to serve, it goes out of print. Thus you get the most advanced construction practice and technology information available from experts who use it on the job.

The Associated General Contractors of America (AGC) statistician advises that the construction industry now represents close to 10% of the gross national product (GNP), some 410 billion dollars worth per year. Therefore, simple, off-the-shelf books won't work. The construction industry is unique in that it is the only one where the factory goes out to the buyer at the point of sale. The constructor takes more than the normal risk in operating a needed service business.

Until the advent of the series, various single books (many by professors), magazine articles, and vendors' literature constituted the total source of information for builders. To fill this need, this series has provided solid usable information and data for and by working constructors. This has increased the contractors' earning capacity while giving the owner a better product. Profit is not a dirty word. The Wiley Series of Practical Construction Guides is dedicated to that cause.

M. D. MORRIS, P.E.

Ithaca, New York
November 1989

v

PREFACE

In the 22 years that followed publication of the First Edition, dramatic advances have been made in the use of prestressed concrete.

Most dramatic have been the marine and offshore structures: the Oosterschelde Storm Surge Barrier, and the offshore concrete oil and gas platforms of the North Sea. Technically most demanding have been the prestressed concrete containments and pressure vessels for nuclear power generation. A number of large floating concrete structures have been built and even more challenging designs are now under construction.

Quality control and assurance have received the attention needed for the technically demanding structures now being constructed.

The use of prestressed concrete in bridges has expanded worldwide. Spans have extended to over 200 meters; if we include cable-stayed bridges, then to over 400 meters.

Prestressed concrete piles, poles, and railroad ties are now commonplace throughout the world.

Comparable advances have been made in materials. Concrete strengths twice those of the 1960–1970's are being utilized; we have new reinforcing details, new approaches to durability. Technologies have been radically improved as experience has been gained.

We have learned a great deal about durability as our concrete structures have been exposed to new and more demanding environments, such as the Middle East and the subarctic.

It's becoming increasingly difficult to classify structural systems: the advent of partial prestressing, and the increased blending of prestressed reinforcement with conventional passive reinforcing bars make any separation into prestressed and nonprestressed categories quite arbitrary. Indeed, the newest codes now recognize a continuum covering the entire range.

So now it appears appropriate to write a second edition, not only to bring the book up to the state-of-the-art of the 1990's but also to expand it to include the major new subjects of high-strength concrete, floating concrete structures, offshore, and arctic platforms, and pressure vessels for industrial processes.

BEN C. GERWICK, JR.

San Francisco, California
April 1992

PREFACE TO THE FIRST EDITION

Prestressed concrete and prestressing have rapidly become of universal importance in construction.

Prestressed concrete has gone through the research and development phase of the 1930's and 1940's, and through the specialized design and specialized construction phases of 1955 to the present. Now it is a tool and technique which every structural designer must possess. It is also a technique in which every good construction engineer and contractor must be versed and competent.

Because of the historical nature of its development, there are many excellent texts on design of prestressed concrete. Unfortunately, few references exist to guide the construction engineer and contractor. Specialized subcontracting organizations do exist but for economical reasons have tended to become over-specialized, restricting themselves to either pretensioned precast manufacture or to posttensioning. A few very large national construction concerns, recognizing both the opportunity and the need, have established prestressing capabilities within their own firms, but these are frequently competent in one phase of prestressed construction only; e.g., pre-tensioned pile manufacture of posttensioned bridge girders.

Serious problems have arisen where lack of full knowledge of prestressed concrete construction engineering has led to disastrous errors, such as errors in installation practice, or errors in concrete technology.

This book provides a basic and inclusive exposition of prestressed concrete construction technology, and a practical guide to the contractor and construction engineer. It is impracticable to cover every detail of every phase, but a serious attempt has been made to set forth construction engineering principles that will be an adequate guide for the majority of prestressed concrete construction, and to indicate, by references, sources of greater detail in specific aspects.

Contractors must perform their work at a profit. Economics are inextricably interwoven with techniques; hence, the emphasis throughout the book on methods of minimizing the cost and time of construction in prestressed concrete.

The orientation of this book is principally to construction; but, in highly technical work, design and construction are increasingly intertwined. Thus, the emphasis throughout this book will be placed on the integration of design with construction. As the demands of society require the construction of structures of increasing complexity and sophistication, the construction engineer and contractor must have the technical resources at his command to enable him to carry out the design, not only in a timely and economical manner, but to assure that the intent of the designer and need of the owner are fulfilled. Then the contractor will have earned his profit; hopefully, it will be a good one.

This book is dedicated to G. W. "Bill" Harker, of Australia, a pioneer in prestressing, and a man of indomitable will.

BEN C. GERWICK, JR.

San Francisco, California
November 1970

ACKNOWLEDGMENTS

As author, I gratefully acknowledge the contributions and assistance of numerous organizations and individuals. Among those many, these are especially noted:

Dywidag International
Freyssinet International
Bureau BBR
VSL International
J. H. Pomeroy and Co., Inc.
Preload Engineering
Norwegian Contractors
DORIS Engineering
Department of Transportation, Oregon
Department of Transportation, Washington
ABAM Engineers
Raymond International
Parsons Brinckerhoff, Quade, and Douglas, Inc.

T. Y. Lin International
John Holland Constructions
Global Marine Development Co.
Buoygues
Herbert Brauner
J. Phillip McQueen
Robert Bruce
Jack Closner
Post-Tensioning Institute
Guy F. Atkinson, Inc.
COWIconsult
Prestressed Concrete Institute
Fèdèration Internationale de la Precontrainte (F.I.P.)

CONTENTS

Construction of Prestressed Concrete Structures

0.1 PRESTRESSED CONCRETE CONSTRUCTION PRINCIPLES

Prestressing is the creation within a material of a state of stress and strain that will enable it to better perform its intended function.

As most commonly applied, prestressing creates a compressive stress within concrete that will partially or wholly balance the tensile stresses that will occur in service. Concrete, having a reserve of strength in compression, is an ideal material for prestressing; it is universally available, low in price, easily molded to the desired form, and provides corrosion and fire protection to steel.

However, prestressing can be and has been applied to other materials: to steel trusses, to stone and ceramics and brick, to timber, and to native rock and soils. Furthermore, prestressing can be utilized to overcome not only tensile stresses due to applied loads, as in a bridge girder, but also tensile stresses and deformations due to dynamic waves (as in piledriving or machinery vibrations), temperature stresses (as in pressure vessels), shrinkage, direct tension (as in tie rods), and shear (diagonal tension).

The prestressing principle is not confined solely to pre-compression. There are some applications under study where the introduction of a tensile prestress at a particular stage is desirable to overcome excessive compression.

Prestressing (pre-compression) is usually introduced by means of internal steel tendons which are stressed and then anchored. While high-strength steel is currently the universally used material for tendons, research and development continues on such promising tendon substances as aramid and alkali-resistant glass. Tendons need not necessarily be inside the concrete. They can be external to the concrete section, as in a cable-stayed bridge, or inside the hollow box of a trapezoidal box girder.

In many applications, the aim is to create, through prestressing, a state of stress that will just balance the stress that will be imposed in the member in service. In other applications, the intent is to overcome deflections, for example, to produce a truly flat floor under normal service loadings.

The construction engineer can utilize prestressing effectively to overcome excessive temporary stresses or deflections that may occur during construction; for example, prestressing can give him the technique of cantilevering in lieu of falsework, or the means of erecting large and unstable elements.

Prestressing is not a fixed state of stress and deformation, but is time-dependent. Both concrete and steel deform plastically under continued stress. This plastic flow is greatly increased by high temperature and decreased by low temperature. Much of the continuing research and development is aimed at finding more stable materials: high-strength steel that is "stabilized" against stress relaxation, and high-strength concrete that has low creep, shrinkage, and thermal response.

About 1890, Henry Jackson, a San Francisco engineer, reportedly "invented" prestressed concrete. He built lintels in which the steel bar was prestressed. But after a year or so, the lintels cracked and collapsed. He did not understand the phenomenon of creep of low-strength concrete and the relaxation of mild steel, which together "dissipated" the prestress. So the technology of prestressing had to await the availability of better quality materials and the technical knowledge. Prestressing in its present form emerged in Europe in the 1930's.

High-strength materials are essential to prestressed concrete for reasons of efficiency and economy of performance, but other properties are essential to the long-term stability and performance. These other properties basically relate to the assurance that the state of stress and deformation of the prestressed concrete will be maintained within certain acceptable limits.

The provision of a "stable" tendon is basically the province of the steel manufacturer. The contractor can contribute to this by protecting it during construction operations, so that it gets installed in nearly the same condition as it was furnished by the steel manufacturer, by the quality of his anchoring and stressing operations, and by the durability of his concrete encasement.

The provision of a "stable" concrete is very much under the influence of the contractor. It is the most critical and sensitive aspect of "prestressed concrete." It is paradoxical that so much attention is given in prestressing specifications and literature to the steel and its stressing, and so little to the concrete! The fabricator or contractor can control the aggregates and the water. He may have considerable say about the cement and admixtures. He controls the batching, the mixing, the transportation, the placement, the consolidation, and the curing. He controls the forms and the accuracy with which tendons and reinforcing bars are placed. Finally, in post-tensioned work, he controls the placing (and maintenance of position during concreting) of ducts and anchorages, and the grouting or other corrosion protection.

With prefabricated members, he may also control the handling, transportation, erection or installation, and the fixing. Small wonder that most of the difficulties that have occurred with prestressed concrete relate to the construction phase.

0.2 BASIC DEFINITIONS AND PRINCIPLES

Prestressing is the imposition of a state of stress on a structural body—prior to its being placed in service—that will enable it to better withstand the forces and loads imposed on it in service or to better perform its design functions.

Prestress requires a pre-strain. It is important to remember that prestress cannot be imparted into a member unless that member is able to shorten.

A beam may be prestressed by pre-compressing its lower flange, so that it can resist positive moment tensile stresses in service without cracking.

A pile may be prestressed (pre-compressed) so as to maintain itself crack-free under shrinkage, transportation, handling, and driving stresses.

A column may be prestressed (pre-compressed) so as to enable it to act as an uncracked homogeneous section of a long-column, and to prevent buckling under accidental or intentional eccentric loading.

A pressure vessel may be prestressed so as to resist the membrane tensile stresses due to thermal gradients and internal pressure.

A thin floor slab may be prestressed so as to remain truly flat under normal service loads.

Tendons are the stretched (tensioned) elements which are used to impart precompression to the concrete. Tendons may be high-strength steel wire, strands made of high-strength steel wire, or high-strength alloy steel bars. As noted earlier, aramid and alkali-resistant glass tendons have been used experimentally.

A state of *pre-compression,* i.e., prestress, may be induced in the concrete by tendons or may be imparted by jacking at the ends of concrete members, reacting against abutments. This latter method has been restricted in usual practice to arch ribs, and airfields or similar pavements.

Pretensioning is the imposition of prestress by stressing the tendons against external reactions before the hardening of the fresh concrete, then allowing the concrete to set and gain a substantial portion of its ultimate strength, then releasing the tendons so that the stress is transferred into the concrete.

In the most common case, high-strength strands are stretched between two abutments and jacked to 70 to 80% of their ultimate strength; the concrete is then poured within forms around the tendons. The cure of the concrete is accelerated by low-pressure steam curing, and the strands are then released, so that the stress is transferred into the concrete by bond. The high-stretched wires shorten slightly, precompressing and shortening the concrete.

Pretensioning is most commonly applied to precast concrete elements manufactured in a factory or plant. Typical products produced by pretensioning are roof slabs, floor slabs, piles, poles, bridge girders, wall panels, and railroad ties. A limited amount of pretensioning has been applied to cast-in-place concrete in the form of pavements and floor slabs.

Posttensioning is the imposition of prestress by stressing and anchoring tendons against already hardened concrete.

In the most usual case, ducts are formed in the concrete body by means of thin-walled steel sheaths. After the concrete has hardened and gained sufficient strength, the tendons are inserted and elongated by jacking, then wedges are seated so as to transfer the load from the jacks to the anchors at the ends of the concrete member.

Posttensioning is most commonly applied to cast-in-place concrete members, and to those requiring a curved profile of the prestressing force. Bridges, large girders, floor slabs, roofs, shells, pavements, and pressure vessels are among the constructions usually prestressed by post-tensioning. Posttensioning is extremely versatile and quite free from limitations on size, length, degree of stress, etc. It may be used for factory-made products as well. However, for mass-manufacture, it is inherently more expensive than pretensioning per unit of prestress force.

Stage-stressing is the application of the prestressing force in a sequence of two or more steps. This is done so as to avoid overstressing or cracking the concrete during the construction phase, before further dead load is applied.

Internal tendons are tendons which are embedded within the cross section of the concrete body or member. They may be either pretensioned or posttensioned tendons. Usually this refers to posttensioned tendons located in ducts cast in the concrete.

External tendons are tendons which lie outside the cross section of the concrete body or member as cast. They may, however, be within the void of a box girder. They may later be bonded to it by concreting or grout. External tendons may be in grooves or channels at the side of the concrete member or may be located inside the open box of a box girder.

Bonded tendons are tendons which are substantially bonded to the concrete throughout their entire length. Pretensioned tendons are almost always bonded. Posttensioned tendons which are in ducts which are grouted after stressing are considered as "bonded."

External tendons may be bonded if stirrups and grout encasement provide full shear transfer along the entire length.

Unbonded tendons are tendons whose force is applied to the concrete member only at the anchorages. Bond throughout the length is intentionally prevented. When the posttensioned tendon is in a duct, the duct may be filled with grease. Some posttensioned tendons are coated with a corrosion-inhibiting grease and then encased in plastic. They are then placed in the forms, the concrete is poured and cured, and the tendons are stretched. Single strands, so encased, are increasingly used for the prestressing of floor slabs.

Cable-stays and ties are generally considered as unbonded.

Partial prestress is a design philosophy in which the degree of prestress is tailored to the service conditions rather than the ultimate loads. The aim usually is to provide a residual compression (zero tension) under normal service loads, but to permit tension and even a minor degree of cracking under occasional overloads. Ultimate strength is then provided by mild-steel (passive) reinforcement or by additional unstressed tendons.

Creep is the plastic change in volume of concrete under sustained stress. It is most marked during the early ages of the life of a concrete member; an old rule of thumb says: "One-third in three days, the second one-third in three months, the last one-third in ten years." Creep occurs in both the aggregate and the paste. The higher the sustained stress, the greater the creep. If stress is applied at an early age, the creep will be greater. Elevated temperatures increase creep. Creep is essentially irreversible; when the stress is removed, the creep stops but does not reverse to any appreciable extent.

Camber is the upward deflection of the member due to prestress.

Shrinkage is a volume change of the concrete due to chemical reaction and the drying-out of the contained water. Some shrinkage occurs at set, but the largest amount occurs during drying. Shrinkage is reversible, and sustained humidity or soaking will cause a swelling, largely offsetting the drying shrinkage. The effect of shrinkage is to reduce the pre-compression in the concrete. Before the application of prestress, shrinkage may produce tensile cracking in the concrete which, although closing again under prestress, may still reduce the inherent tensile strength of the concrete and its ability to remain crack-free under load.

Stress-relaxation is an irreversible plastic flow in the steel under sustained high stress. Stress-relaxation, as the name implies, leads to a reduction in the degree of stress in a tendon, thus reducing the prestress in the concrete.

While the greatest portion occurs during a short time of very high stress, relaxation may continue for very long periods. Since it is a molecular phenomenon, various treatments of the steel can be used to reduce the stress relaxation.

Elastic shortening is the change in dimensions occurring in the concrete as the pre-compression is applied to it during prestressing.

0.3 CONSTRUCTION ENGINEERING

Construction engineering is the art of applying engineering approaches to construction operations. Within the scope of this book, it will encompass such matters as planning, operations, scheduling, and control. Planning starts with an analysis of the work to be accomplished, a selection of methods and techniques, a layout of sequential operations, and assignment of equipment and labor. Scheduling includes the interrelation with other operations at the site as well as with external aspects such as weather, floods, air temperature, and contract requirements. Control involves the assignment of supervision and inspection, the establishment of detailed procedural instructions, the provision for adequate inspection, the maintenance of records, and cost control.

Construction engineering analyses, properly applied, will generally result in considerably more instructions, charts, schedules, etc., than are commonly employed today; but these should not be voluminous and complicated. The objective is strictly practical: to so plan, schedule, and control the work that every worker and every activity is contributing to the accomplishment with minimum waste and interference. The charts and instructions must be clear, concise, and definite, and directed at understanding by those who actually perform the work. There is no need to restate all the manifold considerations that lie behind a set of instructions.

The highly technical nature of many prestressing operations makes it essential that prior planning be carried on in considerable detail. Most problems associated with prestressed concrete could have been prevented had proper planning been given before the time of the actual construction. Conversely, the economic results of prior planning have been very profitable and, in some cases, have actually resulted in substantially reducing the labor costs in a particular prestressed concrete construction operation.

For many construction projects, the prestressing operations must be fitted into an overall critical path schedule. Prestressing is usually on the critical path; therefore, the selection of the method and procedures may be heavily influenced by the time allotted.

Unlike many other types of construction, with prestressed concrete, the details of tendon layout, selection of prestressing system, mild-steel details, etc., are often left up to the contractor (and/or his specialist subcontractor), with the designer merely showing the final prestress and its profile, and setting forth criteria.

Thus, the constructor must understand enough about design considerations to help him select the most efficient and economical system. Such knowledge may often provide him a competitive edge.

Finally, in many aspects of construction, the contractor and engineer are functionally united. Industrial projects and commercial developments are frequently undertaken by a combination of the contractor and engineer, either joined within one firm or as a cooperative effort. The same is true of many overseas contracts where the owner and his engineer merely establish the basic requirements and criteria, leaving the final design to the contractor. Foundation projects in the United States are frequently bid on performance specifications only, or alternates are permitted. In many new areas of construction, such as nuclear reactor power plants, offshore structures, etc., the contractor must bid on a "design and construct" or "turnkey" basis. Therefore the basic principles of design are essential background knowledge for an enterprising construction engineer, even though he himself may not undertake the design task. He must understand the phenomena, the reasons behind the technical requirements, the advantages and the limitations of prestressing.

In all aspects of construction, but perhaps most intensively in prestressed concrete, design and construction must be integrated if the most effective results are to be obtained. The historical separation of the two disciplines, design and construction, is an artificial one, a division of labor in order to enable concentration of skills. The overall endeavor is to build a structure that will serve a specific purpose well. Designer and constructor are thus members of a single team with one goal.

PART ONE

MATERIALS AND TECHNIQUES FOR PRESTRESSED CONCRETE

CHAPTER 1

Materials for Prestressed Concrete

1.1 CONCRETE

1.1.1 General

Concrete is a heterogeneous material composed of aggregates embedded in a cementitious matrix which bonds them together. Most commonly, the aggregates are natural sands and gravels, or crushed rock, and the matrix is Portland cement which has been hydrated by water.

Careful attention is invited to the above definition because, in this technological age, every possible variation is being explored and even commercially applied in an attempt to produce desired properties. Thus we have artificial aggregates, such as expanded shale, as well as pozzolanic and organic cements, being applied on a substantial scale in engineering construction. (Structural lightweight concrete, produced from expanded slate, shale, and clay aggregates, is discussed in detail in Section 4.2.1).

The remainder of this chapter will deal with conventional concrete, consisting of limestone or siliceous coarse aggregates (crushed rock or gravel), limestone or siliceous sand, Portland cement, water, and admixtures. Pozzolans may be introduced to replace or supplement a portion of the cement. This produces concrete weighing 22.4 to 24.0 KN/m^3 (140 to 150 pcf) and, when properly selected, batched, mixed, placed, and cured, capable of strengths up to 40 to 60 MPa (5800 to 8700 psi) in 28 days. Even higher strengths are being achieved in special cases.

Concrete for prestressing is most commonly in the range of 35 to 40 MPa (5000 to 6000 psi). Such concrete can be produced with reasonable economy in most parts of the world, provided proper care is taken in all phases of the concreting operation. The existence of an established concrete industry in metropolitan and technologically sophisticated centers must not be allowed to induce complacency or carelessness, as the uniform high qualities, rather easily obtainable in these centers, were attained only after many years of intensive efforts on the part of all segments of the industry and, even then, there has been a substantial amount of trial-and-error which eliminated the unfit. So, when commencing operations in new areas, it is essential that fundamental considerations be resurrected and applied.

There are adequate and detailed specifications available for the selection and evaluation of concrete materials, such as those of the American Society for Testing Materials (ASTM). All that can be done in this chapter is to highlight certain qualities that can be particularly important or serious for prestressed concrete.

1.1.2 Aggregates

Since we are attempting to get moderately high-strength concrete in order to be efficient in utilizing the prestressing, the maximum size of coarse aggregate should be limited. For most applications, 20 mm ($\frac{3}{4}$ inch) is the optimum maximum size.

Coarse aggregates must not contain clay seams that produce excessive volume change, such as creep and shrinkage.

Both gravel and crushed rock are used successfully. For normal high-strength concrete for prestressed application, gravel will give better workability and compactibility at low water/cement ratios. For extremely high-strength concrete, crushed rock of proper angularity is superior but requires very intensive vibration to achieve proper compaction.

Aggregates to be used in concrete which will be in contact with sulfate ions must be sound under the aggressive expansion which characterizes sulfate attack. The most serious cases of sulfate attack have occurred in hydraulic structures carrying fresh water, and in foundations and underground structures in sulfate-bearing soils. These most often occur in the mountainous deserts such as those of Southern California, Arizona, and Saudi Arabia and the Emirates.

While seawater contains free sulfate ions, the presence of chlorides appears to inhibit damaging attacks, especially when the cement content is relatively high (3.5 $KN/m^3 = 600$ lbs/yd^3 or more), and the water/cement ratio (W/C) is below 0.5.

Pozzolanic additions, such as pulverized fly ash (PFA), are effective in preventing the attack of sulfate ions.

Alkali-aggregate reaction is one of the most insidious forms of attack because of its delayed reaction. In some cases, the cracking and disruption has shown up only after 30 to 40 years. Siliceous aggregates, especially those containing quartz, are most frequently involved in reactive actions but limestone aggregates can also be reactive due to inclusion of nodules of opal and flint.

Several approaches are used to minimize the occurrence of alkali-aggregate reactions.

One is to test the aggregates and eliminate those of high reactivity. A second is to restrict the alkali in the cement to a very limited percentage of sodium oxide and potassium oxide.

A third approach is to supply an excess of active silica in the form of pozzolanic substances such as PFA or microsilica.

Although alkali reactivity first emerged as a hazard in Southern California, widespread damage due to this cause has occurred in the United Kingdom, Eastern Canada, and to a lesser extent in France and Germany.

Because of the high surface areas of fine aggregates, they are most often the culprits in alkali-aggregate reactivity.

Fine aggregates should grade in the coarser ranges since, with the rich cement factors usually employed in prestressed concrete, perfect grading is not necessary and may be undesirable. Gap-grading, properly applied, can often reduce shrinkage and improve strength and modulus of elasticity.

Aggregates must be clean. Even a few percent of silt can make the low W/C mixes for prestressed concrete excessively sticky and difficult to place. Silt reduces strength and increases shrinkage. Silt can usually be removed by rewashing, with very beneficial results.

Aggregates must not contain salt. Salt can be deposited on aggregates, particularly fine aggregates, from seawater immersion, from the upward migration and evaporation of groundwaters in desert countries, or from salt fogs. Even small percentages of salt reduce the corrosion-inhibiting value of the cement and may help initiate electrochemical corrosion. This is particularly dangerous with steam curing.

Aggregates must be of a proper temperature for incorporation in the mix. Since the aggregates are by far the largest component of the mix, it is often effective and economical to cool the aggregates, as by water evaporation, in summer, or to heat them in winter. Water "soakers" running continuously over the aggregate piles will prevent dust and cool the aggregate by evaporation.

Lightweight aggregates are ceramic particles containing numerous air voids. These occur naturally in volcanic deposits, e.g., pumice, but the great majority of applications use manufactured ceramic particles produced in a kiln from shale, clay, or slate.

These ceramic particles may have a hard-burned surface which is relatively impermeable. These aggregates must be formed to the proper size before burning.

Other types are burnt first as relatively large particles, then crushed and screened to the proper gradation.

The air voids are generally disconnected; hence the aggregate particles, while porous, are also impermeable.

Such particles are available both as coarse aggregate and as fine aggregate. Due to size effects, it is the coarse aggregate particles which contribute most to the reduction in the weight of the finished concrete.

In most cases, therefore, a blend of lightweight coarse aggregate and natural sand will give the optimal values of low density, high strength, and impermeability.

Lightweight aggregate concrete has lower thermal conductivity, hence improved resistance to fire, although some precautions have to be taken to ensure against explosive spalling in the case of aggregates with high water absorption.

Entirely satisfactory prestressed concrete of high quality can be attained by careful selection of the lightweight aggregates, proper mix design, and construction techniques. These are more fully described in Section 4.2.

1.1.3 Cement and Cementitious Materials

The almost-universal cementing material for concrete which is to be prestressed is Portland cement. The American Society of Testing Materials (ASTM) Standard, C-150, designates five types. Type I is standard and has relatively wide tolerances in its chemical constituents. Type II is controlled as to the amount of tricalcium aluminate (C3A) and the fineness of grind. It also contains restrictions on the percentage of alkalis. It is moderately low in heat generation, and moderately sulfate-resistant. Type III is high early strength, Type IV is low-heat, and Type V is sulfate-resisting.

Most prestressed concrete employs Types I, II, or III, or a modification of these. The cement is usually selected on the basis of rapid early strength, minimum shrinkage, durability, and economy. Flash set is to be avoided.

Types I and III are most widely used for prestressed concrete in buildings but Type II is preferred for structures in coastal and marine environments. Salt fogs can penetrate 50 miles or more inland, and Type II cement gives the optimum combination for the durability of the concrete, along with good corrosion-inhibiting properties for the steel.

A few Type III cements tend to develop flash sets or have excessive shrinkage under steam curing.

In an effort to obtain the optimum balance of properties, modified Type II cements have been developed. These generally are ground much finer than conventional Type II, for example, to a Blaine fineness of 4000 to 4200 cm^2/g. They have been specifically developed to meet the needs of the prestressed industry and are often marketed as "prestress cements."

Type IV cement, low heat, is for mass concrete, such as dams. It is slow to hydrate and to gain strength, making it unsuitable for most prestressed applications.

Type V cement (sulfate-resisting) is not well suited for most prestressed applications. Type V is very low in tricalcium aluminate (C3A) which, while it gives the concrete itself greater durability under sulfate attack, unfortunately significantly reduces the corrosion protection of the steel.

Blast furnace slag (BFS) cement is being increasingly utilized because it is believed to provide increased durability, reduced heat of hydration, and hence reduced thermal stresses, and greater impermeability, hence protection against corrosion of the reinforcing steel. The ground blast furnace slag is blended with Portland cement, since it takes the heat of hydration of the cement to initiate and support the hydration of the BFS. The blast furnace slag generally constitutes 60 to 80% of the total.

In an effort to accelerate the reaction of BFS cement mixes for building elements, the BFS is often ground very finely, as fine as 5500 cm^2/g (Blaine).

Pozzolanic materials consisting of fine particles of silica were used as cement by the Romans. They are almost entirely silica and as such combine with free lime (calcium hydroxide, gypsum). Their rate of hydration is very slow but, in the presence of moisture, continues to add strength and impermeability over time. Pozzolans increase impermeability and durability. Most importantly, since they hydrate at a slower rate than Portland cement, the heat generation is reduced, resulting in a lower maximum temperature. Typically, a 15% replacement of Type II cement with pozzolans will reduce the maximum temperature by 8°C (15°F).

Natural pozzolans occur from volcanic eruptions, Mt. Vesuvius in the case of the Romans, Mt. Lassen and Mt. St. Helens in the case of the Western United States. Fly ash, a waste product of coal-fired power plants, is the most common pozzolan, but it is important to select fly ash which contains minimal impurities such as unburned carbon.

In the Eastern part of the USA, it is possible to buy Portland cement containing 15 to 20% of pulverized fly ash. In other parts of the country, it must be purchased separately and batched into the mix in a separate operation.

When PFA was first adopted as a partial replacement and supplement to Portland cement, it was customary and convenient to use existing weight batch equipment, batching the cement first, then adding PFA to reach the total amount. This practice led to several very serious errors, including collapses of structures in which the proportions accidentally were reversed.

One pozzolan deserves special mention: this is condensed silica fume or micro-silica. Almost pure silica, with grains $\frac{1}{100}$th the size of the cement grain, like cigarette smoke, it fills the interstices between the cement grains and, with its high surface area, creates a favorable reaction that changes the gel structure, increasing strength, impermeability, and alkali-aggregate resistance.

PFA is normally used in prestressed concrete to replace 15 to 20% of the Portland cement.

Microsilica is normally used in the amount of 3 to 6% of the cement. Higher percentages make the concrete mix "sticky" and hard to place.

1.1.4 Water

Until recently the standard requirement for water was merely that it be potable. However, water for use in prestressed work should be more definitely restricted in salt, silt, and organic contents. Suggested limitations are:

a. No impurities that will cause a change in time of set greater than 25% nor a reduction in strength at 14 days age greater than 5% as compared with distilled water.
b. Less than 650 parts per million of chloride ion. Some authorities permit up to 1000 ppm.
c. Less than 1300 parts per million of sulfate ion. Some authorities limit this to 1000 ppm.
d. Water shall be free from oil.

Water may be added to the mix in the form of crushed ice, in order to reduce the temperature of the fresh concrete mix. This also reduces the maximum temperature during hydration and thermal strains. In extremely hot weather, the mix can be cooled even further by injection of liquid nitrogen.

The optimal temperature of the fresh mix is from 5°C to 10°C (40°F to 50°F).

In cold weather, the mixing water may be heated to raise the temperature of the concrete mix above freezing and to accelerate set and strength gain. The water should not be introduced into the mix hotter than 30°C (90°F).

A newly developed process injects live steam into concrete during the mixing phase, raising its temperature to 65°C to 75°C. The concrete is then placed immediately (within 10 minutes) and maintained at 60°C for three hours. At the end of this period, strengths up to 60% of the 28-day strength are reported.

These processes generally reduce the ultimate strength but that may be acceptable because the design strength required may be adequately met. More serious, but still often acceptable, the permeability of the hardened concrete will be substantially increased. However, if the products are utilized in a controlled environment, e.g., floor slabs in buildings, durability may not be a major concern.

1.1.5 Admixtures

Admixtures to alter concrete properties are a significant contribution to prestressed concrete fabrication and construction.

Water-reducing admixtures are highly effective in reducing the ratio of water to cementitious material. If cement is the only such cementitious material, this is denoted as W/C, whereas if PFA or other pozzolanic materials are incorporated, it is denoted as W/CM.

Low W/C and W/CM not only increase the strength of the concrete but reduce its permeability and hence minimize the probability of corrosion of the reinforcement and long-term deterioration of the concrete itself.

Conventional water reducers usually retard the set of the concrete to some degree. However, this is usually rapidly offset by heat applied during curing, e.g., steam curing.

In recent years, high-range water-reducing (HRWR) admixtures, the so-called "superplasticizers," have been introduced. These make it practicable to achieve the so-called "flowing concrete" with a slump of 200 to 250 mm (8 to 10 inches), thus facilitating placement, while still achieving a W/C ratio of below 0.40.

It is very difficult to obtain a workable mix to which microsilica has been added without use of an HRWR admixture.

Although the HRWR admixtures enable the concrete to flow well, internal vibration is still necessary to ensure thorough consolidation around the prestressing tendons or ducts and the reinforcing steel, especially in the anchorage zones where complete consolidation is essential.

HRWR admixtures tend to experience a sudden loss of slump after about $1\frac{1}{2}$ hours. When this happens with a low W/CM ratio, the continued placement may become impossible. To overcome this, most commercially available HRWR admixtures incorporate a retarder which extends the workable life.

Cold weather may also extend the period over which the mix remains highly fluid. This will cause increased form pressures, especially in the deep webs typical of box girder bridges.

To reduce the time before the prestress can be imposed on the concrete, accelerating admixtures can be employed. These would normally be considered in cold weather construction. Most commercially available accelerators incorporate calcium chloride ($CaCl_2$) which is today prohibited in prestressed concrete construction because of potential corrosion of the steel.

Catastrophic failures have occurred in prestressed concrete water tanks, pressure pipes, and penstocks, where calcium chloride had been used in the concrete mix. The availability of water from the inside, and of oxygen from the outside, acting on prestressing tendons which had been depassivated by the chlorides, resulted in accelerated corrosion and rupture.

Therefore, the accelerating admixture used in prestressed concrete must contain no chlorides. Nitrates are usually forbidden also. There are such nonchloride accelerating admixtures available but they are expensive.

Another type of admixture is that designed to inhibit corrosion. It is used in many structures where chloride corrosion is a potential problem, e.g., the marine or coastal environment.

Most of these are based on nitrites, either sodium nitrite or calcium nitrite. Sodium nitrite has been effectively used in marine structures in the former USSR for 50 years or more. However, calcium nitrite reduces the alkalis and hence is more widely used today. There is some evidence that the nitrite is used up over a period of years; hence the use of such compounds does not obviate the need for impermeable concrete.

Proprietary admixtures are also available which purportedly minimize the diffusivity of chloride ions through the concrete.

For environments and exposures subject to freezing and thawing, the use of an air entrainment admixture is essential. This produces tiny air bubbles in the mix which act to absorb the expansion of moisture and prevent internal fatigue and consequent deterioration under many cycles of freeze and thaw.

Concrete that is saturated in service is much more susceptible to freeze-thaw attack.

It is important that the air entrainment admixture produce not only the proper amount of entrained air (usually 5 to 8% depending on the size of coarse aggregate) but also the proper pore-void characteristics. Whereas air meters can be used to determine the amount of air, they can't distinguish between entrapped air (which gives no benefits) and entrained air. The pore-void characteristics can only be determined by microscopic investigation of hardened concrete cores.

A rational practice, therefore, is to make and examine samples before the start of actual production or construction and at intervals of, say, 2 to 4 weeks thereafter, meanwhile using the air meter to control the daily concrete quality.

One of the properties of conventional concrete is "bleed," in which excess water migrates to the surface, unless it is trapped beneath reinforcing steel or coarse aggregate particles. It reveals itself on the surface as "bug holes" or "worm holes", usually only 15 to 25 mm deep ($\frac{1}{2}$ to 1 inch). Some of the conventional water-reducing and retarding admixtures promote bleed.

Recently developed thixotropic admixtures inhibit bleed. While currently expensive, they produce better surface characteristics as well as reduce internal micro-cracking due to bleed water trapped under the steel or aggregate.

Microsilica also tends to minimize bleed.

1.1.6 Storing of Aggregates and Cement

Much contamination and deterioration of aggregate quality can arise from improper storage. Stored on the ground or on a slab at grade, the aggregate can be contaminated by dirt, ice, and undersized particles. Stored in bins, it is much better protected, except from snow and ice. In either event, considerable dust and chips can collect at the bottom. The only way to positively eliminate this last problem is rescreening above the batcher. The need for this depends on the character of the rock, the abrasion in transport and handling, and the quality of concrete desired.

Aggregates exposed to the summer sun can become overheated. They can be shaded by galvanized or aluminum corrugated roofing. Soaking or spraying of the aggregate pile will cool it very effectively through evaporation. Vacuum-cooling and the injection of liquid nitrogen are two of the newest techniques.

Cement must be stored and used in such a manner that none of it is left to age excessively. Heat accelerates aging and loss of strength, so insulation of bins and/or use of reflective surfaces such as galvanized sheet metal are advised. Obviously, the cement must be completely protected from moisture; a tropical or a humid atmosphere may present difficulties in this regard and may require dehumidification of the storage shed in the case of sacked storage.

1.1.7 Batching, Mixing, and Transporting

Approved procedures are set forth in the American Concrete Institute's Recommended Practice ACI-614. Accuracy of batching is essential for production of consistent high-quality concrete. Therefore, batching should be by weight and preferably by automatic rather than hand controls. A continuous correction must be made for water contained in the aggregates.

Mixing must be thorough, especially with low-slump mixes. The turbine mixers are especially adapted to this need. The more recently built ready-mix truck mixers have been improved so as to handle a fairly low-slump mix, but the blades must be kept in good condition. The horizontal turbine mixer is definitely preferable, therefore, for the highest quality mixes. Adequate mixing time improves the uniformity, strength, and impermeability of the concrete. Mixer blades must not rotate too fast.

Transporting may be successfully carried out in a number of ways. Mixes may be transported dry (dry-batched), with water added and the mixing at the point of use by either a concrete mixer or paver. Wet mixes are transported in ready-mix trucks or in hoppers (concrete buckets, "dumpcrete" bodies, rail-mounted hoppers, or scoop-loaders). The essential cautions here are to prevent segregation and premature set during transport. A low W/C ratio and a set-retarding admixture are helpful. Vibration or mixing during transport is of considerable value in preventing segregation and premature set. Remixing can also be done at the point of discharge.

1.1.8 Placing and Consolidation

This subject is set forth in detail in the *ACI Manual of Concrete Practice*. With the generally dry mixes employed in prestressed concrete, intensive vibration is necessary to insure complete filling of the voids, especially in congested areas, and to thoroughly consolidate it. Internal vibration is the most effective method for the great majority of cases, as it serves to ensure compaction around the tendons, embedded steel, anchorages, etc. Frequencies of 9000 rpm are most commonly employed. As noted earlier, while highly workable mixes are practical with the use of HRWR admixtures, internal vibration is still essential.

External vibration can be used very effectively with thin products, particularly precast elements cast in heavy steel forms. Such vibration definitely helps placement and produces excellent surfaces on the vibrated face, eliminating most of the entrapped air voids (bubbles). Vibrators placed opposite each other tend to cancel out. It is usually best to stagger their location. External vibrators can usually consolidate only to a depth of 150 to 200 mm (6 to 8 inches); hence on thick products, a combination of internal and external vibration will prove most satisfactory.

In placing low-slump or no-slump concrete in forms, it is best to dump it on the advancing face of the concrete where it will get the full effect of the vibration. This will speed concreting and produce better consolidation.

External or form vibration places the forms themselves under high stresses and may even develop fatigue in the connections, etc. Therefore forms must be specially designed when form vibration is to be employed.

Although vibration is the most widely employed method for consolidating concrete, other processes are employed in special product fabrication processes.

Centrifugal spinning is widely used in the production of poles and pipes. The intense acceleration forces the material to the outer surface, allowing excess water to drain out from the central void.

This process may, unless carefully controlled, produce an outward deflection of reinforcing or prestressing bars, or may result in a small gap on the outside face of prestressing strands or wires. These difficulties have generally been overcome by concrete mix design and control of the spinning process.

The compaction can be increased by introducing steel balls into the void after the initial spinning. These balls roll as the tube turns, exerting pressure on the inside face.

Pressure can also be used to consolidate concrete slabs. In the former USSR, after the concrete has been placed and screeded, a steel pad, with an absorbent form liner, is pressed down to firmly compact the concrete, and expel excess water. This process has been combined with vacuum processing so as to both apply pressure and suck out excess water.

A process used extensively in architectural precast concrete fabrication is that of dynamic shock, in which the heavy timber pallet molds, filled with concrete, are repeatedly "shocked" by means of eccentric rollers on which the pallet is seated. This process would appear difficult to combine with prestressing.

Prestressed railroad sleepers (ties) are pressed out in an adaptation of the concrete block process. They use a very dry "zero slump" concrete mix and the sleeper is stamped or pressed out. The sleeper contains duct formers for subsequent insertion of post-tensioning bars.

When relatively stiff mixes of concrete are placed, there is a definite tendency for water and air pockets to form on vertical and overhang surfaces. Entrapped air and excess water try to escape from the mix under vibration, and are trapped under the overhang and, to some extent, along the side of a vertical surface.

There appears to be no practical technique available for completely eliminating these when dry mixes are used, but they can be minimized by the following steps:

a. Selection of aggregates for cleanliness. Minimization of the silt content. This reduces the tendency to entrap air.

b. Selection of a type of form oil suited to the surface that reduces capillary attraction.

c. Use of HRWR admixture to reduce the amount of water in the mix.

d. Use of a thixotropic admixture that prevents bleeding and promotes workability. Air entrainment may frequently be beneficial—entrained air is not the same as entrapped air. Addition of microsilica to the mix.

e. Thorough internal vibration.

f. External (form vibration) is especially valuable if continued after internal vibration is completed.

g. Spading along the form sides, where accessible, following vibration.

Absorptive and porous form liners will absorb air and excess water from the surface skin, but pockets may still be located a fraction of an inch below the surface. Porous forms are being increasingly used for architectural concrete.

Vacuum processing will remove entrapped air and excess water to the depth to which the vacuum is effective, but is generally not economically practicable.

Strangely enough, these surficial problems may not be so noticeable with some wet mixes, as the pockets and voids are then distributed as pores in the mix. Of course this is generally not a satisfactory solution at all, as it results in low strengths and low durability.

Experience has shown that the previously cited techniques of placing and consolidation of low W/C ratio mixes will produce the strongest and most durable concrete, even though minor surface blemishes still occur. For architectural or exposed surfaces, rubbing and sacking will produce the desired uniformity. Properly done, this finished surface is durable from both an architectural and structural point of view.

For concrete exposed in the splash zone or other region of attack, the "bleed" holes should be filled if they are deeper than a nominal amount, say 10 to 15 mm ($\frac{3}{8}$ to $\frac{1}{2}$ inch) deep. Depth can be determined with a wire. Filling should be with a suitable grout or dry pack or epoxy mortar.

1.1.9 Curing

Immediately after pouring, a fresh concrete surface exposed to hot sun or drying wind may lose so much water that it sets and shrinks, even while the body of concrete is still plastic. This can be prevented by one of the following means:

a. Fog spray, which is especially adaptable to large, flat surfaces. The amount of water must be minimized so as not to increase the W/CM ratio of the surface.

b. Covering with an impervious blanket. Polyethylene is preferred by many, as the moisture steams in the sun and improves curing.

c. Covering with wet burlap.

d. Covering and injecting low-temperature, low-pressure steam.

e. Spraying on a membrane curing compound which prevents loss of moisture. In hot climates where the curing compound evaporates rapidly, a second coat should be applied after about eight hours.

Curing of concrete should provide sufficient moisture to allow the completion of the chemical reactions that produce strong, durable concrete. Some of these reactions take place in a short time; others require a much longer period. The exact time required is highly temperature-dependent, with different reactions responding differently to the temperature. For example, compressive strength typically develops more rapidly than tensile strength. It is essential to continue curing sufficiently long to enable all the desired reactions to be completed, not just until the concrete has reached a minimum compressive strength.

When precast concrete products are removed from steam curing, especially when dry cold winds are blowing, the moisture extraction is accelerated due to the change in temperature and humidity. Cracking frequently occurs under this situation, so the product should be covered or blanketed so as to protect it from the external environment.

Similarly, for prestressed concrete structures with thick members, 500 mm (20 inches) or greater in thickness, once forms are stripped, blankets should be draped so as to insulate the surface from sudden drops in temperature which will cause thermal cracks. This applies both to winter and cold, windy situations and also to environments where the temperature drops rapidly when heat radiates to the night sky. The desert, with its severe changes from day to night, is especially vulnerable.

Concrete, especially dense concrete, will continue to cure internally due to the excess water of the mix. It is the surface layers for which moisture must be contained or supplied.

Curing is accelerated by heat. Practicable means of supplying heat include low- and high-pressure steam, radiant heat from hot water or hot oil pipes, and electric-resistance heating. In these latter cases, humidity should also be introduced, as by fog spray.

1.1.10 Steam Curing

Steam curing at atmospheric pressure is widely employed in precast prestressed concrete manufacture to accelerate gain of strength. A great deal of attention has been directed at the effect of this accelerated curing on long-term ultimate compressive strength, durability, shrinkage and creep, loss of prestress, etc. Other studies have been directed to the determination of the optimum cycle for the steam-curing process. Finally, the interactive effects of cooling, release of prestress, and subsequent thermal effects must be considered.

As a general statement, it may be said that properly applied, low-pressure steam curing improves the quality of the concrete product. Some recent research indicates an increased permeability but this may be overcome by inclusion of pozzolans in the mix.

Steam curing at atmospheric pressure has been one of the important techniques which has made it possible to attain economical production of prestressed concrete elements, permitting daily turnover of forms. It has also made it practicable to have a short cycle between manufacture and erection, thus eliminating in large part the need for stockpiles and inventory. Steam-cured prestressed piles have actually been successfully driven when only one day old.

The adoption of a proper cycle of steam curing is essential. The generally accepted optimum cycle is (Fig. 1.1):

- *a.* A delay period of three to four hours until concrete has attained initial set. Concrete should be protected from drying out during this period.
- *b.* A heating period, with a temperature rise of 20°C to 30°C (40°F to 60°F) per hour to a maximum temperature of 60° to 70°C (140°F to 160°F).
- *c.* A steaming period of six hours at 60°C to 70°C (140°F to 160°F).
- *d.* A cooling period (steam tarpaulins or hoods still covering concrete). During this period, the exposed portions of the strand cool faster than concrete, thus pulling on the concrete. Similarly, steel forms cool faster, as do the outer portions of the concrete. Thus tensile stresses may be set up in the concrete element. For this reason, with many products, especially massive ones where the inner core of concrete may retain its heat for a considerable period of time, it is necessary to release all or a portion of the prestress into the member during the cooling period, and then re-cover the units to permit a slower and thus more uniform rate of cooling.

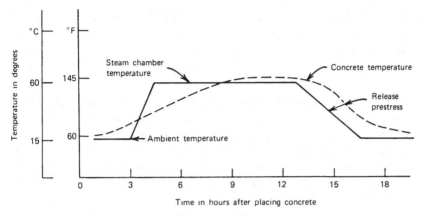

Figure 1.1. *Typical steam curing cycle (atmospheric pressure).*

e. An exposure period (steam hoods removed from the concrete, which is now exposed to the atmosphere). This may be a critical period insofar as surface shrinkage, crazing, and ultimate durability are concerned. The concrete is warm and moist. In winter the concrete surface may be subjected to a substantial temperature differential and to very drying winds. The inside core is still warm. A combination of thermal and drying shrinkage may cause crazing and hairline cracks. This is often referred to as "thermal shock." In summer there may be hot drying winds, which again cause drying shrinkage on the surface. Also, the inner heat of the concrete tends to accelerate evaporation of the water from the surface. Therefore, in temperatures above freezing, it may be desirable with certain products to apply supplemental water cure immediately after exposure. The first few hours are the critical ones.

Prestressed concrete cylinder piles, with their relatively thin walls, are typical of a product for which this supplemental water cure is recommended.

For temperatures below freezing, two methods have been employed. The cooling period has been programmed over a longer period, so as to permit the concrete to attain a rather uniform low temperature before exposure to the atmosphere. This can, for example, mean merely storage inside at room temperature for two or three hours before exposure. The second method is to apply membrane sealing compounds after the conclusion of steam curing, in order to prevent rapid drying shrinkage of the surface, and then to cover the product with insulating blankets for 24 hours or so.

Some plants apply membrane sealing compound before steam curing. Although the compound evaporates during the steaming, it provides protection against moisture loss during the concrete placement process, until the tarpaulins or covers are placed and the steam is introduced.

Steam curing has the following effects on precast products as compared with field water cure.

a. Shrinkage may be reduced, depending on presence of gypsum in cement.

b. Creep may be reduced if prestress is released into concrete when concrete strength is greater.

c. Relationship between compressive strengths at various ages are shown in Fig. 1.2.

d. There is some indication that the rapid rise in compressive strength by steam curing is not matched by the rise in tensile strength. At ages of a few days, steam-cured concrete may not have the tensile strength that would be expected of a water-cured concrete of the same compressive strength, and thus may be more susceptive to drying shrinkage cracking. Hence the employment of supplemental water cure will improve the tensile strength of steam-cured concrete.

e. With proper cycling and supplemental water cure, steam-cured concrete is approximately as durable as fog-chamber concrete. Its durability is usually better than that of field water-cured concrete because of the difficulty or impracticability of truly maintaining all surfaces in a moist condition by field water curing. This applies to both seawater durability and freeze-thaw durability, with the supplemental water cure being particularly desirable for the latter.

f. When the steam cycle is commenced, the metals inside the steam chamber respond much more rapidly to the heat than does the concrete. Steel forms expand while the concrete has low tensile strength; if there are changes in cross section, or fins, etc., the forms may crack the concrete. Internal ducts of metal should be plugged; otherwise they may expand under steam heat and cause longitudinal cracks in the concrete.

Satisfactory solutions to the form problem have been obtained by careful detailing, use of transverse elastic strips (neoprene or sponge rubber), elimination of fins,

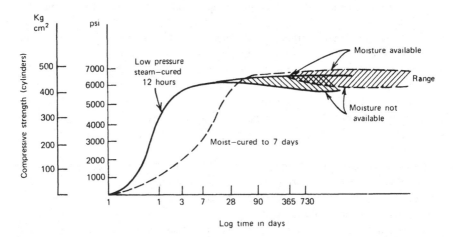

Figure 1.2. *Typical relationships between strengths at various ages and type of cure.*

at the joints in the forms, by sealing with tape, and by raising the temperature slowly.

Other methods of accelerating the initial cure through heat have been applied. The beds or forms may be heated electrically or by pumping hot oil or hot water through them. Covering the surface with black nylon will cause radiant heat to be absorbed from the sun. As noted earlier, moisture should be introduced to ensure 100% humidity.

Field-placed concrete may be similarly subjected to accelerated cure. With segmental construction (see Section 12.13), it may be particularly desirable to accelerate the rate of gain of strength of concreted joints. This may be done by steam jackets placed around the joint concrete or by electrically heated forms.

Heating the concrete internally by electrical resistance methods has been used for concreted splices in prestressed concrete piles, using embedded copper wires. The potential problem of electrolytic corrosion between copper and steel prestressing tendons must be considered because this matter has not yet been fully investigated to determine methods of positively insuring against such corrosion.

Concrete may also be heated internally by steam through piping or ducts. Steam should not be allowed to come in contact with tendons as it will cause rusting.

Autoclaving (high-pressure steam curing), which has been successfully applied to concrete block production, has also been applied in Germany and Japan to prestressed precast concrete elements.

In Japan, a standardized and highly mechanized system for prestressed concrete pile segments 12 and even 15 meters (40 to 50 feet) in length includes the use of pretensioned high-strength bars and autoclaving. The standard pile sizes range from 600 mm (24 inches) to 1 meter (3.28 feet) in diameter. The piles have hollow cores. Strengths up to 70 MPa are attained and shrinkage is virtually eliminated.

Stress relaxation of the pretensioned bars during autoclaving is about 20%. Creep of the concrete is accelerated during the curing. Thus, a significant portion of the prestress is lost but this is compensated for in design.

While concerns have been expressed about the potential for corrosion initiation during autoclaving, this does not appear to have occurred.

1.1.11 Hot Weather Concrete Practice

Refer to American Concrete Institute Standard ACI 605. Coarse aggregates should be cooled by evaporation, letting soakers run over the stockpiles. The stockpiles may also be shielded from the direct sun by aluminum roofing or galvanized sheet metal. Blowers or vacuum can be used to intensify the evaporation.

Ice can be utilized in lieu of mixing water.

In extremely hot climates, for example, desert, it may be preferable to pour at night or in the very early hours of the morning. This will help prevent flash set.

The forms should be cooled prior to placing concrete by shielding from the direct sun by means of aluminum roofing and/or by a fog spray. After placing concrete, cover promptly with wet burlap or polyethylene prior to start of steam cure.

Since finely ground cements develop greater heat of hydration, it may be preferable to use a standard or coarse grind of Type II cement (Blaine fineness of 2200 to 3200 cm²/g) in very hot weather.

Curing is very important and it should be recognized that extreme temperature changes may take place in the desert environment.

1.1.12 Cold Weather Concreting Practice

Refer to ACI 306. Aggregates may be preheated by steam. The stockpiles should be kept free from ice or snow by providing coverage and drainage.

The mixing water may be preheated to 60°C (140°F) to heat the concrete to 30°C (90°F).

After placing and finishing, cover the concrete promptly. Place steam covers and apply low-pressure, low-temperature steam. Keep inside temperature about 20°C (65°F) until time to start the temperature rise in the steaming cycle.

After steam curing, protect concrete with blankets or tarpaulins or insulated forms, and bleed steam into them so as to keep them warm and moist.

Structural lightweight aggregates should not be intentionally pre-saturated prior to mixing when they are to be used in winter concrete, unless they can be protected from freezing. Pre-saturation also reduces freeze-thaw resistance. Preheated lightweight aggregates hold the heat better for winter concreting than sand-and-gravel aggregates.

Type III cement and finely ground Type II cement develop more heat of hydration and thus may be preferable in cold weather.

1.1.13 Testing

Most design specifications are very strict as to the required cylinder strength at the time of introduction of prestress, and at 28 days. In addition, there may be strength requirements for removal from forms, for cessation of supplemental water curing, for transportation, and for erection or driving. It is important that attention be paid to the techniques of testing to eliminate radical and erroneous results.

Poor testing procedures almost always result in lower indicated strengths than the true values.

Test cylinders should be produced from concrete samples taken during the middle of any flow (as down a chute), so as to eliminate the effect of segregation. They should then be well-rodded, or vibrated, using a small pencil vibrator, not a large vibrator. Cover cylinder tightly with a polyethylene bag to prevent surface evaporation. Burlap is not satisfactory; it acts as a wick to draw the water out. The cylinder should then be placed under the steam hoods or in the curing chamber at about the same height as the members themselves, so as to receive the same heat cycle. On the removal from the steam hoods, the cylinder should be carefully transported (not bounced around in a truck) and immediately placed in a fog chamber at 23°C (73°F) or in a tank of warm water at 23°C (73°F). It should be continuously stored until the day before the test, for example, 27 days in the case of a 28-day test.

Add lime (in a porous sack) to the water in the tank to prevent leaching of lime from the cylinder. The tank should have thermostat control to maintain temperature accurately.

On removal of the cylinders, allow them to dry for one day and to cool. If testing a cylinder taken directly from steam hoods in order to determine strength for release, allow it to cool before testing.

Prepare a cylinder for testing as follows:

a. Rub ends with a stone to remove any projections, as from date markings.

b. Use new capping compound for highest results; keep cap as thin as possible. If capping job is not perfect, scrape it off and start again.

Warning: *With high-strength concretes, fracture may be explosive. Be sure to provide a wire screen protector.*

When testing cubes cut from a prestressed concrete member, cut cube between strands, polish ends and sides with a stone so they are perfectly plain and square. Throw away imperfect cubes. Test without capping.

1.2 PRESTRESSING TENDONS

1.2.1 General

The most prevalent means for inducing a compressive stress into concrete is by stretching a tendon and anchoring it to the concrete. This tendon may be located inside the concrete cross section, either directly embedded or in a duct. Alternatively, it may be located external to the concrete cross section, either immediately adjacent, or at some finite distance away from it (e.g., inside the box of a box girder).

The tendon must have a very high tensile strength and an ability to sustain indefinitely a high state of stress, with little loss due to relaxation, corrosion, or fatigue. Cold-drawn steel wire and alloy steel wire and bars have these attributes and are the most common materials for tendons.

However, there are other materials (carbon and boron filaments, and nonmetals, such as glass fibers) which have potentially even better strength properties. Their practical use depends on technical aspects, such as chemical stability, practical aspects such as means of anchoring, and the economic aspects of cost of material and installation.

Cold-drawn steel wire is produced in diameters up to 7 mm (0.276 inches). It has ultimate strength ranging from 1700 to 2100 MPa (250,000 to 300,000 psi) and moduli of elasticity about 20,000 MPa (29,000,000 psi). However, the bond between smooth wires and concrete is low. This requires the use of large numbers of small wires in pretensioning, or a means of mechanical anchorage such as button heads for post-tensioning.

Hot-rolled alloy steel wires are extensively employed in Europe, especially in Germany. They are frequently rolled with a flattened or oval cross section, and may have a pattern or "profile" of indents rolled onto the section. This greatly improves the bond in pretensioning, enabling fewer larger wires to be employed.

By weaving several wires into a strand, a tendon of substantial capacity can be produced with excellent bonding properties. There is a slight reduction in modulus of elasticity (5 to 10%). Seven-wire strand is widely used both in pretensioning and in most systems of posttensioning. Strands are manufactured in diameters of 9 to 17.5 mm (.375 to 0.7 inch) although the great preponderance of prestressing employs 15 mm (0.6 inch) strand. (Fig. 1.3) Three-wire strands are used in some lightly stressed, mass-produced pretensioned floor slabs. Nineteen-wire and thirty-seven wire strands, formed similarly to high-strength wire rope, are also available.

The wires used in stranding are usually round; thus there are interstices between the wires of a strand. A recent development has been to produce shaped wires in which the wires completely fill the cross section. This concentrates more force within a given cross-sectional area and is particularly valuable for highly stressed members where space for tendons is limited.

Wires can also be drawn with an indent, so as to increase bond.

In pretensioning, individual strands are most commonly used, although two or three may be bundled. In posttensioning, individual strands are used in floor slabs of buildings and in tanks and pipes, but in most other work, e.g., bridge girders and marine structures, the tendons are grouped in multistrand assemblages. A typical posttensioning assemblage will consist of 19 strands or 37 strands of 15 mm (0.6 inch) diameter (Fig. 1.3).

Even larger assemblages are occasionally employed, for example bundles of 3 × 19 15-mm strands were used in the prestressed concrete pressure vessel for the nuclear power plant of Wylfa, Wales (Fig. 1.4).

Figure 1.3. *Typical posttensioning tendon assemblage of 37 15-mm (0.6 inch) strands. (Courtesy of VSL)*

Figure 1.4. *Bundles of 3 × 19 strands for Wylfa Nuclear Power Plant, Wales.*

Alloy-steel bars are extensively used for post-tensioning. The bars generally have ultimate strengths ranging from 1000 to 1200 MPa (140,000 to 165,000 psi). Bar diameters range up to 35 and even 44 mm (1.375 and 1.75 inches). They may be smooth or have spiral corrugations rolled on, so that they have excellent bond characteristics. These bars can then be spliced or anchored by means of coarse-threaded couplers or nuts (Fig. 1.5). Smooth bars are anchored by wedge-type grips. While alloy-steel bars are less efficient in themselves than wire, they offer practical advantages in many installations because of their ease of coupling. Since the end anchorage can be screwed up tight after jacking, there is no seating loss, which makes them very efficient for short tendons.

By stress relieving, the internal stresses of cold-drawn steel are relieved and the elastic properties of the tendon are rendered more uniform. Initially the relaxation or creep is reduced; however, the ultimate relaxation may approach that of nonstress-relieved steel.

Recently, cold-drawn steel strip has been produced for use in external winding

Figure 1.5. *High-alloy bars for posttensioning, showing dead-end anchor and splice sleeve.*

applications, e.g., tanks and pressure vessels. This strip has approximately the same properties as cold-drawn wire.

Cold-drawn wire, alloy bars, and prestressing tendons of other types can be galvanized, which provides corrosion protection during the transport, storage, and installation phases, and also during the service life, or at least the early years. There is some indication that over the long term, corrosion losses merge with those of nongalvanized steel.

To prevent hydrogen liberation which, theoretically at least, could potentially lead to hydrogen embrittlement of the prestressing steel, the steel should be passivated by a chromate wash after galvanizing. Only hot dip galvanizing or sherardizing should be employed, never electrogalvanizing.

Prestressing tendons may also be epoxy coated. In this case, to prevent excessive bond slip, a fine sand is blown onto the wet epoxy. This appears to enhance the bond properties, even beyond those of uncoated steel tendons.

Strand and bar tendons may be sheathed in polyethylene, with a corrosion-inhibiting grease coating over the steel. Such a tendon has no bond, so the anchorages must sustain the entire force at transfer. See Section 2.3.6.

1.2.2 Stress Relaxation

Tensioned steel tendons undergo a stress loss due to plastic flow or creep of the steel. For tendons in common use, stress relaxation losses of 6 to 13% are to be expected.

TABLE 1.2   Properties of Commonly-Available Strand, Wire, and Bars

A. Seven-wire strand – ultimate strength =

Ultimate strength (fpu) MPa	1860	1860
Ultimate strength (fpu) psi	270,000	270,000
Nominal diameter mm	12.5	15
Nominal diameter inches	0.5	0.6
Area (Aps) mm²	97	129
Area (Aps) inches²	0.153	0.215
Weight kN/m	7.8	10.9
Weight plf	0.53	0.74
Ultimate capacity kN	186	262
Ultimate capacity lbs	41,300	58,100
Effective capacity kN	112	157
Effective capacity lbs	24,800	35,000

B. Prestressing wire

Ultimate strength (fpu) MPa	1650	1620
Ultimate strength (fpu) psi	240,000	235,000
Nominal diameter mm	6.25	6.9
Nominal diameter inches	0.250	0.276
Area (Aps) mm²	31.6	38.7
Area (Aps) inches²	0.049	0.06
Weight N/m	2.5	2.95
Weight plf	0.17	0.20
Ultimate capacity kN	53	63.3
Ultimate capacity lbs	11,780	14,050
Effective capacity kN	31.8	38
Effective capacity lbs	7070	8420

C. Smooth prestressing bars

Nominal diameter mm	25		35	
Nominal diameter inches	1"		1¾"	
Area (Aps) mm²	491		962	
Area (Aps) inches²	0.785		1.485	
Weight N/m	39.4		74.3	
Weight plf	2.67		5.05	
Ultimate strength MPa	1000	1100	1000	1100
Ultimate strength psi	145,000	160,000	145,000	160,000
Ultimate capacity kN	512	565	970	1070
Ultimate capacity lbs	114,000	125,600	217,000	237,600
Effective capacity kN	307	338	580	640
Effective capacity lbs	69,000	76,000	129,000	142,000

In a process known as "stabilization" the steel wire is given a treatment combining temperature and stressing, which reduces the long-term stress relaxation substantially. Stabilized strand is also known as "low-relaxation" strand.

Overstressing a tendon (to 75 or 80% ultimately) during the stressing operation and holding it for one or two minutes at a normal temperature of about 20°C (70°F) will induce performance similar to stress relieving.

Some European authorities prefer to limit the initial stress to 50% of ultimate, which practically eliminates stress relaxation. Others take the opposite approach, with initial stressing to 80%. Normal practice for initial stress in the USA is 70%.

1.2.3 Ductility

Ductility is an essential requirement for tendons in order to prevent brittle failure during installation and during service. Ductility is usually measured by bend tests and by elongation tests. Elongation tests must be conducted on substantial lengths of tendon to overcome the influence of the necked-down area. Most specifications require an elongation of 2% or more. Most tendons in practice actually have about 3% elongation.

1.2.4 Corrosion

It is essential that tendons be protected from substantial corrosion. Corrosion may affect the ductility of the tendons or may simply reduce the cross section of tendon and thus reduce both the prestress and the ultimate strength. Corrosion may also reduce the fatigue strength.

Hydrogen embrittlement, while extremely serious, appears to be rare. Stress corrosion is likewise of rare occurrence. Atmospheric corrosion prior to installation is common but, fortunately, is not serious. Salt cell corrosion, and its related electrochemical types of corrosion (chlorides, etc.) are often a serious matter, and are primarily related to the concrete protection over the tendon.

All these forms of corrosion and means of protecting against them are discussed in detail in Chapter 5, Durability.

The intent is to place a tendon in service and maintain it in service in substantially the same condition as when it left the manufacturer. This period extends during transport, storage at the site, handling, installing, stressing, and concreting or grouting. The following "rules" are recommended to give the best assurance of attaining the objective. Not all are currently followed as standard practice, but this does not necessarily mean that excessive corrosion will result. Corrosion is a matter of probability—by following the listed rules, the probability of corrosion will be reduced to a minimum.

For shipment, tendons should be protected by water-soluble oil, wrapped with waterproof paper, with vapor-phase-inhibiting crystals (VPI) inside the packaging. They should not be coiled too tightly; coils of very small radius produce microcracks in the steel surface, leading to stress corrosion (Fig. 1.6).

Storage should preferably be enclosed and heated to maintain a relative humidity of less than 20%. Never store in mud or on a bench. Avoid open storage near refineries and industrial plants, especially those burning coal or oil and thus emitting sulfides into the air.

In handling the tendons, they should not be dragged across the ground. They should not be subjected to sharp bends.

Pretensioned tendons should not be exposed to the atmosphere more than 24 hours before concreting. In the vicinity of refineries or other plants emitting even

Figure 1.6. *Coils of strand are shipped wrapped in heavy waterproof paper, with corrosion-inhibiting substances inside.*

minor amounts of sulfide gases (H_2S), expose tendons in a stressed condition for the minimum time practicable.

In posttensioning, be sure that ducts have been protected by caps from water entry prior to installation of the tendons (Fig. 1.7). Avoid abrasion at point of entry into duct by providing, if necessary, a funnel entry piece (Fig. 1.8). After installation, proceed with stressing within 48 hours, and grouting within 24 hours thereafter.

Where stressing is not possible within 48 hours, tendons should be swabbed with water-soluble oil or thoroughly dusted with VPI powder during threading. (VPI powder is positively effective for only about 150 mm (6 inches) so dusting must be thorough). Seal ends of ducts with vapor-tight plastic covers (or tape) prior to stressing.

Where stage stressing is required, it is better to stress some tendons to full value, and grout, leaving the remainder in an unstressed or low-stressed condition. These latter, of course, should be protected as noted above.

Pretensioned tendons are protected by the quality of the concrete placed and compacted around them. A mix of high plasticity, rich in cementitious materials, with a low W/CM ratio, consolidated thoroughly by vibration is the best protection.

Figure 1.7. Posttensioning ducts are capped with highly visible red caps to prevent entry of water and debris.

Figure 1.8. Roller guides funnel the tendon group into the duct without abrasion.

Posttensioned tendons are protected by injection of cement grout. The mix and placement are intended to ensure full embedment of the strands and to penetrate the interstices between adjacent strands. Detailed provisions are given in Chapter 5.

Posttensioned tendons may alternatively be given corrosion protection by injection of corrosion inhibiting grease. Care must be taken to exclude all water pockets and lenses.

1.2.5 Fatigue

Since prestressing tendons undergo only a very small range of stress change while the live load is varied from zero to maximum, fatigue is generally not a problem with commonly used prestressing steels. Extensive tests by the American Association of Railroads and others have verified this. Fatigue is normally only investigated for girders and other members under railway tracks, or in machinery foundations, both of which may be subject to millions of cycles.

However, there may be some reduction in bond between the tendon and the concrete due to a lesser number of cycles. The reduction in bond will increase the transfer length and may be a problem with very short pretensioned members, such as railroad ties and overhanging cantilevers.

Fatigue is also a phenomenon for concern in those structures which are saturated and then subjected to cyclic loading. Offshore structures and floating structures, such as floating bridges, must consider fatigue in their design.

Fatigue problems can be minimized by providing the best possible bond (e.g., good internal vibration) and the most impermeable concrete possible.

The endurance of prestressed concrete subject to cyclic loadings, both dry and wet, can be significantly enhanced by the inclusion of microsilica in the mix. Lightweight concrete reacts very positively to the inclusion of microsilica as the lightweight aggregate develops a secondary bond with the cement paste.

1.2.6 Bond

Bond, and transfer, or development length, are the result of complex mechanisms which determine both the static behavior of a prestressed concrete structural element and its behavior under dynamic and cyclic loadings. Bond is developed as a result of autogeneous shrinkage of the concrete around the tendon, the swelling of the tendon due to Poisson's effect, mechanical interlocking of deformations or strand twists, and adhesion. The adhesive effect is not yet completely understood but is highly dependent on minute variations in the surface. A highly polished surface will give much reduced bond. Similarly, any drawing lubricant left on the wires will reduce the adhesive bond. Epoxy-coated bars have essentially no adhesive bond and must rely entirely on deformations and anchors.

Deformations normally do not come into action until after the adhesive bond has partially failed.

Lateral binding, such as spirals or stirrups, increases the bond, and is especially required for those structures exposed to dynamic and cyclic loads. Wire and strands generally available have a surface which is satisfactory but, from time to time, changes in manufacturing processes have resulted in excess drawing lubricants or other bond reductions. Once again, this is usually only critical in very short members and short cantilevered overhangs.

The use of microsilica in the mix appears to enhance bond.

A slight coat of surface rust will improve bond significantly, but this is normally not to be encouraged. Attempts to purposely attain a rusted condition will inherently be nonuniform, leading to danger of pitting corrosion in some zones. In coastal environments, the rust may become contaminated with salt from the saline atmosphere.

Any dirt, oil, or grease causes a significant decrease in bond.

Decreased bond (increased transfer length) may actually be of benefit in girders, piles, etc., by reducing the end-zone tensile stresses. Means of intentionally obtaining reduced bond are the following:

a. Enclosure of the end portions in a plastic sheath.
b. Coating of the ends with wax or bitumastic and wrapping with paper.
c. Greasing of the ends. A proprietary substance known as "Lubabond" is often used for this purpose.

1.2.7 Very Low Temperature

At very low temperatures, such as those associated with Liquefied Natural Gas (LNG) and other cryogenic storage, prestressing steels in common use, such as cold-drawn wire, still maintain excellent properties. They are not subject to brittle behavior nor to a transition zone effect. Strength and modulus of elasticity increase while ductility decreases, but only to a minor degree.

1.2.8 Elevated Temperatures

As the temperature rises, the rate of stress relaxation increases substantially. Ductility increases. Above a critical temperature of 750°C to 900°C (1400°F to 1650°F) there is a substantial loss of stress.

1.2.9 Fiberglass and Polymeric Tendons

A considerable amount of research has been expended in the United States, Great Britain, Germany and the former USSR in an attempt to find practicable means of employing fiberglass tendons. Long term chemical stability is the major problem. At present, this seems to be best provided by encasement of the fiberglass filaments in an inert matrix, thus making up a rod. A second problem is that of finding a practicable means of anchorage. One system employs a wedge driven into the center of a cluster, which in turn is locked within a sleeve.

Progress has been made in these areas and several small experimental structures have been constructed in Europe. They have been thoroughly instrumented so that long-term properties may be monitored.

Aramid tendons, commonly known as Kevlar, consist of aramid filaments encased in a matrix of epoxy resin. The characteristic axial tensile strength is 2800 N/mm^2 (400,000 psi). These tendons are noncorrosive, nonmagnetic, and nonconductive. Anchorage systems have been developed based on wedges forced into the end of the tendon, and experimental small projects have been carried out.

"Arapee" consists of aramid filaments embedded in epoxy resin. It has a uniaxial tensile strength of 3000 MPa (420,000 psi) but stress relaxation under sustained loading limits the usable stress to about 50% of these values. Among other nonmetallic tendons under research and development are carbon fibers, made into strands and encased in epoxy resin.

Fiber tendons possess attractive long-range potential because they are non-corrodible and non-magnetic.

1.2.10 Special Tendons

Stainless steel wires have been used for prestressing where nonmagnetic properties are required.

Titanium bars have been used for the prestressing of some historical monuments, such as the Parthenon, where exceptionally long life is desired and initial costs are not controlling.

1.3 ANCHORAGES AND SPLICES

1.3.1 General

Anchorages are mechanical devices used to transmit the tendon force to the concrete structure. They include the means of gripping and securing the tendon and the bearing plate or reinforced cone or other means by which the concrete reacts against the tendon's force.

Practically all anchorages are part of proprietary systems for post-tensioning. For competitive reasons, most European contractors found it necessary to have a system of their own, which explains the great number of devices available. Certain of the systems have now emerged as dominant, usually because of more intensive and thorough engineering and more aggressive marketing.

Anchorages may grip the tendon by means of mechanical wedges. These may have serrations that dig into the tendon to grip it, or may be smooth, with friction furnishing the necessary grip (Fig. 1.9).

Anchorages may also employ the wedging action of a swaged fitting, with molten zinc, epoxy, or cement mortar gripping the tendon through a combination of friction and adhesive bond.

Another group of anchorages depends on upset enlargements that bear directly on bearing plates. Buttonheads may be cold-formed on the ends of wires (Fig. 1.10).

Figure 1.9. *Machined anchorage for 37-strand tendon. Wedges will be hydraulically forced into tapered holes to anchor each strand.*

Deformations may be hot rolled on bars. Upset threads may be rolled on bars on which nuts are screwed.

Because of the tremendous forces involved, it is essential that the anchorage perform properly and safely. A failure may produce a serious or even fatal accident. Therefore only thoroughly developed and tested systems should be employed, and

Figure 1.10. *Preformed buttonheads anchor multi-wire tendons.*

these should exactly follow the manufacturer's recommendations. Only anchorages approved by the responsible engineer should be used; serious slippages have occurred when minor changes in specifications, metal composition, or tolerances have been made (Fig. 1.11).

Anchorages and the tools must be kept clean and free from corrosion during storage.

After installation, anchorages should be properly protected from corrosion or fire in accordance with the design engineer's details.

It is important that the anchorages are aligned axially with the tendons. Most anchorage systems are equipped with special chairs and jacks to accomplish this.

In some systems tendons may also be anchored at one end by looping around steel or concrete, by flaring or kinking, or by simple embedment in concrete for a substantial length. These are the so-called "dead-end" anchorages (Fig. 1.12). For looped anchorages, the specified bending radii must be closely adhered to in order to prevent failure or rupture during tensioning. A special flattened and curved steel tube will provide more uniform bearing for multistrand tendons (Fig. 1.13).

After jacking the tendon, the anchorage must be seated. This usually involves a small inward movement or "set" of the tendon, losing a slight amount of stress. This is usually of little importance, except on very short tendons such as those used for connections; on these, "set" may be serious. Certain types of anchorages provide for shimming or locking to overcome "set."

Before grouting, stressed tendon anchorages are under very high stress and,

Figure 1.11. Three different anchorage systems were employed to provide triaxial prestress of the Oosterschelde Storm Surge Barrier structures.

Figure 1.12. Dead-end anchorages are formed by kinking of the ends of the tendons.

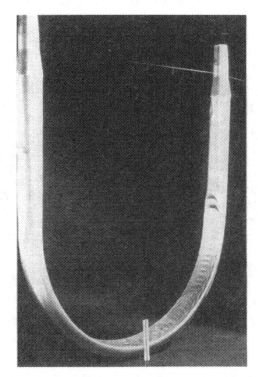

Figure 1.13. Special flattened tube for "looped" anchorages where tendons are turned through 180° Flattened tube aids uniformity of stress.

therefore, are vulnerable to accidental blows, accidental heat from a welding rod, etc. As such, they can become deadly missiles, shooting off with all the energy stored in the tendon.

1.3.2 Splices in Prestressing Tendons

Since bars for prestressing are made in lengths only up to 20 meters, splices are frequently required. Special bar splices have been developed which enable full transfer of force and which have high endurance under cyclic loading. Spliced bars are frequently employed in segmental construction and whenever incremental installation is required.

The most commonly used splice involves a coupler with coarse female threads that screws onto the cut ends of corrugated prestressing bars. It is important that the ends of both bars be engaged the specified length. This coupler has both right-hand and left-hand threads enabling the easy coupling of the bars, since only the coupler has to be turned.

In the case of "blind" anchorages or wherever the coupler cannot be properly inspected, so that the second bar is turned, the very act of screwing it on may simultaneously unscrew the first bar.

To prevent this, the following rules should be followed:

1. On all bar-coupler splices, mark on each bar the proper distance so it can easily be verified after coupling that the proper lengths are engaged.
2. On blind couplings, especially when the second bar is turned, daub the threads of the first bar with epoxy so that when it is engaged, and the epoxy has set, it cannot be further moved.

For splicing of strand tendons, various devices have been perfected. These are essentially a double anchorage. Since after seating, the wedges are not locked under stress until later when the entire length has been stressed, it is important to precisely seat the wedges.

Since couplers and splices are inherently larger in diameter than the bars or strands they join, a special enlarged duct section must be fitted, of adequate length to enable the splice to travel when it is stressed. Not giving adequate length led to the failure of a bridge in Germany.

1.3.3 Precautions for Anchorages and Splices

Use only anchorages of the selected system, in strict accordance with manufacturer's recommendations.

Store so as to keep clean and dry.

Use manufacturer's recommended practice and tools for installation.

Take necessary steps to insure axial alignment.

Protect workers during tensioning; protect anchorage from accidental blows, heat, and electric arcing (as from welding).

Take special pains in applying protective cover (to protect against corrosion or fire) to anchorage. This is a highly stressed critical mechanism and deserves full protection.

1.4 REINFORCING STEEL

Mild-steel reinforcing bars play an essential role in most prestressed concrete construction by resisting secondary tensile stresses and by confining the precompressed zones. Such bars are classified as "passive" steel, as opposed to prestressed steel. However, in actuality, the application of prestress to the prestressing steel, for example, longitudinally, will produce a small compression in the "passive" reinforcement. The unstressed reinforcement also provides resistance to transverse and torsional shear. It also may be used as additional primary steel to give greater ultimate capacity or to control behavior.

Reinforcing steel bars are available in sizes up to 35 and even 50 mm (1.4 to 2 inches) in diameter. Grades of steel, according to their yield strength, range from 300 to 500 MPa (45,000 to 70,000 psi). Grades 550 and 600 MPa (80,000 to 87,000 psi) are occasionally available but generally are not economical in design.

As a general statement, prestressed members will behave as intended only if the reinforcing steel is properly detailed and placed. Adequate reinforcing will serve to confine the member and, in effect, force it to function the way it was designed. The proper installation, positioning, and securing of reinforcing bars becomes of fundamental importance.

Because in prestressed members, the reinforcement frequently consists of closely spaced small bars, and because the concrete mixes used require heavy vibration, the reinforcement is often dislocated either by concrete placement or by the vibrators. This can be prevented by proper tying and "dobe" blocks.

Tack-welded cages have been extensively used in precast members, but the location of welds must be carefully detailed since a reduction in strength may be induced. Tack welding is generally prohibited in members subject to dynamic or shock loading and those subject to cyclic loading, because of the danger of brittle fracture at the weld.

To prevent corrosion, the specified cover must be maintained over the reinforcement. Plastic chairs, or concrete "dobe" blocks, should be used at close spacing.

Reinforcing bars are almost always deformed, so as to achieve more effective and reliable bond. These deformations have been designed so as not to reduce the bar's static ultimate strength. There is some evidence that the deformations reduce the fatigue strength under a high number of cycles of loading.

Mild-steel reinforcement, as used in the United States, is generally Grade 60, that is, nominally of 60,000 psi yield strength, whose equivalent in Europe is Grade 400 (400 MPa) and in Japan, 40 Kg/mm^2.

However, in recent construction, higher grades of reinforcement have been used, e.g., with yield strengths of 500 MPa (75,000 psi), and very occasionally 600 MPa (85,000 psi). In order to properly develop such high yield strengths, adequate transverse and spiral confinement is required, especially in the development zones. Crack-width criteria usually limits the utility of very high yield reinforcement.

Unstressed prestressing strand or unstressed alloy-steel post-tensioning bars may be used to provide ultimate strength in special cases. Unstressed strand was so used in the Gladesville Arch Bridge in Sydney, Australia.

Reinforcing bars may be used in compression as well as tension. Especial care has to be taken at splices and ends of large compressive bars in order to prevent a bearing or punching shear failure: this requires confining spiral or milled ends, so as to butt the bars. Mechanical splices are increasingly used for large bars under high compression.

Splices in small and moderate size reinforcing bars are usually "lap splices," in which the ends of the bars overlap. This lap length has to be increased when the loads are dynamic and cyclic. Tying both ends of splices is effective in ensuring proper splice behavior.

This tying is especially critical with compression splices. Although for static loads, the codes permit shorter lap lengths than for tensile splices, inadvertent spreading of the bars may lead to splitting.

Tension lap splices should be staggered and confined, especially when bars over 25 mm (1 inch) diameter are lapped. Spiral or a grid of transverse steel is the best way to confine such regions.

Mechanical splices are used with larger reinforcing bars. There are a number of types, threaded bars with couplers, hydraulically forged couplers, and swaged couplers. As in all mechanical items, their use requires care in storage, protection in transport and erection, cleanliness, and precision in installation. Because of the reduction in congestion and assurance of continuity, the use of mechanical couplers is becoming increasingly common.

Full-penetration butt welded splices are rarely used because of the danger of accidentally heating the prestressing strand, which would cause it to lose its properties. When welding is employed, it should be carried out before insertion of the prestressing tendons.

Stirrups are required in most prestressed concrete members to resist shear and transverse tensile forces. For example, bridge girders usually require closely spaced stirrups in their end blocks.

Stirrups, because of their sharp bends, are especially difficult to install, particularly in congested zones. In order to bend them to short radii so they can be fit around the primary steel bars, stirrups are usually limited in size to 12 mm ($\frac{1}{2}$ inch) or less, most often 8 or 10 mm ($\frac{3}{8}$ inch).

The "tails," the bent ends of the stirrups, should tuck into the confined core of the member for highest performance. This is especially required for structures in earthquake zones or where the loading is dynamic. This is extremely difficult to do in congested areas, so in the preponderance of cases, the "tails" are simply bent parallel to the primary steel. They should be tied at their ends.

Recently, in order to ease installation of transverse and confining steel and to reduce the congestion that impedes concrete placement and vibration, mechanically headed bars (T-headed bars) have been introduced, in which an enlarged head is anchored behind a bar of primary steel. Installation requires only a simple insertion and twist.

Because of the better anchorage, larger diameter T-headed steel bars, up to at least 20 mm ($\frac{7}{8}$ inch) may be used. The number of bars to be placed can often be reduced by a factor of four to six.

1.5 DUCTS (SHEATHS)

"Ducts" and "sheaths" are terms used to describe the conduit through which the post-tensioning tendons run. In its broadest sense, the term "duct" may include a formed void, or a conduit of any material.

For reasons of economy, practicability, and electrochemical compatibility, the majority of ducts are of bright steel.

Flexible metal ducts are spiral coiled, with crimped but nonwatertight joints. They can be formed of very thin metal, easily shipped in coils, and are draped in the forms. However, their very flexibility is their greatest drawback; they may have excessive "wobble" (erratic misalignment), thus requiring very frequent support by tying or other means. One of the best solutions when using flexible metal ducting is to insert a length of electrical conduit in the flexible metal sleeve. This can be sized to provide the proper degrees of flexibility for curvature and rigidity for support. During concreting this electrical conduit will hold the duct in an acceptable alignment, and after concreting, it is easily withdrawn.

Flexible metal conduit, being usually very thin, is subject to mechanical damage and to rusting. It must be protected by boxing during shipping and stored in a dry place till actual use.

Because it is nonwatertight, blockage sometimes occurs when grouting an adjoining duct if a duct or joint has been ruptured. When several ducts are in a vertical line, one above the other, and touching (e.g., at the point of maximum negative moment over a support), the act of stressing tendons may cut through or otherwise damage the conduit, allowing the tendon to pull through.

Semirigid metal conduit, although more expensive (because of thicker gauge) and more difficult to ship and handle, is gaining popularity. It requires little tying or support, has a low friction factor (because of low "wobble"), and is grout-tight (provided that joints are properly sealed) (Fig. 1.14).

To protect flexible and rigid conduit from corrosion and to reduce friction during stressing, coatings are sometimes employed. Galvanized coating has been widely and successfully used. A few warnings have been raised about the possibility of free hydrogen being liberated from the reaction of cement and zinc; however, such hydrogen embrittlement of tendons has not been duplicated in laboratory tests and the reported field instances are extremely scattered and not completely convincing.

The possibility of hydrogen evolution can be minimized by passivation of the zinc after the galvanizing process by a chromate wash. This is common practice in most galvanizing process plants. Cold-drawn steel wire and strands are relatively unsusceptible to hydrogen embrittlement.

Figure 1.14. *Semirigid spirally crimped metal conduit.*

At present, in the United States, galvanizing is generally considered a beneficial means of reducing friction and providing corrosion protection, and the potential for hydrogen embrittlement is considered negligible.

Epoxy coated ducts are often used for the transverse prestressing of bridge decks. The coating is only applied on the outside: the inside is still bright steel.

Plastic conduit has been used from time to time; it can be formulated so as to be inert in concrete. Corrugated polyethylene drainpipe and specially fabricated spirally corrugated duct are available. They have the proper rigidity with flexibility for normal curvatures and are easily joined (or spliced) in the field. However, with curved tendons, the tendons bite into the walls of the plastic conduit and friction factors may be excessive. The tendons may even cut through the corrugations, allowing the grout to leak. For straight tendons, such as the transverse ducts in bridge decks, plastic ducts appear to be very satisfactory.

Bond values (plastic to concrete and grout to plastic) are generally very low; thus there may be insufficient adhesion for fully bonded behavior. This places increased responsibility on the long-term behavior of the anchorage. The heat resistance of plastic is generally substantially less than that of the tendons so, in a fire, damage may occur to the sheath. However, in the course of many actual fire tests, this was not found to be of significance, nor did it lower the fire (heat) resistance of the member.

Where voids are formed in the concrete by removable ducts or forms, abrasion may take place while inserting and tensioning the tendon. Friction losses may be very high. If grease or a similar substance was used to permit removal of the duct form, bond may be reduced. The most serious problem, however, is with grout leakage through cracks, honeycomb, or rock pockets, resulting in blockage of adjoining ducts. Therefore formed voids are generally not used except for very short, straight ducts, as for connecting tendons.

Splices in ducts should be suitable to the material and the need for water-tightness. Wrapping overlapping sleeves with waterproof tape has proven moderately effective. Sleeves, plus taped joints, are widely used in segmental construction. Plastic sleeves have similarly been used, but have been knocked askew during concreting. In recent tests, heat-shrink taping of sleeved joints was found to give the best performance of all when using thin metal ducting.

For the highest integrity, as in pressure vessels, long-span bridges, and offshore structures, rigid ducts, with belled female ends, and taped joints have been used. The ultimate, of course, is to use threaded joints and screwed-on couplers. Care must be taken to prevent a kink in the duct profile due to the extra rigidity at the sleeve. Welded and brazed joints have also been used. Welding or brazing should never be used to splice a duct when the tendon is in that duct.

1.6 EMBEDMENTS

Embedments are, of course, common to all concrete construction. With prestressed construction, certain precautions are necessary.

Prestressed members shorten in both elastic and plastic movement. This may change the dimensional relationship of critical embedments. It often leads to cracking at the corners of rigid embedments. These cracks can usually be eliminated by affixing soft wood chamfer strips or sponge rubber strips.

The high-strength, low W/C ratio concrete used in prestressed work generally requires more vigorous compaction from vibrators, etc., which may disturb inserts and cause their dislocation.

Steam curing with metal forms produces an expansion of the form before that of the concrete, while the concrete is still weak. This may cause a loosening of inserts. It also causes the inserts themselves to expand, producing cracks at the corners. As noted earlier, softwood or sponge rubber strips will usually solve this problem.

Electrochemical compatibility must be assured where an electrolyte (e.g. saturated concrete) can electrically connect an insert and prestressing tendon. For this reason, copper and aluminum inserts are very suspect and can be used only where special insulating or other protective steps, such as epoxy coating, are employed.

Concern has often been voiced over the possibility of corrosion proceeding up an insert, particularly an insert of strand or steel bar, to a tendon. This fear is not borne out by experience. Atmospheric corrosion is generally limited to $\frac{1}{8}$ to $\frac{1}{2}$ inch in depth. However, if the concrete around an insert is shattered or cracked in service, e.g., by freeze-thaw attack, then corrosion may proceed more deeply.

It is the exposed insert itself that will usually corrode since it becomes the anode, with the embedded steel tendon or bars being the cathode.

Rust staining from inserts (and exposed strand ends) can be detrimental for architectural concrete and even in engineering structures. An epoxy coating may be applied to the ends. Alternatively, the inserts may be galvanized or cadmium-plated. Sacrificial anodes are usually affixed to important underwater inserts of offshore structures.

Stainless steel embedments are sometimes used, for example in prestressed concrete cooling towers. Care must be taken to select the correct stainless alloy to ensure against electrochemical corrosion of the dissimilar metals also employed, for example, bronze bearings.

Where high prestress crosses an opening, the hole should be formed in such a manner that the form for the hole will not be locked in by the distortion due to prestress. The form can be designed to collapse or sponge rubber or polyurethane strips affixed.

Generally, tendons can just be spread around penetrations or openings or inserts without adverse effect on the stress pattern. As in all prestress work it is the concrete that is under stress; a plasto-elastic flow net is automatically formed, regardless of the exact location of the tendon.

1.7 BEARINGS

Prestressed concrete girders and beams are subject to considerable volume change in service. They continue to shorten over many years under the influence of creep. Moisture differentials between top and bottom flanges may cause differential shrinkage. Temperature differentials may cause increase in camber. Service loads cause changes in camber (deflection), with accompanying rotation of the ends and change in length.

Although all these phenomena are to some extent true of conventional reinforced concrete, they are aggravated in prestressed concrete because of the generally thinner sections and the effect of continuously maintained prestress.

Bearings must allow for longitudinal movement and rotation, while maintaining adequate vertical support.

Steel plate bearings, even stainless, are likely to "freeze" in service and become inoperative, due to minor corrosion. Lead plates have been used but are subject to plastic deformation in themselves under continued service.

Neoprene bearings are by far the most widely used type. They are now readily available with suitable durometer hardness, and of varying thicknesses to accommodate the anticipated total movement. For larger bearings a sandwich of neoprene and steel plates is frequently used. Neoprene permits rotation as well as movement.

Teflon provides an almost frictionless surface, hard and durable. Teflon, combined with neoprene and stainless steel plates, is being increasingly used on long-span girders and bridges.

Movement may also be provided for by designing appropriate flexibility into the columns or supports. Where dowels are used to secure the ends of girders, it is recommended that one end have an oversize hold and that it be filled with bitumastic, not grout.

Although it is the designer's responsibility to ensure that the connections and bearings provide for adequate movement, nevertheless, the construction engineer must be aware of the dangers of too rigid connections and the factors causing and influencing movement. When, for example, both ends of prestressed building slabs or girders are welded to fixed plates, or embedded in cast-in-place concrete, distress will show up as cracks and spalling at the girder ends and in the seats. Usually the cast-in-place concrete seat shows the earliest and most severe distress. In extreme cases, either the seat or the girder end may fail in shear. The fabricator may then be charged with faulty construction practice, resulting in long and costly disputes. Probably more dissatisfaction has arisen over this matter of time-dependent movement than any other aspect of prestressed concrete construction.

Therefore it is recommended that, as a matter of practice, careful check be made by the construction engineer of the connection and support details to be sure that there is adequate provision for time-dependent shortening of girders, beams, and slabs. If provision appears inadequate, he should then notify the client and/or the engineer in writing, so that the matter can be properly resolved before construction.

SELECTED REFERENCES

1. Libby, James R. *Modern Prestressed Concrete*, Third Edition, Van Nostrand Reinhold Co., 1984, New York.
2. Mehta, P. K., *Concrete—Structure, Properties, and Materials*, Prentice-Hall, Englewood Cliffs, New Jersey, 1986.
3. *FIP State of Art Report*, "Condensed Silica Fume in Concrete," Thomas Telford Ltd., London, 1988.
4. ACI State of Art Report on "Polymer-Modified Concrete," ACI 548.3R-91, American Concrete Institute, Detroit, Michigan, 1991.
5. *ACI Manual of Concrete Practice*, American Concrete Institute, Detroit, Michigan. ACI 212, Admixtures; ACI 305, Hot Weather Concreting; ACI 306, Cold Weather Concreting; ACI 308, Curing Concrete. (Latest edition)

CHAPTER 2
<hr>

Prestressing Systems

2.1 INTRODUCTION

Prestressing denotes the intentional creation of a favorable state of stress or strain in a structural element. In its usual form, it consists of pre-compression of those cross sections of a member which will later see tensile forces in service. Because of the pre-compression, all or most of these tensile forces will be countered by reductions in the pre-compression.

Since concrete typically used in prestressed concrete construction has an uniaxial compressive strength of 35 to 60 MPa (5000 to 8500 psi), but a tensile strength of only 3 to 6 MPa (450 to 850 psi), it is obvious that the efficiency of the section will be greatly enhanced by pre-compressing it to some degree so that the concrete will be restrained from cracking until well beyond the design load. Prestress may also be used to create a favorable state of strain, e.g., an upward camber, so that future loadings will reduce the camber but not create a sag. Prestress may be used to prevent cracking and to enhance fatigue endurance, aesthetic appearance, and durability.

However, this whole process is complicated by factors such as shrinkage and creep of the concrete, friction in stressing, and stress relaxation of the steel as well as other effects of loading history. It is the aim of the fabricator and constructor of prestressed concrete structures to so carry out his operations as to minimize the losses and problems and to produce efficient structures and structure elements whose behavior over time will be in accordance with that specified during design.

A state of "prestress" or "pre-compression" may be introduced into concrete by internal or external means. External means include superimposed weights and external jacking or thrusting. Internal means include pretensioning and post-tensioning. Composite means are those where internal expansion is resisted by external abutments or internal anchorages.

It is normally the responsibility of the fabricator, in the case of precast prestressed concrete, or the contractor, in the case of field prestressed concrete, to prepare shop drawings. These are very demanding and must be done with care in order to prevent conflicts with other embedded items due to the fact the prestressed tendons cannot be adjusted in location in the field (other than to a very small degree, typically about 6 mm ($\frac{1}{4}$ inch).

The purpose of the shop drawings is to eliminate conflicts with conventional reinforcing steel, ducts for telephone and electricity, plumbing and air conditioning ducts, and other embedments.

The post-tensioning systems further require detailing of the end anchorages including trumpets, anchorage plates, spiral confinement, and grout tube inlets and outlets. For major girders and beams, large scale drawings of the end blocks should be prepared, in order to reflect tangible physical dimensions of bars, anchor plates, trumpets, and spiral.

2.2 PRETENSIONING

Tendons are stressed and anchored against external abutments. The concrete is poured and cured so as to have substantial strength in compression and bond. The tendons are then released from the abutments and the "prestress" is transferred into the concrete member (Fig. 2.1).

Prestress can be transferred only by elastic shortening of the concrete. Until the concrete has actually shortened, it is not prestressed.

In manufacture by the pretensioned method, the frictional weight of the concrete member, or binding in forms, may prevent longitudinal shortening until the member is lifted. During this period, shrinkage or thermal cracking may occur, unresisted by prestress, since the member has not yet shortened. Similarly, in picking from the forms, the initial lifting force may cause cracks, particularly at the point of pick, since the member has not yet shortened and become prestressed.

Pretensioning requires adequate transfer length to transfer the stress from the tendon to the concrete. Tendons for pretensioning are, therefore, designed to make the transfer in as short a distance as possible. Smooth wire is used only in smaller sizes, usually 3 to 5 mm diameter ($\frac{1}{8}$ to $\frac{1}{5}$ inch). Above this size, strands of 3, 7, or 19 wires are employed. The 7-wire strand is now almost universally applied in pretensioned manufacture in the United States, in diameters of 9 to 17.5 mm (.375 to 0.7 inch). Strand with a diameter of 15 mm (0.6 inch) is perhaps the most widely used.

Indented strand is manufactured in Japan for use in those cases and products, such as railroad sleepers (ties), where bond must be developed over a short length.

In Europe, similarly deformed ("profiled") wires are available in 5-mm and 8-mm sizes. Cross sections are frequently oval, and the deformations provide mechanical bond at frequent intervals. Careful design of these ribs or protuberances has eliminated notch sensitivity and fatigue susceptibility.

Recent tests have indicated that the "adhesive" influence plays a very important role in transfer bond, and that this adhesion is largely determined by surface charac-

Figure 2.1. *Typical pretensioning facility manufacturing prestressed concrete piles.*

teristics of the wire. Slightly rusted wire or strand has several times better bond than bright wire; however, the degree of rust is difficult to control and the dangers of pitting corrosion generally rule against intentional rusting. Current best practice requires that the strand be in covered, protected storage until not more than 24 hours prior to concreting.

However, manufacturers have widely adopted wire-drawing practices and lubricants which improve the drawing process but adversely affect bond.

Short transfer length is desirable when high moment must be resisted a short distance from the end; for example, in cantilevered beams, railroad ties, and columns at rigid connections, for which, special efforts must be made to obtain the shortest possible transfer length. Such steps include:

1. Thorough consolidation of concrete at ends.
2. Higher compressive strength of concrete at release.

3. Reduced stress level in tendons obtained by increasing steel cross-sectional area.
4. Increased surface area of tendons obtained by using smaller wires or strands.
5. Gradual release, that is, by hydraulic de-jacking.
6. Adequate lateral binding, by spiral or stirrups, at ends of members.
7. Use of indented strand or wires.
8. Cleaning of the lubricant from the strand used in drawing.

On the other hand, for girders, piles, beams, and slabs—in fact, the great majority of all pretensioned concrete—a relatively long transfer length is desirable. The short transfer length causes high transverse tensile stresses in the end block and may even lead to cracks between strand groups, etc., in the end of the girder. With a slightly longer transfer length, these internal stresses are distributed over a longer zone.

To intentionally increase transfer length, the ends of tendons may be sheathed in plastic or coated with a grease. One such product, Lubabond, is especially designed to minimize early bond but not ultimate bond.

After the concrete member has cured and attained a specified strength, e.g., 60 to 80% of its 28-day design strength, the tension in the strands is released and the strands at the ends are cut. For bridge girders and major beams, this release is usually required to be by hydraulic jack, but for lightly stressed members where a longer transfer length is permissible, the tension may be released into the member by burning the strands. This burning should be performed for each strand by heating with a low-temperature flame (low oxygen) until the strand stretches, then by burning through. A careful sequence should be followed so as to effect the release of prestress equally over the cross section of the member. If the member is free to shorten, the last tendons may break before burning. Therefore this writer prefers to heat all the strands at one end with a low oxygen (yellow) flame before cutting any. "Shock" release automatically destroys the bond over a distance of 1 meter or so (3 to 4 feet).

Because of the uncertainties associated with release by burning, most authorities now prefer or require gradual release, and the practice of release by burning is diminishing.

Pretensioned tendons, by the nature of their being stressed before concreting, must be straight between points of support. On the other hand, for many girders, etc., the specified tendon profile is a parabolic curve.

To achieve an approximation of such a curve, the strand may be stressed up over chairs and down under "hold-downs." These hold-downs can be attached to the soffit (Fig. 2.2), or thrust down from above.

During stressing, a considerable longitudinal movement and force is developed. Later, on release, the longitudinal prestress must be transferred into the concrete before the hold-downs are released.

Provision, therefore, must be made for small movement and angular rotation of the hold-down devices as the longitudinal prestress is released.

The most accurate production method is that in which the strands are stressed as

Figure 2.2. *To achieve the most efficient stress profile, strands are deflected at several locations in the middle of the span.*

straight strands to a pre-computed value, then pulled down (or up) to match the desired parabolic chord profile. This movement, of course, increases the stress in the tendons. Then the longitudinal pretension is checked and adjusted to the design pretensioning value.

Some plants use the alternative method of stressing the strands in their deflected profile. Rollers or shoes are provided to minimize frictional loss. Variations of about 5% have been measured on strands deflected by this method. It should not be used for sharp profiles.

Since the vertical forces at hold-downs are high, it may be more economical and satisfactory to increase the number of hold-down points, thus distributing the vertical reaction along the bed. This will also minimize the actual combined stresses in the steel wires across their cross section, where bending and axial stresses are combined. Note that at these points of deflection, the steel wires are also subject to transverse bearing stresses. Fortunately, unless the curvature is more than 15°, the combined stresses are sufficiently low that rupture does not occur and, due to stress relaxation at high values, the stress is later readjusted.

Deflection of strands is a critical construction procedure. Should a hold-down fail, the stored energy in the strands is like a bowstring and the hold-down becomes a missile. Positive means must be taken so that this work can be done from the side, so that a worker does not have to expose any part of his body above the hold-down.

Remote hydraulic control is a preferred method for pulling up or pushing down tensioned strands.

During concreting, a vibrator may hit a hold-down device. Hold-down positions

Figure 2.3. *Pretensioned bridge girder with deflected strands shown in Fig. 2.2 has optimum prestress.*

should be carefully marked so the vibrator man does not expose himself. A protective device of steel and/or timber should be slipped in place above the point and fastened to the forms.

With deflected tendons, if the girder does not have its longitudinal prestress transferred into it, the release of the hold-down will cause the girder to crack. The frictional force of the girder deadweight, plus the hold-down forces themselves, may prevent movement and transfer of prestress even after release. Therefore longitudinal prestress should be released first. The hold-down devices should be fastened in such a way as to permit the necessary longitudinal movement and angular rotation due to shortening of the girder under release of prestress (Fig. 2.3).

Since the concrete products will usually be subjected to either heat curing or steam curing, the effect of that on exposed lengths of strand between the concrete elements needs to be considered. Long lengths which are not embedded will expand quickly, thus reducing the prestress in the adjoining products and possibly causing permanent reduction in bond. Therefore, the total length of exposed strand that will be heated should be minimized.

2.3 POSTTENSIONING

2.3.1 General

Posttensioning is the system of prestressing by which the tendons are tensioned against the concrete after it has hardened and reached a substantial portion of its design strength. The tendons are typically inserted through the concrete element, through holes which have been formed by ducts. Then the tendons are stressed by hydraulic jacks, and anchored against the concrete. The duct is then filled with cement grout or other substance designed to prevent corrosion of the tendons.

In some cases, the tendons are inserted into the ducts before casting. This poses the hazard of accidental grout leakage into the ducts, which would impede the later tensioning. It also increases the exposure to corrosion.

Posttensioning tendons are sometimes also pre-encased in plastic sheaths, with a lubricant, and installed before concreting, then stressed after the concrete has hardened.

Posttensioning is used to create the desired state of initial stress in cast-in-place concrete structures. This requires that the structure be supported in such a way that it is free to shorten, either by sliding on its bearings or by deflection of the supports.

Posttensioning can also be utilized to join precast segments while at the same time creating the desired prestress in the structure. For such operations, accuracy is required in the assembly of the precast members, especially to ensure the continuity of ducts.

2.3.2 Systems

Posttensioning systems consist of the following assemblages:

Single wires or single T-wire strands.

Groups of wires or multistrand tendons.

Single wires or single strands pre-encased in a sheath of plastic or paper with corrosion-inhibiting and lubricating material between.

Single bars.

Anchorages for these systems include:

Buttonheads for the wires.

Wedge anchors for the strands.

Threaded nuts for the bars.

2.3.3 Installation

While inserting the tendon, swabbing on of water-soluble oil or dusting on of a vapor-phase inhibiting (VPI) powder, followed by sealing of the ends of the duct, will prevent corrosion prior to grouting. The minimum practicable time should ensue between stressing of tendons and grouting; many specifications limit this to 48 hours, but longer periods have been satisfactory when the above special precautions were employed (Fig. 2.4).

Tendons may be installed in the ducts after the concrete has hardened by pushing in one wire or strand at a time or by pushing in a group, with a guiding "nose" attached.

Figure 2.4. *Unreeling seven-wire strand from a pack for insertion into duct. (Courtesy of VSL)*

Tendons may also be pulled through a duct, using a "Chinese-finger grip" and a pulling wire. The wire may be attached to a rubber ball and blown through the duct by compressed air. Then it can be used to pull in a heavier pulling wire, which can then pull in the strands or wires.

Accurate alignment of the anchorages normal to the tendon is essential. The bearing plates must be accurately aligned and have complete bearing on the concrete beneath. The fixing of these anchor plates in the end forms so as to prevent dislocation during concreting is extremely important (Fig. 2.5). If an anchorage plate tends to deform during stressing, it usually indicates defective concrete beneath the anchor plate. The stressing should be stopped and the plate removed. After investigation, the defect can be repaired as necessary.

With some large, heavily stressed girders, the end blocks are separately precast in a horizontal position, with bearing plates, etc.; then these precast end blocks are set in the main girder forms and the ducts connected, and the girder concreted. This procedure assures highest quality concrete under the bearing plates and accuracy of their alignment.

2.3.4 Stressing and Anchoring

Friction during post-tensioning can be verified by stressing first from one end, then from the other (Fig. 2.6). Excessive friction can sometimes be minimized by mov-

Figure 2.5. Bearing plates are fabricated with initial section of trumpet so as to facilitate assembly.

ing the tendon at the start of the jacking operation a few feet one direction, then back. Stressing from both ends is recommended for all double-curved tendons and for all single-curved tendons over 30 meters (100 feet) in length, and wherever sharp bends are involved. Where this is impracticable, calculations and tests must be made to determine friction losses and adequate compensation must be provided.

Figure 2.6. Stressing a multistrand tendon. (Courtesy of VSL)

Excessive friction can also be overcome by injection of a water-soluble oil or lubricating grease. This must, however, be thoroughly flushed from the duct after stressing. Fortunately, water-soluble oil can be flushed out with the initial injection of cement grout.

When multiple strands or bundles of wires are inserted into a single-curved duct, binding of the strands and wires on one another may occur during stressing. This may be minimized by pulling the tendons back and forth a meter or so at moderate stress before final stressing. Flared ducts can be used at sharp bends, and of course, stressing from both ends will help to assure that the tension is equalized.

Large tendons must be supported along the desired profile by wiring or blocking of the duct. The duct can be tied to the reinforcement or supported on it. Depending on the thickness and rigidity of the duct, a steel softener or saddle plate should be used to prevent local denting during concreting.

Personnel should be kept clear of the space directly in back of the stressing as equipment failure or human error may cause the anchorage to shoot rearward like a projectile.

Walkways or platforms should be provided at stressing locations to permit the operators to carry out the jacking and later grouting.

In the posttensioning of large members, such as girders and beams and engineering structures, multiple tendons are involved. The stressing of these should be carried out in a carefully predetermined pattern so as to prevent eccentricities in the concrete structures.

The stressing of the later tendons in the sequence will further shorten the member and relieve some of the tension imposed by the first tendons. In some cases this may be compensated by overstressing in first tendons but this expedient is limited to about 5% of the initial force so as not to run the risk of fracture. In most cases it will be necessary to re-tension the early tendons so as to achieve the design force in all.

In stressing, the theoretical and actual elongations should be checked against the jack gauge readings and calculated friction losses. Discrepancies of over 5% should be carefully investigated, the cause determined, and necessary corrective action instituted.

In anchoring tendons, a small seating loss of "set" will occur. With long tendons, this set, typically 3 to 5 mm ($\frac{1}{8}$ to $\frac{1}{4}$ inch), is of no importance, but with short tendons this loss may result in significant loss of tension. In the latter case the anchor may again be re-stressed against the concrete and shims inserted. Some types of anchorages are designed so as to eliminate "set." At least one post-tensioning bar system permits the taking up of the set by a screwed nut.

Stage stressing may be employed to keep stresses at proper levels during all stages of erection and superstructure construction. In this process, some tendons are stressed on erection, while others are left until a later stage of erection or until part or all of the deck is poured. Tendons which are to be stressed at different stages should be in separate ducts in order to minimize problems of friction and corrosion protection. In some instances a tendon may be partially stressed in one stage, then re-stressed to a higher level at a second stage. In such a case the anchorages should be designed to facilitate this re-tensioning without damage to the tendon.

Warning: *A stressed and ungrouted tendon is very susceptible to corrosion.*

When multiple ducts are spaced closely above one another and the radius of curvature is small (as in negative moment areas), there is the possibility of the tendons on top pulling through into the tendons below. It is desirable therefore to have at least 50 mm (2 inches) or so of concrete between adjoining ducts and to stress and grout the lower ducts first. Heavy-walled ducts (e.g., thin-walled rigid pipe) may be used in the critical areas and/or mild-steel reinforcement bars run between the ducts.

2.3.5 Wire-Wound Systems

Posttensioning tendons may also be external to the concrete cross section during the time of stressing. A long-practiced application of external posttensioning is that of wire-wound concrete pipe and tanks. The concrete core is wrapped with wire under tension, then encased in a concrete mortar. The same principle has been applied to beams and other structural members, particularly ring beams and ring girders of domed roofs.

Posttensioning tendons (wires, strands, or strips) are sometimes positioned in grooves in the sides of girders or walls of tanks; the grooves are later filled with cement mortar. The use of grooves reduces the friction during stressing (as compared with ducts) and makes it easier to ensure full compaction of the mortar encasement.

Shrinkage, however, and the lack of bond between mortar and concrete may present a durability problem. For this reason, application of a rich cement flash coat over the wires, followed by a mortar application, is generally recommended.

The application of mortar is usually by rotating brush, in the case of pipe, and by pneumatic means (shotcrete, gunite) in the case of tanks. The difficulty with shotcrete is that varying degrees of porosity may occur, particularly if the operator allows rebound to be trapped. With a salt-fog or salt-spray atmosphere, electrolytic salt cells may form, and corrosion ensue. Therefore care must be taken to ensure the density of the coat and its water impermeability. The wet process of shotcrete appears to produce more uniform density and impermeability.

Use of a poured and vibrated concrete protective layer, in lieu of shotcrete, may often prove economically feasible and technically superior, especially when the tendons are placed in grooves or recesses in the concrete. For tanks, the placing of the wires under tension is generally accomplished by drawing through a die (the Preload method), by use of a differential-winding system, or by a braked drum system.

Pretensioned wires and strands for slab and girder manufacture may also be laid out under stress, using sophisticated mechanical systems.

In the former Soviet Union, wire and strand have been laid out under full tension in large slab manufacturing plants, by use of the "turntable" and by winding around pegs or dowels. These procedures are probably the most rapid means for placement of tensioning force in a two-dimensional pattern.

For external tendons, the mortar cover is not prestressed and therefore is theoretically subject to shrinkage cracks and tensile cracks in service. Actually some prestress does enter these coatings through the creep of the main concrete member. Favorable behavior of thin mortar coats is also assured by the continuous bond with the main concrete. When thick or large coatings are involved, consideration must be given to provision of mesh reinforcement or of other special protective means, for example, dowels or studs to tie the concrete to the structure.

It is especially important with external systems which are later fully bonded to the structure, such as tanks, pipes, and some girders, to prevent delamination at the plane of the tendons, with the potential for in-leakage of water and corrosion.

2.3.6 Unbonded Posttensioning, External Tendons

This is widely employed in floor slab construction of buildings. The tendons may be pre-wrapped in plastic or fiberglass sheaths, after coating with an appropriate corrosion-inhibiting grease or bitumastic compound. The sheathed tendon is then placed in the forms, the concrete poured and hardened, and the tendon stressed. This is very adaptable to thin flat slabs, where ducts would take up an appreciable portion of the cross section. It facilitates the draping of the tendon into an up-and-down profile characteristic of flat slab moments, and thus is very economical. Many of the difficulties that have arisen with such systems are a result of this very ease of installation; it has led to carelessness. The tendons should be supported at 1- to 1.5-meter (3- to 5-feet) spacing so as to maintain the design profile within a tolerance of 3 to 6 mm ($\frac{1}{8}$ to $\frac{1}{4}$ inch). It is particularly important to insure adequate cover, through chairs or dobe blocks, and to insure thorough consolidation of the concrete under the anchorage bearing. The placement and support of the anchorage must be accurate and rigid enough to prevent displacement.

Obviously, the integrity of the sheath must be assured. Repairs to damaged sheathing must be completed before concreting. Fiberglass (FRG) tape is often used to patch damaged sheathing. Excessive wobble and sharp bends must be eliminated.

In order to ensure shear transfer to columns, the tendons in the slab are often deflected sidewise. Some tendons may be run through the columns. There are a number of differences in behavior under static and dynamic loads between unbonded tendons and fully-bonded tendons. These must be considered by the designer. The main concern of the construction engineer utilizing unbonded prestressed tendons must be to ensure their proper installation, their integrity during construction and durability thereafter, and the proper placement of the anchorages and construction of any intermittent deflection saddles. Many of the past problems with unbonded tendons are really due to careless and poor construction practice unrelated to the matter of "bonding" or "not bonding." See Section 5.5.3.5 and Reference 7 to Chapter 5.

Large external tendons have been used for long-span bridges and in other engineering structures. These uses are increasing because they eliminate the need to thicken the concrete webs to accommodate internal tendons. Hence the deadweight is reduced. The practice of external posttensioning also reduces friction during stressing. In this case, the tendons are outside the concrete cross section and only attached to the concrete at the ends (through anchor blocks) and intermittently along the member (with saddles or deviators). These external tendons, while external to the concrete cross section, may be contained inside a box or space frame, for example, inside the box of a box girder.

Such tendons are usually threaded through polyethylene tubes which are later injected with cement grout. Where the tubes curve at saddles, the tube may be of steel, or the polyethylene duct may be run continuously through the steel tube.

Saddles over which these external tendons run must be carefully designed to minimize friction and to transmit the high forces involved. This latter has usually been done with concrete bolsters, tied to the web of the girder by multiple stirrups, with consequent congestion at just the location that requires the most thoroughly consolidated concrete. This is where T-headed bars could be beneficially employed.

To minimize friction at the saddles, sliding blocks of stainless steel, coated with Teflon or with an inserted sheet of Teflon, have been employed.

Use of external tendons permits future inspection, verification of stress level, and, where necessary, replacement.

2.4 FLAT JACKS

This generic term is used to describe all types of external jacks which prestress the member against abutments. Flat jacks may be thin neoprene bags, which are inflated with air or water to erect a high force working over a very small distance. Several of these can be superimposed to gain greater movement. Since the prestress is an external, not an internal, force, buckling is definitely a consideration and must be guarded against, particularly with thin slabs.

Use of water under pressure is the safest means of inflation (jacking). Then grout can be pumped in under high pressure to displace the water and to fix the jacking distance. A special valve is used that will permit the water to be ejected by the grout without loss of pressure. The new thixotropic grouts with antiwashout admixture would appear to be well suited to this task.

Steel plates are used to contain multiple flat jacks in batteries so that their expansion will be uniform.

Flat jacks are extremely useful for de-centering of arches. By inserting them at several locations near the crown, the arch can be raised off the falsework and forced to act in true arch fashion. Flat jacks have been used to prestress pavements; they are very effective for a limited time, although the losses of stress in the pavement due to creep and temperature are irrecoverable.

Flat jacks have also been used to raise extremely heavy weights, such as prestressed concrete nuclear reactor pressure vessels, from temporary supports, and to balance large reaction forces, which are then fixed by grouting. Flat jacks were used to equalize the eight support reactions for the superstructure of the Ninian Central Offshore Platform.

A number of types of mechanical and hydraulic jacks have also been developed for utilization in the same manner.

SELECTED REFERENCES

1. *FIP Guides to Good Practice,* "Recommendations for Acceptance of Post-Tensioning Systems," Thomas Telford Ltd., London, 1992.
2. Freyssinet International, "Prestressing Systems," Paris, current issue.
3. VSL International, "Post-Tensioning Systems," Berne, Switzerland and Los Gatos, California, current issue.
4. Bureau BBR Ltd., "Prestressing Technology," Zurich, Switzerland, current issue.
5. Neturen Co. Ltd., "Prestressing Steel," Tokyo, Japan, current issue.
6. DSI, "Post-Tensioning Systems," Dywidag Systems International, USA, Inc., Long Beach, California, current issue.
7. *Post-Tensioning Manual,* Post-Tensioning Institute, Tucson, Arizona.

CHAPTER 3

Applicable Reinforced Concrete Practice—A Background for Prestressed Concrete Structures

3.1 GENERAL

Prestressed concrete structures, the theme of this book, are an outgrowth of conventional reinforced concrete structures and the new and advanced technologies depend heavily on the basic principles of reinforced concrete construction. So in this chapter, a summary of reinforced concrete construction will be presented, with the explicit understanding that it does not pretend to cover the entire field nor delve into the myriad facets of this widespread construction system.

The basic material constituents of reinforced concrete, concrete and reinforcing steel, have already been dealt with in Chapter 1.

Only basic principles will be presented here, and these will be set out in brief, prescriptive presentation.

3.2 FORMING SYSTEMS

The forms for reinforced concrete construction are intended to confine it to the desired dimensions, and to minimize local distortions such as bulging under the hydrostatic pressures of fresh concrete and the local effects of vibration.

Where soffit forms are used, they and their supporting scaffolding, falsework, or shoring, are intended to support the weight of the concrete, including the impact of placement and the concreting equipment, with minimum sag.

The scaffolding and shoring must support these vertical loads, with particular attention to prevention of buckling and lateral failure due to sidesway. The lateral forces may be external environmental forces such as wind and water current, but most often are dynamic, due to swing of a crane on top, accidental impact, and unbalanced loading.

Forms moreover must provide a surface that meets the architectural requirements. Plywood surfaces may be oiled to prevent adhesion, while absorbent polyurethane liners may be used to absorb excess moisture from the surface. Fiberglass (FRG) forms produce ultrasmooth architectural surfaces, whereas rough-sawn lumber may be purposely used to give a desired texture.

The construction engineer must of course meet the specification requirements but in addition must concern himself with the time and cost of assembling and "buttoning-up" of the forms, and the subsequent stripping. Hence the use of wedges, strips, drop-outs, and sheet metal at reentrant angles and corners is appropriate.

To prevent spreading under the hydrostatic pressure of fresh concrete, form ties or similar restraints are used. The ACI Manual on Formwork gives appropriate design pressures for fluid concrete.

Form ties extend through concrete walls and therefore must be removed and patched after the concrete has gained sufficient strength to stand on its own.

Cantilevered forms use inserts in the top of the preceding pour. They often also use ties over the top of the current pour.

Slip or sliding forms rise up continuously during the concrete placement which goes on 24 hours a day, continuously until completion. The forms are typically jacked up on climbing rods embedded in the preceding concrete pours, and are guided from the concrete section previously placed.

Rates of rise up to 200 mm/hour (8 inches /hour) are standard. With accelerating admixtures, these may be extended to 300 mm/hour (12 inches/hour) or, with retardants, slowed to as little as 50 mm/hour (2 inches/hour).

With the use of internal "spider" frames, walls may be tapered inward or outward.

3.3 REINFORCING STEEL SYSTEMS

The reinforcing steels available and their importance to prestressed concrete have been briefly described in Chapter 1.

Assembling the reinforcing steel into meaningful and structurally useful patterns in complex structures is an extremely important aspect of design and construction, often overlooked because of the tendency in US practice to turn this over to a specialty subcontractor.

Mild-steel reinforcement is a highly important component of prestressed concrete design, for it can furnish the required ultimate strength, ductility, and confinement capabilities.

Because of the high performance requirements of advanced concrete design, a number of techniques have had to be adopted.

Where more steel area per square meter is required and there is limited space, bars may be bundled in groups of two, three, or four. ACI-318, the Building Code of the American Concrete Institute, gives applicable rules for bundled steel.

The typical high performance structures, subject to demanding loading conditions, require that the steel be placed to close tolerances. This is relatively straight-forward in the in-plane dimension, but more difficult in the through-thickness and the linear dimensions.

The through-thickness dimension can be controlled by dobe blocks against the forms or by ties. The linear dimension depends on the accuracy of the cut length of the steel bar. Relatively easy to maintain on straight planes, it is very difficult to achieve with curves and bent bars.

Stirrups provide the through-thickness steel reinforcement. The problem that must be met is how to anchor them. The usual rectangular stirrup has two "tails." Ideally these would be bent back into the core for anchorage; this requires field bending after placement, which is extremely difficult and time-consuming, hence costly.

Other alternatives have been adopted: overlapping U stirrups and stirrups configured in J and Z shapes. Yet stirrups are not ideal from the point of efficiency since they all require tight bends. Numerous tests have shown that they do not develop their full yield strength due to crushing under the bends.

Tolerances, especially of cover, are also difficult to maintain as a result of the bends. Concreting is rendered difficult by the congestion (Fig. 3.1).

The mechanical-headed (T-headed) bars described in Chapter 1 are more efficient and more easily placed but for best results require locking behind the in-plane reinforcement (Fig. 3.2).

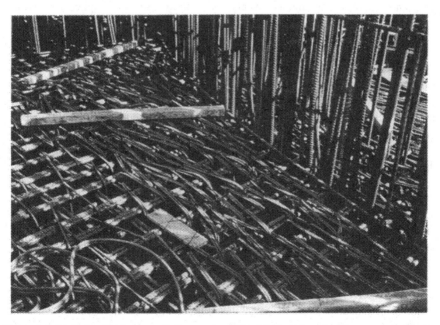

Figure 3.1. *Multiple closely spaced stirrups are difficult to install and make concrete place-ment difficult in base slab of offshore platform.*

Figure 3.2. *Mechanically headed "T-bars" lock behind the in-plane reinforcement.*

Typing of reinforcing steel bars is highly important, not only to hold bars in position relative to one another but also to ensure continuity in splices.

3.4 INSERTS

Inserts, whether piping or plates, are common in all concrete construction. They must be accurately positioned in the forms and held there as the concrete is placed and vibrated.

Some inserts, such as electrical, require protection against concrete paste leakage. Seals must be provided: usually watertight tape will suffice.

Consideration must also be given to stripping of the forms since the inserts may tend to bind them.

Care must be taken, when metals other than mild steel are used, that electrolytic corrosion cannot occur. One positive means of insulation is to coat the embedded portion of the insert with epoxy or polyethylene.

Steel embedments which have exposed surfaces, e.g., plates, will tend towards accelerated corrosion since they become the anode and the extensive cage of reinforcement becomes a large and powerful cathode. Separating the anchor bolts from the reinforcing bars by plastic spacers is one solution. Epoxy coating of the anchors is another.

3.5 CONCRETE PLACEMENT

Concrete mix design has been presented in Chapter 1. This section addresses the placement.

Concrete placement may be by pumping, conveyor belt, buckets, chutes, or even wheelbarrow. The method should be selected which is appropriate to the actual placement, as to rate of placement, continuity, prevention of segregation, and number of men required, this latter being the prime determinant of cost.

Considerations that should be given involve the quality of the concrete delivered into the forms. With air entrained concrete, pumping will decrease the size of the air voids and may decrease their efficacy. Pumping may also reduce the workability.

Curing has been presented in Chapter 1. An additional caution must be noted that elevated temperature curing, such as steam curing, may produce early expansion of large and heavy concentrations of reinforcing bars prior to expansion of the concrete, thus causing microcracks that may lead to future corrosion.

3.6 VOLUME CHANGES IN STORAGE AND SERVICE

Reinforced concrete is subject to significant volume changes, especially in its early life.

The matter of thermal strains due to exposure to cold air and cold winds has been described in Chapter 1. Protection in the form of shielding or insulation is often required. Steel forms may be insulated by sprayed-on polyurethane. It is remarkable how often the simple draping of a tarpaulin over an exposed concrete surface will prevent cracks that would otherwise have arisen.

Drying shrinkage is due to moisture loss and can be prevented by supplying or retaining the moisture. It is generally a surficial phenomenon. When excessive, it results in surface cracks or crazing.

Both thermal and shrinkage strains are resisted by the reinforcement, which as a result incurs a moderate initial compressive stress.

Creep is especially a concern at the early ages and may result in permanent deformation of structural elements which were initially cast to extremely close tolerances. Proper storage, including blocking, shoring, and strutting as appropriate, is essential. Creep results from a state of sustained stress. Concentrated loads may cause local deformation. A sustained bending moment may cause a permanent sag or deflection.

SELECTED REFERENCES

1. Bresler, B., *Reinforced Concrete Engineering*, Vols. I and II, Wiley-Interscience, New York, 1974.

2. ACI 318, "Building Code Requirements for Reinforced Concrete," American Concrete Institute, Detroit, Michigan, current issue.
3. *ACI Manual of Concrete Practice,* American Concrete Institute, Detroit, Michigan. ACI 315, Concrete Reinforcement; ACI 439, Mechanical Connections of Reinforcing Bars, current issue.
4. Berner, D., Gerwick, B., Hoff, G. "T-Headed Stirrup Bars," Concrete International, May 1991, American Concrete Institute, May, 1991, Detroit, Michigan.

CHAPTER 4

Special Techniques

4.1 HIGH-STRENGTH CONCRETE

4.1.1 General

In the early days of prestressed concrete, it was quickly realized that high-strength concrete was essential to the technical and economical success of this new technique. During certain stages of its service life, the concrete must resist a combined compressive stress due to prestress and dead-load stress. Higher strength produces a higher modulus of elasticity, reducing prestress loss. Higher strength usually is concurrent with reduced creep and shrinkage. Higher strength also means higher bond of paste to aggregate and to steel. Therefore cylinder strengths of concrete in the range of 35 MPa (5000 psi) were utilized.

The advent of commercial precast pretensioned plant production required high strengths at early ages, often one day. It was quickly discovered that the techniques for producing high early strengths automatically led to concrete with 28-day strengths of 40 to 50 MPa (6000 to 8000 psi). These strengths were achieved by using conventional concrete materials and practices, with emphasis on the reduction of the W/C ratio. In addition, the following practices were adopted:

1. Use of coarse aggregate with a maximum size of 20 mm ($\frac{3}{4}$ inch).
2. High cement factor 350 to 400 Kg/m^3 (600 to 700 lb/cy).
3. Water-reducing admixture, as necessary to achieve a water-cement ratio of 0.40 to 0.42.
4. Thorough mixing.
5. Thorough vibration.
6. Excellent curing: low-pressure steam and/or hot water curing.
7. In hot climates, pre-cooling of aggregates.

There are many advantages through the use of moderately high-strength concrete.

These moderately higher strengths give many direct and indirect advantages to prefabricated products as well as to cast-in-place concrete.

In the last decade, a second surge in concrete strengths has taken place. The high-range water reducers (super-plasticizers) have allowed reduction in practicably obtainable water-cementitious material ratios (W/CM) to 0.35 to 0.38 for field cast concrete and to 0.32 for precast concrete. Even lower values have been obtained in special instances.

Such concrete, moreover, is flowable, permitting placement among dense patterns of reinforcing steel and prestressing ducts or tendons.

A second advance came with the advent of microsilica (condensed silica fume), which has particle sizes about $\frac{1}{100}$ the size of cement grains. These fit into the interstices of the crystal-gels of the cement, where they have both a pozzolanic effect, combining with the gypsum in the cement, and a mechanical effect due to their extremely small size. The result, when used in proportions of 4 to 10% by weight of cement, along with control of the water-cementitious material ratio, can add significantly to the strength of the concrete.

With concentrated efforts on selection of aggregates and their cleanliness (freedom from extreme fines such as silt), on cooling of the mix, and on internal vibration, cylinder strengths up to 70 to 80 MPa (10,000 to 12,000 psi) have become commercially attainable in practice, and for highly demanding uses, strengths as high as 130 MPa (19,000 psi) have been attained.

Such ultra-high-strength concrete has a stress-strain curve that is much steeper than that of more conventional concretes: a curve which continues in almost a straight line up to ultimate strength. Thus, it has a lower plasticity and the failure mode is brittle.

The use of ultra-high-strength concrete, that is, concrete with strengths above 80 MPa, requires three-dimensional confinement, in the form of heavy spirals, stirrups, mechanically headed bars, or even an enclosing steel tubular sleeve.

Use of ultra-high-strength concrete requires very close adherence to dimensional tolerances, since there is less margin for adjustment. It means less elastic and creep deformation under concentrated and uneven bearing loads.

With some members, for example, thin, high girders, more temporary bracing may be required to prevent buckling.

Returning to the properties of high-strength concrete, the following may be noted:

The modulus of elasticity of concrete increases approximately as the square root of the strength; thus deflection characteristics are improved. The tensile strength of concrete also increases approximately as the square root of the strength, thus improving behavior under overload or partial prestress conditions. Durability is generally proportional to strength in that the same factors that improve strength also improve durability by reducing porosity and permeability. An exception is air-entrainment, which reduces strength while improving freeze-thaw durability.

As experience has grown with the use of higher concrete strengths, research has been intensified. The most successful practical means have been the reduction of the water-cementitious ratio (W/CM) through the use of high-range water reducers (HRWR) and the inclusion of moderate amounts of microsilica. However, these immediately successful means should not be allowed to overshadow the fundamentals. These include:

1. Selection of aggregate for high inherent strength.
2. Use of crushed rock instead of gravel.
3. Maximum size of coarse aggregate to be kept small. For normal structural members, 20 mm ($\frac{3}{4}$ inch) maximum size is used, but for very high strengths, 10 mm ($\frac{3}{8}$ inch).
4. Gap grading, especially elimination of fines at the lower end of the grading curve.
5. Cleanliness of aggregate, removal of silt and dust by rewashing and re-screening.
6. Cooling aggregate by water soaking and evaporation, shielding from sun, or by refrigeration. Further cooling may be accomplished by batching of ice, instead of water, or injection of liquid nitrogen directly into the fresh concrete mix.
7. Rich cement factor. Although up to 600 Kg/m^3 (1000 lb/yd^3) has been used in extreme cases, present practice favors a maximum of 400 Kg/m^3 (680 lb/yd^3).
8. Selection of cement for high strength, with attention to grind. This may often involve a compromise, since a fine grind may be desired for high early strength and a somewhat coarser grind for high late strength.
9. Replacement of 15 to 20% of the cement by pozzolon (PFA).
10. Addition of 4 to 10% of microsilica (condensed silica fumes). Too high a percentage will make the concrete "sticky" and hard to place. Six percent appears to be a practicable compromise.
11. Low W/CM ratios (0.30 to 0.38). To obtain a workable mix, water-reducing admixtures are necessary. These may be HRWR (super-plasticizers) or high doses of conventional water reducers.
12. Heavy vibration to consolidate the concrete and drive out entrapped air.
13. Special curing, that is, continuous ponding or soaking in water, or steam curing followed by water curing.
14. Adopting a two-stage batching-mixing process so that the sand is coated with cement before the coarse aggregate is added.

Careful attention to the above, using skilled and experienced personnel, has enabled top concrete producers, especially those in precast plants, to consistently and reliably produce concrete to design strengths of 55 to 77 MPa (8,000 to 10,000 psi). One of the most impressive of such projects was that for the prestressed girders for O'Hare Airport, near Chicago. On this project, the manufacturer paid particular attention to rewashing of the aggregate to remove dust and silt particles (which raise the water requirement and thus the W/C ratio), to the cooling of the aggregate piles by pre-soaking and evaporation, and to careful grading to obtain a maximum density. Actual strengths averaged 70 MPa (10,000 psi) with some values ranging to 12,000 psi.

Maximum aggregate size was ⅜ inch, a gap-graded mix was adopted, a high range water-reducing (HRWR) admixture employed, and the W/CM ratio was 0.35. Compaction was by intense vibration.

Precast tunnel liner segments for the English Channel Tunnel have been produced with a W/CM ratio of 0.32 and have achieved 90-day cube strengths of 70 to 80 MPa (10,000 to 12,000 psi). Norwegian Contractors are achieving 90-day characteristic cube strengths of 80 MPa (11,000 psi) with cast-in-place concrete on current large offshore platforms. The 90-day period for strength evaluation seems to be appropriate where PFA is included in the mix, since such a period allows time for complete hydration.

For the columns of high-rise buildings, higher and higher strengths are being specified and achieved. Values of 80 MPa (12,000 psi) are achieved in reinforced concrete columns in Chicago buildings, and up to 120 to 130 MPa (18,000 to 19,000 psi) in composite columns of 90-story buildings in Seattle, Hong Kong, and Taiwan. These composite columns consist of high-strength steel tubes, with studs to transfer shear, filled with the high-strength concrete. Some relatively minor degradation of ultimate strength over time has been noted in these latter cases, also some autogeneous shrinkage. Nevertheless, these cases, substantiated by work in defense-related projects, show that design cylinder strengths of 100–120 MPa are practicably attainable on real projects.

It is important that elements and members of high-strength concrete be well-confined by reinforcement.

Measuring and testing techniques must be compatible with the strengths to be tested. Attention is directed to Section 1.1.13, Testing. Correlation between test cylinders and actual concrete must be achieved; use of a testing hammer and extraction of cores may prove valuable for such verification.

Other studies have been directed towards increased cohesion, increased compaction, triaxial stress, fiber concrete, and polymer concretes. These are briefly described as follows.

4.1.2 Cohesion

For conventional concrete of 60 to 70 MPa (8,500 to 10,000 psi), it is the loss of adhesion between cement matrix and aggregate which initiates failure. Use of cementitious aggregates, such as Portland cement clinker and aluminous cement clinker, would provide improved bond. In England there is commercial production of prestressed concrete beams of design strength of 12,000 psi (cube), using alumina cement, fine aggregate, and a high-quality granite coarse aggregate. E is 600,000 to 700,000 kg/cm² (9,000,000 to 10,000,000 psi).

Another approach exploits the silica-lime bond. Structures from ancient China, such as the Great Wall, have shown constant improvement in strength through the centuries. To obtain this reaction within a practicable time period, autoclave heating to 200°C or more is required. From a present-day practical viewpoint, it would seem that potential use would be limited to small-size products. One possibility, however, that is being studied, is the production of high-strength segments which could be prestressed together to form heavily loaded and long-span beams.

Microsilica appears to form a chemical bond with lightweight aggregate that enables high-performance concrete to be achieved.

Considerable research and development has been directed towards the use of epoxy-resin and polyester-resin binder to provide greater bond between aggregate particles. Problems are lack of fire-resistance, low E, and high cost. A variation is to precoat the aggregate with epoxy-resin, then incorporate it in a concrete mix having a conventional cement matrix.

Research in Japan has shown improved cohesion and higher strengths when the sand particles are precoated with cement paste, before the coarse aggregate is introduced into the mix.

4.1.3 Compaction

Present day vibration frequencies compact the coarse aggregate only. Ultrasonic frequencies could be used to compact the sand, and even the cement particles. The practical problem has been to develop vibrators of such frequencies that have sufficient power. However, such vibrators do exist, and commercial availability appears entirely practicable. These will undoubtedly develop very high form pressures, necessitating more rigid forms and design of details to prevent fatigue failures of the forms. Frequencies of 5,000 to 10,000 cps have been used in experimental work with great effectiveness in obtaining compaction and high strength. Commercially available vibrators have frequencies up to 3000 cps and only moderately high force.

Experimental work at Ohio State University has shown promising results when very high-frequency transponder vibrators are combined with conventional low-frequency, high-powered vibration.

Increased compaction and activation of the cement may be obtained by the electrohydraulic impulse method. This latter, still in the development stage, consists of interrupted high intensity electric discharges, which create a plasma sphere within the grout. Collapse of this sphere causes a breakdown of the cement particles, formation of water films on the grain surface, and ionization of the grout. Gel formation is increased. The resulting grout is both stronger and more dense.

Revibration of the concrete mix has been found to be effective in achieving greater compaction and strength with some mixes. A retarding admixture will make revibration more practicable. Recent work in France has shown the benefit of previbration of the concrete mix, followed by vibration and pressure after placement.

Compaction can also be increased by pressure. This method has been applied for many years in France for manufacture of prestressed concrete poles. A combination of pressures and vibration appears most effective. Use of autoclaving or low-pressure steam curing appears to enhance the effects of pressure.

Pressure can be applied by the following:

a. Spinning.
b. Use of a strong, fully enclosed form, with an inflatable rubber tube which is expanded.
c. In the former USSR, a steel pad, to which a rubber pad is attached, is placed over a slab and locked to the soffit. By removal of air below the membrane, a vacuum is created which exerts the compacting pressure.
d. Ramming. This may be accomplished by shock dropping of the forms (as in the Schokbeton process) or by mechanical ramming. In the spinning process, a ramming ball may be used to increase compaction.

Strengths up to 140 MPa (20,000 psi) [and even to 250 MPa (35,000 psi)] have been achieved with pressures of four atmospheres combined with vibration. These are accompanied by values of E of 60,000 to 70,000 MPa (8,500,000 to 10,000,000 psi) and densities of (2900 kg/m³; 180 lb/foot³). Poisson's ratio is 0.28.

Finally, compaction can be improved by modification in the mix design. In some cases careful gradation has been adopted to ensure the maximum aggregate content, with the minimum requirement for cement paste. In what seems like a paradoxically opposite approach, high strengths have been achieved with mixes in which a small size of coarse aggregate [e.g., 9 mm (⅜ inch)] has been used with no fine aggregate whatsoever and a high cement content.

4.1.4 Triaxial Prestress

If confinement in the form of high-strength steel wire spirals is used to restrain the concrete normal to the applied load, substantial increases in strength can be achieved.

The problem is that the concrete cover over the binding spalls off long before ultimate strength is reached. Since the modulus remains constant at the level of the concrete itself, initial strains are large. Preloading, however, will stiffen such a member, improving the apparent modulus.

Enclosure in a steel pipe shell has a similar effect and is practicable in many actual applications (e.g., compression members, such as piles or columns). Concrete heavily compacted inside high yield strength steel cylinders has developed apparent strengths up to 700 MPa (100,000 psi).

4.1.5 Fibers

Use of random fibers, such as finely divided wires and polypropylene and asbestos fibers, has been studied for many years. Such research and development is still continuing. Tensile strengths are greatly improved; compressive failure is delayed. Strains remain large.

4.1.6 Polymerization

Recent research has shown that substantial increases in strength and other properties can be achieved by impregnation of dried concrete with a monomer followed by treatment by irradiation or thermal processes. The monomers which have proven most effective are methyl methacrylate and styrene. Vinyl acetate and ethylene gas dissolved in sulfur dioxide have also been used. The concrete is usually thermally dried, and subjected to a vacuum of about 3 inches of mercury. Then the precast concrete element is soaked with the monomer, under a nitrogen blanket (excess pressure) of one-third atmosphere. This is followed by an irradiation treatment of Co^{60}, or more practically, thermal treatment at 75°C, for about four hours, which converts the monomer to a polymer.

Such polymerization of conventional concrete of 35 MPa (5000 psi) improves the properties as follows:

Compressive strength, three to four times, to a maximum of 140 MPa (20,000 psi).

Tensile strength, three to four times.

Modulus of elasticity, two times.

Freeze-thaw resistance, three to four times.

Permeability—decreases almost to 0.

Water absorption—decreases to about $\frac{1}{20}$.

Corrosion under sulfates, acids, and hot brines—decreases toward 0.

Bond strength, two to three times.

Creep, negative (slight expansion under continued load).

Hardness, two times.

Such treatments appear to be economically feasible for mass production of standardized precast segments. For cast-in-place concrete, such as bridge decks, polymerization by heat appears to be practicable, although it is an expensive process. It has recently been applied for the decks of bridges in the Florida Keys.

Examination of polymerized concrete specimens shows substantially increased bond between the matrix and aggregates, as well as essentially complete filling of the voids in the matrix.

Normal concrete (35 MPa) will absorb about 6% of monomer by weight.

Experiments continue on the incorporation of monomers in the mixing water; to date, results have been only about 50% as effective as with impregnation of dried concrete.

Provided suitable caution is used by designing engineers to prevent the development of such phenomena as buckling and vibration, previously known only to steel structures, and as long as dimensional tolerances and consistent quality are maintained, the use of higher strength concrete offers greatly improved performance and economy in a wide variety of applications.

4.2 PRESTRESSED LIGHTWEIGHT CONCRETE

4.2.1 General

Structural lightweight concrete is being employed very extensively in prestressed applications. The production of high-quality lightweight aggregates is well established in the United States, Germany, Canada, Russia, Japan, and Australia, and is being developed elsewhere. The techniques of mix design and placement (vibration, finishing, etc.) have been developed to the point that lightweight concrete may be used with confidence both for highly technical construction, such as bridge girders and offshore structures, and also for mass production, such as roof and wall slabs. However, such high-quality structural lightweight concrete can be obtained only if these techniques and controls are carefully utilized by trained and skilled personnel. Failure to take this care may result in excessive creep and shrinkage, segregation, and nonuniform deflections.

Prestressed lightweight concrete is best considered as a unique material in its own right. Its particular properties can then be fully utilized and properly employed. Design and construction procedures can be adopted to give the desired results.

Structural lightweight aggregate concrete contains expanded clay, shale, or slate as the coarse aggregates. It may utilize expanded aggregates for fines, or natural sand, or a combination. For the higher strengths required for prestressed applications, natural sand is generally used. The unit weight of the fresh concrete will run from 1400 to 2000 Kg/m^3 (85 to 120 lb/foot^3) with the heavier weights being associated with higher strengths. The strength is quite dependent on the W/C ratio and cement factor, but design strengths of 30 to 65 MPa (5000 to 9000 psi) are obtainable by following proper procedures. These strengths, and the associated properties, render lightweight concrete entirely suitable for prestressed applications.

The higher strengths are attainable with the use of HRWR admixtures and the inclusion of microsilica (condensed silica fumes) in the mix. Strengths are then limited by the aggregate crushing strength. There are several manufacturers of high-strength lightweight aggregates throughout the world. These high crushing strengths are obtained by longer, hotter kilning, which gives a strong ceramic sealing "skin" over the particles' surface.

Laboratory tests and scanning electron microscopic views show that microsilica forms a chemical gel bond with the lightweight aggregate which increases bond, tensile, and compressive strengths substantially.

Thus 28-day cylinder strengths as high as 65 MPa (9000 psi) are practicable, although very strict controls are essential.

Such mixes, with microsilica, also exhibit high bond with the reinforcement, leading to better fatigue performance and better abrasion resistance.

Conventional concrete, made with sand and gravel (or crushed rock) aggregate, exhibits a broad spectrum of properties. Similarly, the properties of structural lightweight aggregate concrete can best be represented by a band, rather than a numerical figure or straight line.

Since structural lightweight aggregates are basically a manufactured product, it is to be expected that even higher quality and more closely controlled ceramic materials will become available in the future. This will undoubtedly make available higher strengths and reduced creep and shrinkage.

A parallel and recent development has been the exploitation of natural deposits of pumice. While the majority of natural aggregates are unsuitable for high-strength and prestressed applications, deposits do exist in Hawaii, Nevada, Japan, and the Caucasus which have very satisfactory properties and which are being used for prestressed panels, slabs, and even small bridge girders.

Some interesting experiments are being conducted on the use of small glass balls as lightweight aggregate. However, before commercial use can be undertaken, matters such as chemical stability and aggregate-paste bond must be thoroughly evaluated.

4.2.2 Properties of Structural Lightweight Aggregate Concrete

Comparisons are between high-quality structural lightweight concrete, made from expanded clay, slate, or shale aggregates, and conventional (normal weight) concrete made from sand and gravel aggregates. These comparisons are purposely general in nature and apply to the typical mix of lightweight concrete suitable for structural elements for buildings.

Unit Weight. In air, 60 to 80% of normal concrete. Submerged in water, 50% of normal concrete.

Compressive Strength. With a 5 to 10% increase in cement content, strengths are about equal up to a maximum of about 40 MPa (6000 psi). Above that, the special mixes described earlier are required.

Shear (Diagonal Tension). Ultimate strength 65 to 80% of normal weight concrete.

Tensile Splitting Strength. Dry conditions, somewhat lower than normal concrete. Moist conditions, the same.

Modulus of Rupture. Dry conditions, somewhat lower than normal concrete. Moist conditions, the same.

Bond. Slightly less than normal concrete, unless microsilica is added, in which case bond is increased by 50 to 100%.

Transfer Length and Development Length. Same or slightly greater than normal concrete.

Modulus of Elasticity. Approximately 50 to 60% of that of normal concrete.

Creep. (Defined as the time-dependent deformation of concrete under sustained loading.) Usually slightly greater than normal concrete of same strength (may run from 0 to 100% greater, depending on aggregates). Creep is also dependent on relative humidity under service conditions. For the usual prestressed lightweight concrete, 40 to 50% greater. The creep is proportional to the ratio of applied stress to the strength at time of loading. Thus, with a given prestress level, the creep may be reduced by having the concrete at a greater strength at the time of transfer of prestress.

Shrinkage. For best lightweight aggregates, slightly equal or greater than normal concrete. (For some low-quality lightweight aggregates, 50 to 100% greater.) Steam curing reduces shrinkage. Replacement of a portion of lightweight fines with natural sand reduces shrinkage.

Total Prestress Loss. This loss is 110 to 115% when both are subjected to water cure; 120 to 125% when both are subjected to steam cure. (*Note:* Lightweight concrete subjected to steam cure has approximately the same total prestress loss as normal concrete subjected to water cure.)

Thermal Conductivity. Forty percent of normal concrete.

Thermal Transmittance. Fifty percent of normal concrete. This much greater thermal insulation has a decided effect on prestressing applications, such as the following:

a. Greater camber when one side is exposed to sun.

b. Slower response to steam curing and in service.

c. Greater fire resistance.

Permeability. Same, since structural lightweight aggregates are basically impermeable in themselves, i.e., their pores are sealed. With microsilica, impermeability may be greater.

Water Absorption. Runs 10 to 22% by volume as compared with 12% for normal concrete. However, the highest grade lightweight aggregates have only 6 to 10% absorption. This is of importance for freeze-thaw resistance under saturation conditions.

Abrasion Resistance. Lower, where abrasion is local and concentrated since lightweight coarse aggregate particles may be "plucked" out. Same for general wear, as in pavement. Recent tests in the Swedish Baltic Sea of high-quality lightweight concrete show resistance to ice abrasion is approximately the same.

Freeze-Thaw Durability. Equal or superior in most installations, especially where deicing salts are employed. Air-entrainment is recommended. Aggregates must not be pre-saturated. This may make placement by pumping difficult. Saturated aggregates of high absorption are subject to damage by deep freezing.

Marine Durability. Experience appears to clearly show superior durability for high-quality lightweight aggregate concrete.

Atmospheric Durability. Equal.

Corrosion Protection to Reinforcement. Approximately equal, although with high-quality lightweight concrete, experience seems to indicate better performance. The cover is generally specified the same as for normal concrete. (For lightweight aggregate concrete, the cover should always be at least $1\frac{1}{2}$ times the maximum size of lightweight coarse aggregate since carbonation take place along the surfaces of externally exposed coarse aggregate particles.)

Fire Resistance. Better by 20 to 50%. Sealed-surface aggregates should have low water absorption to prevent explosive spalling as absorbed water turns to steam. Alternatively use crushed lightweight aggregate so as to allow steam to escape.

Fatigue. Equal or better, as shown by tests of railroad girders, etc., provided adequate shear reinforcement is specified. When microsilica is added to the mix, fatigue endurance is up to twice that of normal concrete, both in air and under water.

Coefficient of Thermal Expansion. With all lightweight coarse and fine aggregates, the coefficient is 80% of normal concrete. With lightweight coarse aggregate and replacement of 30% of fines by natural sand, the coefficient is 90% of normal concrete. While thermal expansion is slightly lower than normal concrete, there have been no reported difficulties in service with reinforcement. Nor have problems been reported when lightweight aggregate concrete has been used in composite action with normal concrete.

Poisson's Ratio. From 0.17 to 0.21, with an average of 0.20.

Alkali-Aggregate Reaction. Apparently nonexistent.

Dynamic Behavior Under Impact Stresses. Stress wave velocity is about 20% less than normal concrete. Length of stress wave, same. Period of vibration, longer. Vibration damping greater; probably due to the interface condition between paste and coarse aggregate. Global shock and energy absorption, believed to be substantially greater than for normal concrete. Spalling under impact is a different mechanism than with normal concrete, exhibiting a delamination and ejection of flat plates.

Ultimate Strain. Research to date indicates greater ultimate strain under sustained loading.

4.2.3 Utilization of Prestressed Lightweight Concrete

Prestressed lightweight concrete has been utilized as a lightweight substitute for prestressed normal concrete in almost every type of application. The extent of such substitution has varied rather widely, depending on the immediate economic advantages obtained. Therefore, by far the largest use to date of prestressed lightweight concrete has been for the roofs, walls, and floors of buildings.

The lower weight has also been the determining factor in the selection of lightweight concrete for prestressed bridge girders, particularly the suspended span of cantilever-suspended span bridges and for bridges erected as cantilevered segmental girders.

The reduced weight leads to a saving in the structural frame itself and particularly to a saving in foundations. In good soils, settlements can be reduced, or the size of footings reduced. In poor soils the number of piles may be reduced. In in-between soils it may be possible to use spread footings or a mat in lieu of piles. Savings in foundations have been cited as a major economic advantage for use of lightweight concrete.

The reduced weight also reduces the seismic loadings for design; this reduction in lateral force requirement can be of importance, not only in the structural frame, but also in the connection details and the foundations.

The reduced weight of lightweight concrete means reduced draft for floating structures such as floating bridges and tension-leg platforms (TLP). Structures designed to be towed to the arctic have employed prestressed lightweight concrete in order to reduce draft.

Reduced weight also facilitates transportation of prefabricated elements. In a number of instances studied in California, the saving in transport cost just about balanced out the increased cost of the lightweight aggregate when the transport distance was 100 miles; beyond that, the use of prestressed lightweight concrete offered a savings.

Moreover, since the maximum hauling weight is often limited, use of lightweight concrete may make it possible to haul a larger unit. Similarly, in erection, where crane capacities are limited, use of lightweight concrete permits larger single elements to be erected.

A rather dramatic extension of this latter advantage was realized in the reconstruction of a railroad trestle across San Francisco Bay. Because of restricted access on the tidal flats and the proximity of the deteriorated timber trestle, which had to be kept in full use, the weight of individual units was limited to about 50 tons. Use of lightweight concrete enabled the employment of prestressed cylinder piles of that weight, capable of taking lateral loads in bending. This, in turn, enabled direct structural framing with precast pile caps, eliminating the need for cast-in-place pile caps. Thus the structural system was greatly simplified and a significant saving achieved in the time and cost of reconstruction.

The applications of prestressed lightweight concrete which offer the greatest ultimate or long-term benefit are those in which the unique properties of prestressed lightweight concrete are fully utilized, and it is selected not merely as a lightweight substitute but as a new material in its own right.

Foremost among these is the property of thermal insulation. Reduced heat transmission reduces the heat loss in cold climates and, conversely, reduces air-conditioning costs in hot climates. Condensation problems also are reduced.

Prestressed lightweight concrete has been shown to be an outstanding material for the containment of materials (cryogenics) at very low temperatures, down to $-170°C$ ($-340°F$). Prestressing steels, being cold-drawn, are not subject to transition to a brittle state. The concrete itself maintains integrity, even under the thermal shock when liquid nitrogen is poured onto a slab at normal air temperature.

The insulating qualities of structural lightweight concrete produced with aggregates of low water absorption and with low moisture content (no pre-saturation) are superior to those of normal concrete. However, with high moisture content, sealed-surface lightweight aggregate may "explode" during fire due to the formation of steam within the pores.

In this respect, crushed lightweight particles have behaved better in fire tests than sealed-surface particles, since the steam was able to escape from the open pores.

Improved thermal insulation, combined with a slightly lower coefficient of thermal expansion, reduces temperature stresses, especially where these are occasioned by short-term temperature cycles (e.g., daily). This is of direct importance where the major stresses are those due to temperature (e.g., pavements). This may also permit a reduction in the number of expansion joints, which are costly in both construction and maintenance. Future applications may well include furnaces, stacks, and nuclear reactor pressure vessels, where the major stresses are those due to temperature gradients.

The low modulus of elasticity is a property of lightweight concrete that, although presenting problems in some applications, can be utilized advantageously in others.

Lower modulus increases the prestress loss. It leads to increased camber and increased deflections under load. Both of these can, and should, be recognized by the designer and properly provided for in the design.

Where the design stresses are induced by known strains (e.g., temperature stresses or differential settlement), then the lower E reduces the stresses significantly.

Prestressed lightweight concrete is an ideal material for barges, ships, drydocks, etc. Minimum wall thicknesses are generally established by the requirements of rigidity, safety against buckling, space for reinforcement and prestressing tendons, need for cover for durability, and ability to place the concrete. Thus, with an essentially fixed minimum wall thickness, the use of prestressed lightweight concrete offers a significant reduction in dead weight of the vessel. The drafts are frequently reduced by 20% or, conversely, for a given maximum draft, the cargo capacity may be increased by 20%.

As offshore platforms are usually constructed in a construction basin or graving dock, the depth of this dock is determined by the draft at launching. Use of prestressed lightweight concrete for the base raft of offshore platforms permits float-out at a lesser draft.

This can be an especially important asset where completion of the construction afloat is also limited by the available water depths. Use of lightweight concrete, especially in the upper portions of the structure, can significantly enhance stability while afloat.

Of great importance for waterborne structures is the high fatigue endurance of lightweight concrete containing silica fume, which can be twice that of submerged concrete.

The excellent performance of prestressed lightweight concrete at low temperatures and under thermal shock loadings, combined with its inherent suitability to floating structures, makes it of great potential use for floating LNG (liquefied natural gas) production and storage.

Another property of potential interest in special applications is that of low submerged or buoyant weight. Obviously this is directly attributable to low unit weight, but its field of application is quite different. Whereas, in air, prestressed lightweight concrete weighs 75 to 80% as much as prestressed normal-weight concrete, when submerged, the ratio is only 50%.

This lower submerged weight has an application in extremely large and heavy piles and caissons, such as those for major bridge foundations. When these are

TABLE 4.1 Use of Prestressed Lightweight Concrete

Type of Application	Present Volume of Use	Form or Section Employed	Properties and Advantages Leading to Selection	Adverse Properties
Flooring	Substantial	Precast flat slabs. Precast cored slabs. Precast tee slabs, single or double tee. Precast joists with cast-in-place slab. Lift slabs. Cast-in-place composite deck slab on prestressed lightweight joists. Cast-in-place post-tensioned slab or slab and beam floor.	Lower deadweight. Reduced seismic loads. Reduced structure and foundation loads. Easier handling and erection. Better fire resistance. Lower transport costs.	Slightly greater camber and deflection. Reduced shear strength at bearings.
Roofing	Very large and widespread	Precast tee slabs, single and double tees. Precast and cast-in-place shells and folded plates. Cellular composite slabs. Precast slabs for cable-suspended roofs. Canopy roofs for grand-stands and aircraft hangars. Roof trusses. Precast flat slabs, solid or cored.	Lower deadweight. Reduced seismic loads. Reduced structure and foundation loads. Better fire resistance. Better heat insulation. Reduced condensation. Lower thermal movements. Lower transport costs and erection costs.	Slightly greater camber and deflection. Reduced shear strength at bearings.
Walls for buildings	Substantial	Solid panels. Tee panels. Sandwich panels. Cellular-composite panels with special architectural finish.	Prestressing of wall panels prevents cracking, makes thinner sections practicable, and facilitates handling and erection. Lower deadweight. Reduced seismic loads. Reduced structure and foundation loads. Better freeze-thaw durability. Lower thermal movements. Easier handling and erection. Better fire resistance. Less condensation.	

Bridge girders	Some, including a few large and important structures.	Spans erected by segmental cantilever method. Suspended spans. I-girders with composite cast-in-place deck. Hollow-core slabs.	Reduced deadweight. Reduced seismic loads. Reduced structure loads. Reduced foundation loads. Easier handling and erection. For segmental girders, low modulus of elasticity permits better distribution of bearing pressures on adjoining faces when using "dry" joints. Better fire resistance (some installations in fire hazard areas).	Low modulus of elasticity. Greater camber. Reduced shear and diagonal tension strength.
Bridge decks	Substantial	Precast slabs. Cast-in-place composite deck slab on prestressed girders.	Lower deadweight. Better resistance to freeze-thaw and deicing salts.	
Foot bridges	Considerable	I-girders. Tee-girders. Box beams.	Low deadweight, of particular importance where deadweight is a major portion of total design load and where design load is seldom realized in service. Easier erection.	Liability to local "plucking."
Piling	Some	Bearing piles. Fender piles. Batter piles.	Reduced submerged weight means greater capacity for design loads; also less bending stress in batter piles. Reduced weight for transport and handling, especially important with very large piles. Reduced modulus allows greater deflection, hence greater energy absorption.	More subject to spalling from abrasion and local impact.

(continued)

TABLE 4.1 (Continued)

Type of Application	Present Volume of Use	Form or Section Employed	Properties and Advantages Leading to Selection	Adverse Properties
Industrial building elements	Some	I-girders. Rectangular beams. Box sections.	Lower deadweight. Reduced structure loads. Energy absorption. Fire resistance.	Greater camber. Lower diagonal tension. Greater deflection under live load.
Seismic-resistant and shock-absorbent structures	A few	Girders. Beams. Framed structures.	Energy absorption. Lower modulus of elasticity. Lower deadweight means lower seismic loads.	
Guard rails and fenders	A few	Rectangular beams.	Energy absorption. Greater deflection under impact. Fire resistance (as compared with timber).	Spalling under local impact and abrasion on edges.
Poles	Experimental and some production	Tapered. Constant section. Hollow core.	Reduced weight for transport and erection. Reduced seismic loads.	Greater deflection under load.
Railroad sleepers (ties)	Experimental		Reduced weight for handling and installation. Greater deflection reduces "center binding."	Inserts require somewhat larger embedment or anchoring. Lighter weight reduces anchoring effect against thermal expansion of rail.

Floating structures	Some	Precast elements, cast-in-place.	Reduced draft. Reduced weight where wall thickness is determined by cover over steel and by placement, water-tightness, and stiffness requirements. Greater durability anticipated. Reduced permeability. Submerged weight of lightweight aggregate concrete is only one-half that of normal concrete. Much greater fatigue endurance when microsilica is added.	Reduced shear strength. Above-water tensile strength is reduced.
Offshore platforms			Reduced draft. Less draft during launching. Increased fatigue endurance when submerged. Increased durability.	Spalling under local impact. Decreased shear resistance.

driven through weak soils, such as mud, to bear in lower soils, such as sand, the dead load of the pile becomes a significant factor. Assuming a "fluid" density of the soil equal to a specific gravity of 1.8, the use of lightweight concrete results in essentially no added load, whereas a normal-weight pile has an effective or net specific gravity of 0.7 to 1.0. In some soil conditions, this may permit a significant reduction in pile length, or alternatively, in the number of piles.

The improved durability and resistance to corrosion of the reinforcement has led to the adoption of prestressed lightweight concrete for a number of major bridges, where, of course, advantage was also taken of its lower deadweight. An example is the San Diego–Coronado Bridge approach structures and the San Mateo–Hayward viaduct across San Francisco Bay, both projects being in California in areas where chloride corrosion is known to have attacked earlier structures.

Some of the most useful applications of lightweight concrete are in composite action. All possible combinations have been employed successfully. These combinations are the following:

a. Prestressed lightweight concrete joists or girders with a cast-in-place normal-weight composite slab (Fig. 4.1).

Figure 4.1. Prestressed lightweight concrete roof girder combines both pretensioning for erection with a second stage of post-tensioning after the roof slab of normal weight concrete has been concreted.

b. Prestressed lightweight concrete joists or girders with a cast-in-place lightweight concrete slab.

c. Prestressed normal-weight concrete joists or girders with a cast-in-place lightweight concrete slab.

All of these combinations, especially *c,* require that the designer carefully consider the differential moduli of elasticity and shrinkage.

Composite construction has also been employed rather extensively in the former USSR where cellular (aerated) concrete is joined with ribs or slabs of prestressed normal concrete. Cellular concrete, whose structure is produced by gas or foam, contains no coarse aggregates. Its unit weight ranges from 600 to 1200 kg/m³ (35 to 75 lb/foot³). It is not suitable for direct prestressing; hence, the need to join it in composite action with ribs or slabs of conventional concrete or structural lightweight aggregate concrete. These ribs are highly stressed and provide the tensile element. The cellular concrete is poured around or over the ribs, then the entire unit is autoclaved. Elements of this type are utilized for roof and wall panels where the thermal insulating values are of greatest importance. The use of prestressed ribs or slabs gives greater strength and durability than is possible with mild-steel reinforcement.

4.2.4 Manufacture and Construction

The techniques for manufacture and construction of prestressed lightweight concrete are essentially the same as for precast normal-weight concrete. The following specific recommendations are made:

a. The substitution of natural sand to replace 30 to 100% of the lightweight fines will generally give more consistent results with only a small increase in unit weight. The mix design must be modified accordingly so that the volume of coarse aggregates in a cubic yard remains approximately the same. With sand replacement, the drying shrinkage is usually reduced and the tensile splitting strength in the dry condition is improved.

b. Use of air-entrainment is always recommended for lightweight aggregate concrete to improve workability, reduce segregation, and increase freeze-thaw durability. Air-entrainment values of 4 to 6% are generally employed. For some applications where the lowest possible unit weight is required and medium strength of 25 to 30 MPa (3500 to 4500 psi) is acceptable, air-entrainment up to 8% may be introduced.

c. Use of water-reducing admixtures is essential to assure adequate workability at low W/C ratios and in achieving higher strengths at all ages.

d. Careful selection and handling of lightweight aggregates to prevent crushing and contamination and to ensure uniform moisture content and uniform unit weight.

e. Presaturation of lightweight aggregates is widely practiced in order to facilitate placement by pumping. Otherwise, the aggregate will absorb water from the mix. Saturation renders lightweight concrete very susceptible to freeze-thaw degradation. Further, in a fire, the generation of steam within the saturated pores will cause explosive spalling. Use of pumping-aid admixtures will reduce the degree of presaturation required.

Some sealed-surface aggregates have a very low moisture absorption, only 6 to 10%.

Pre-saturation increases the unit weight of the aggregate by 10 to 15% and the unit weight of fresh concrete by approximately half that amount.

If, because of durability and fire considerations, no pre-saturation is permitted, then placement must be by conventional bucket or other mechanical conveyence.

Japanese manufacturers have produced a paraffin-type coating which significantly reduces the 24-hour absorption of the aggregates without apparently adversely affecting aggregate-paste bond and overall performance. It permits pumping of the concrete, without further water absorption.

f. Design of the mix should provide for continuous control for variations in unit weight and moisture content of aggregates.

g. Avoid excessive overvibration as this will cause the coarse aggregate particles to float to the surface.

h. Curing by steam at atmospheric pressure (or at high pressures) generally improves properties by reducing drying shrinkage and creep.

The optimum cycle for low-pressure steam curing is the following:

Delay (holding period)—three to five hours.

Rise in steam temperature—22°C (40°F) per hour.

Maximum sustained temperature—60 to 70°C (140 to 160°F).

Cure at maximum temperature—8 to 14 hours.

Shorter delay periods and higher maximum temperatures may be employed to achieve a shorter production cycle, but generally reduce the ultimate strength, durability, and other properties.

Lightweight concrete gains heat more slowly and holds it longer, due to its insulating properties. Presaturated lightweight aggregate makes water available for internal curing.

Steam curing at higher pressures (autoclave) is very beneficial in eliminating volume change but has so far not proven practicable for prestressed applications other than very small roof slabs. Electrothermal curing is practiced extensively in the former USSR. The cycle is a two- to four-hour rise in temperature, followed by 12 hours of cure at 70 to 95°C. The higher temperatures are employed when the tendons are stretched directly against the forms.

i. Supplemental moist curing after steam curing will reduce drying shrinkage and thermal strains.

j. Due to the longer transmission length on release of pretensioning, slow release by hydraulic jacks is preferable to sudden release, as by burning.

k. With post-tensioning, particular care must be taken to achieve dense concrete in the anchorage zone. Local reinforcement should be provided to prevent spalling and tensile cracking in the end block. In some special cases (e.g., long-span bridge girders) the end blocks have been precast in high-strength normal-weight concrete, then set in the forms and the lightweight concrete poured as the main body of the girder. Similarly, in large diameter precast concrete pipe, the body may be formed of lightweight concrete and the bell and spigot of normal concrete, both placed as fresh concrete and blended together during placement. The heads of lightweight concrete piling have been similarly blended into normal-weight concrete to prevent spalling under pile driving.

l. Lightweight concrete is somewhat more apt than normal concrete to form air and water bubble pockets on vertical and return (overhang) faces. External vibration will minimize these, as will the addition of microsilica.

m. Surface or deck finishing should follow closely behind the placement, and care must be taken so as not to "drag" the coarse aggregate out of the mix.

n. Avoid excessive soaking of surface of concrete element during storage as this may cause drying shrinkage problems after erection.

o. Protect edges and corners from impact as these are more subject to spalling.

p. Protect thin elements (such as roof slabs) from differential exposure to the sun in prolonged storage as this may cause differential camber.

4.2.5 Architectural Treatments

Lightweight concrete wall panels for buildings are frequently given architectural finishes. Prestressing eliminates cracks, at least those normal to the prestress, and often permits thinner panels, with greater dimensional stability in the out-of-plane direction, that is, flatness.

Long-term creep must be considered in respect to its effect on joints and connections.

Detailed discussion of finishes is given in Section 8.2.

4.2.6 Actual Performance of Prestressed Lightweight Concrete

Actual fires have demonstrated that the better thermal insulation does provide greater fire protection for the same cover. See, however, the prior comments on the explosive spalling of saturated aggregate.

Experience from the northern United States, the former USSR, and Canada indicates better freeze-thaw resistance than normal concrete. Researchers in the former USSR state this is due to the porous nature of the aggregates. Experience in New York has shown that, where deicing salts are applied to bridge decks, the lightweight concrete has shown markedly increased durability.

Experience with structural lightweight aggregate concrete (conventionally reinforced) in ships built in both World Wars I and II shows generally excellent marine durability despite the small cover. One especially well-documented case is the ship Selma, which has been beached since World War I at Galveston, in semitropical seawater. Despite only 1 cm ($\frac{3}{8}$ inch) cover of expanded shale aggregate, substantial portions of the ship remain in excellent condition.

Prestressed lightweight concrete piles up to 40 meters (132 feet) in length have been installed in a number of marine structures since as early as 1955. There have been no reports of durability problems with any of these, over 35 years of service.

During installation, these piles behaved generally similarly to piles of normal weight aggregate, except for greater tendency for spalling of thin slabs outside the spiral at the head of the pile. Proper head cushions and driving heads have been able to prevent this.

Recent inspections of three major bridges in California in which prestressed lightweight concrete was used—the San Diego–Coronado Bridge, the San Mateo–Hayward Bridge Viaduct, and the reconstructed portions of the San Francisco–Oakland Bay Bridge—have shown no deterioration nor corrosion in this marine environment. It is generally believed that this is due to the reduced permeability to chloride ions, due to better aggregate-paste interface bond.

Prestressed lightweight concrete has given satisfactory performance in a number of long-span bridges which have been constructed in the USA, Canada, and Europe. In one such bridge, the Parrott's Ferry Bridge in California, excessive creep deformations led to a noticeable sag in the center of the span. The problem was identified as due to use of erroneous long-term shrinkage and creep coefficients. These had been determined by laboratory tests in high humidity while the actual service was in very low humidity. Even the short-term deflections were greater than anticipated and should have been corrected.

This problem could have been readily overcome had means for adjusting prestress during and after construction been incorporated in the design, e.g., providing for additional internal tendons during construction, and for additional external tendons for the long term.

Obviously, shrinkage and creep tests should take the actual service environment into consideration.

Prestressed semi-lightweight concrete was employed for the Tarsiut Caissons in the Canadian Beaufort Sea of the Arctic Ocean. Although these have proven durable in this extreme environment for some eight years, two of the caissons have shown thermal cracking patterns on their faces. These are believed to have been initiated during construction. There have also been several small areas where spalling due to impact from heavy steel barges has occurred.

The CIDS (Global Marine Beaufort Sea One), an offshore concrete platform, which has been working for five years in the Alaskan Arctic, was constructed of prestressed lightweight concrete in Japan. Great care was taken to limit the water content of the lightweight aggregate particles to less than 6% in order to prevent freeze-thaw damage. To prevent thermal cracking upon cooling from the heat of hydration, external insulation was applied during the curing period. The contractor experienced some difficulties in placing the relatively dry, harsh mix in heavily reinforced walls, but did succeed by use of intensive internal vibration.

The structure was towed from Japan to the Beaufort Sea, where it has given five years of highly successful service in sea ice with no durability problems.

Volume-change problems have occurred in a number of buildings, principally with prestressed lightweight concrete floor and roof slabs. These are similar to the problems that occur with prestressed normal-weight concrete, but are aggravated because of lower modulus, lower dead load to live load ratio, greater shrinkage, and increased thermal insulation.

Because of lower modulus and lower dead load to live load ratio, design camber is often greater. (This can, of course, be compensated for in design.)

After installation, several factors may cause the camber to grow. The sun's heat on the top produces a thermal gradient, with the top fibers expanding more than the lower ones. The camber grows. At night, friction and the rotation of the ends prevent shortening so the effect is cumulative, not reversible. This effect is aggravated when the roof slab is covered with dark roofing, such as asphalt, which absorbs the sun's heat. Eventually, the supporting beams may be cracked.

Similarly in northern climates, the top flange may be exposed to cold drying winds while the undersurface and legs of the tees are kept warm and moist. This may eventually offset the camber and produce sag.

Use of neoprene bearing pads to allow rotation and longitudinal deformation, plus proper shear reinforcement of the end zone of the legs of the slabs, or joists, and of the beam, as well as proper curing time (to eliminate creep and shrinkage before erection) will eliminate most problems.

In some cases, it may be possible to spray-paint the roofing with a reflecting paint; this greatly reduces the thermal effect. Insulation may be provided. Most importantly, consideration must be given in the design to these problems and appropriate steps taken. These may include deeper sections, partial prestress, and, of course, proper detailing of bearings and adjacent zones.

Lightweight concrete has proven entirely satisfactory in resisting abrasion and wear. A most notable example is the roadway deck of the San Francisco–Oakland Bay Bridge. Decks of ships built of lightweight concrete have similarly withstood wear, vibration, and fatigue. However, lightweight concrete is sometimes subject to local "plucking" of coarse aggregate. Edges of curbs, rails, etc., are subject to greater spalling.

Tests in the Baltic for ice abrasion have shown performance equal to that of normal-weight concrete.

When concreting in winter, with heated aggregates, lightweight concrete has proven to retain its heat longer and more effectively. Structural lightweight aggregate concrete will be used for the Second Benecia-Martinez Bridge in California, in order to minimize inertial effects under earthquake.

Prestressed lightweight concrete appears destined to play an ever-increasing role in construction. Constant improvement in quality of aggregates, combined with the performance and behavioral advantages of prestressing, present opportunities for increasingly sophisticated design and new applications. The present differential in raw aggregate cost will probably narrow and disappear as close-in-gravel deposits and quarries are depleted. Greater use will lead to a more widespread recognition of lightweight concrete's many concomitant advantages.

The construction engineer has a major part to play in this development. By applying proper techniques and procedures, he can ensure consistent high quality. By his selection of prestressed lightweight concrete for those applications under his control, the direct and indirect economical and structural benefits will be made more obvious. Prestressed lightweight concrete should be treated as a new material in its own right, possessing many advantages when wisely and properly employed.

4.3 PRECASTING

4.3.1 General

Precasting is one of the most important techniques by which prestressed concrete has been made practicable and economical. Precasting is so intimately bound up with pretensioning that often, within segments of the industry, the two terms are used synonomously. From a factual and technical point of view this is, of course, not true; precast concrete can range from unreinforced concrete to conventionally reinforced (mild steel) concrete to prestressed concrete; prestressing can be applied to cast-in-place concrete as well as precast concrete, and can be either pretensioned or post-tensioned. Precasting, however, is the technique by which mass-produced elements of prestressed concrete are constructed. It offers all the advantages of prefabrication: quality, mechanized production, speed, and economy.

Precast prestressed concrete is applied to virtually all types of structural elements for every structural application. Among the widest applications of precast prestressed concrete are the following:

Buildings: Floor slabs, roof slabs, wall slabs, piles, columns (Fig. 4.2).

Bridges: Piles, girders, deck slabs, superstructure segments, pier segments.

Marine structures: Piles, deck slabs, sheet piles, fender piles, also shells and slabs to serve as soffit forms for cast-in-place concrete.

Railroad ties (sleepers).

Poles.

Tanks: Staves, roof slabs, columns.

Pipes.

Precast members may be cast either at a permanent plant or at a temporary jobsite plant. They are cast and cured at this plant, then transported, erected, and joined into the completed structure. Joining may be accomplished by welding, bolting, or post-tensioning, or by cast-in-place concrete connections. Many efficient designs utilize precast and cast-in-place concrete in monolithic action.

Figure 4.2. Precast concrete columns.

Precast members may be pretensioned or post-tensioned. Post-tensioning can be utilized to make individual precast elements work together monolithically, or to increase the prestress values after erection (stage-stressing).

There are two main uses of precasting. The first and most extensive use is in the mass-production of standardized elements.

The second is the casting of unique and complex units in the position and under conditions whereby the necessary quality and tolerance control and economy may be best achieved.

Standardization is an essential consideration for mass-production. Many building elements, bridge girders, piles, poles, railroad ties, pipes, etc., have been standardized by national associations, such as the Prestressed Concrete Institute, American Association of State Highway Officials, American Railway Engineering Association, and American Concrete Pipe Association.

4.3.2 Specific Advantages

Some of the specific advantages of precasting are the following:

1. Control of Shrinkage. Shrinkage may be reduced through lower W/C ratio and through steam curing. By proper aging and drying before erection, the majority of shrinkage may be made to take place before erection.

2. Reduction of Creep. By proper curing and aging before erection, the strength and maturity of the element and its modulus of elasticity are all at higher levels at the time of loading, thus reducing the creep under prestress. Higher strength concrete of lower W/C ratio may be utilized in manufacturing precast elements, thus producing concrete of reduced creep characteristics. By aging for two to three months, 50 to 60% of the creep may be dissipated before erection.

3. Control of Dead-Load Deflection. Precast concrete segments may be erected and adjusted to exact position on the theoretical profile, including an allowance for elastic and long-term deflections, before final jointing and fixing. This is of particular importance on long-span bridges.

4. Quality Control. Precast concrete elements may be manufactured under the best conditions for forming, placement of reinforcement, placement and vibration of concrete, and curing. Uniformity and high quality may thus be obtained practically and economically.

Precasting makes it possible to assure greater durability through accuracy and uniformity of cover. Because it permits improved placement and vibration, while maintaining a low W/C ratio, a more permeable cover may be achieved. Precasting also facilitates a smooth surface finish and proper curing.

Precasting also permits the application of special manufacturing techniques, such as combined vibration and pressure, vibro-stamping, extrusion, spinning, and steam curing, which, in general, cannot be applied to cast-in-place concrete. Similarly, precasting permits the use of pretensioned reinforcement.

Dimensional tolerance control is facilitated by precasting. Rigid forms, often of steel, prevent excessive deflection under vibration. Precast members may be checked for tolerance before shipment and erection. Prestressing steel, mild steel, and inserts may be accurately positioned and held during concreting.

5. Timely Availability. Many standardized, mass-produced elements can be furnished to a construction site on very short order. While some standardized elements are carried in inventory, most plants are able to produce to order at the desired rate and even to meet accelerated schedules; therefore, producing for inventory with its costs of rehandling, storage, and interest, is not a major practice in the industry.

Precasting makes it practicable to utilize steam curing and thus to produce on a 1- or 2-day cycle, with full design strength generally achieved within 7 to 14 days. (However, for some very long, thin members, such as long-span roof slabs, storage at the yard before shipment for a period of 30 to 60 days before erection may be desirable to minimize both shrinkage and creep.)

6. Erection Over Existing Traffic and Minimization of Falsework. Precast concrete girders may be erected without falsework over traffic, railroads, or waterways. (See Fig. 4.3.) Longer spans may be erected by the progressive cantilever scheme or cantilever-suspended span scheme.

7. Reduction in Site Construction Time. Actual erection time and site construction time may be significantly reduced through the use of prefabrication of members or segments. In favorable circumstances, the floor of a building may be erected and jointed in two days. The 50-meter-long approach spans of the Columbia River Bridge were completed at the rate of one complete span per week. Currently, at the Great Belt Western Bridge in Denmark, one 110-meter length of bridge is being completed every two to three weeks, including the piers, highway spans, and railroad spans.

Over water and other site conditions present difficult and costly access and work conditions which can be obviated by prefabrication.

Where falsework is required to temporarily shore or support precast segments, the extent required may be reduced (although total loads may remain the same as for

Figure 4.3. *Nighttime erection of precast concrete girder over existing roadway.*

cast-in-place construction). Since the time it must remain in place under any one area is reduced, maximum reuse of falsework is facilitated.

8. Economy. Precast concrete schemes offer substantial economies in construction. Mass production of standardized elements reduces the cost of forms and manufacturing labor. Site labor can be kept to a minimum. Erection may be performed during the most favorable seasons and conditions. By using higher strengths and thinner sections, etc., deadweight may be reduced, resulting in overall savings of concrete and steel.

Labor and equipment may be utilized with maximum efficiency. Forms may be reused a maximum number of times, and the advantage taken of mechanization in manufacture.

9. Suitability for Composite Construction. Precast segments may frequently be combined with cast-in-place concrete to act in composite action, as a monolithic structure. Provisions must be made for shear transfer. The precast units may serve as partial forms, and also as a support for forms for the cast-in-place concrete (Fig. 4.4).

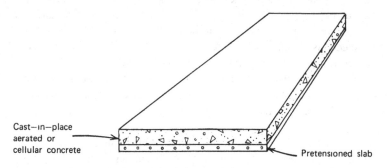

Cast—in—place aerated or cellular concrete

Pretensioned slab

Figure 4.4. *Composite construction for industrial buildings.*

Consideration must be given to the various stages of loading to prevent over-stressing of the precast elements while the cast-in-place concrete is still wet. The designer must consider the effect of differential shrinkage and of different moduli of elasticity.

4.3.3 Dynamic Loading and Long-Term Considerations

Consideration should be given to dynamic loading in all joints of precast concrete elements, especially to welded reinforcing steel connections and to welded structural steel connections. The connection bars should be adequately embedded within the concrete to develop the full connection strength under dynamic loading wherever seismic forces or other dynamic forces may occur. Usually this will require an embedment length almost double that required for static loading. Bars embedded in very thin flanges may tend to break out at right angles to their direct load stress due to eccentricities; these bars should be extended to attain sufficient embedment to hold even if local flange failure takes place under dynamic loading.

The long-term creep deformations of prestressed precast members may impose severe stresses at joints, especially where the connecting member or frame is rigid. Provisions must be made to accommodate such deformations.

4.3.4 Manufacture

The manufacture of precast concrete girders, elements, and segments should follow applicable provisions of ACI Standards and Recommended Practices for Concrete Construction. Chapter 7 discusses the organization and layout for manufacture of precast pretensioned concrete, and lists special considerations, many of which are applicable to all precast concrete construction regardless of whether pretensioned tendons or conventional reinforcing are used.

The manufacture of precast elements basically employs: (1) the assembly and installation of reinforcing and prestressing steel, (2) the production, under control, of concrete and its subsequent placement, and (3) the lifting, storage, and load out of the completed element.

Addressing these in more detail, the concrete is a heterogeneous mixture of a number of components, aggregates with coarse and fine cementitious materials, water, and admixtures. As the industry moves to higher strengths and durability requirements, ever more emphasis has to be put on the control of the basic materials for such aspects as cleanliness, temperature and moisture, surface characteristics, and sequence of addition and mixing. All batching should be done separately, by weight, since accurate and consistent proportioning are essential for control. The thoroughness of mixing becomes of critical importance. Turbine mixers emerge as the most effective, even when discharging to a ready-mix truck for delivery (Fig. 4.5).

Delivery of the concrete must be selected that will ensure that the mix does not segregate nor gain excessive heat, due to external sources or internal hydration, and does not lose workability (slump). In very hot climates, it may be useful to paint the trucks with reflective paint.

Figure 4.5. *Placing low-slump concrete in precasting operation.*

In very cold climates, insulation may be required during transport to prevent freezing.

Forming for precast elements follows standard practice, with these special considerations:

- Forms which are used many times are subject to local and global distortion, from vibration, the release of prestress, and steam curing.
- Local deformations restrict movement as the prestress is released into the member and may lead to local buckling.
- As forms are used and reused, thin wafers of cement paste get into joints in the forms and prevent their tight closure; hence the problem gets worse each use.
- Under steam curing, the steel forms expand more rapidly than the concrete. Any change in configuration along the member, e.g., at an end block, may lead to the local cracking of the concrete member, which has little tensile strength when steam is first applied. Sponge rubber gaskets may eliminate this problem.
- Similarly, embedment plates in precast members may expand under the steam curing more rapidly than the concrete into which they have been installed. A "reveal" of wood, or a sponge rubber gasket usually solves this source of cracking.
- When members are steam cured, they must not be suddenly exposed to cold dry air. In several cases, on the morning after casting and steam curing, the steam hoods have been thrown off in order to obtain the cylinders stored

alongside, so that they can be tested. Failure to re-cover before the member has cooled down and prior to the release of prestress has led to cracking.

- A member is not "prestressed" until it has actually shortened. In removing an element from the forms, it should be done in such a way as to permit the member to shorten before maximum moments are developed. This may require lifting one end slightly before the other or, in some cases, jacking out of the forms. A typical adverse case is that of a pile cast in old forms which have seen much abuse due to vibration, etc. The pile has not been able to shorten due to the restriction by the forms, even after release of prestress. The picking point has the dead load of the pile plus the suction in the forms. Negative moment may well exceed the member's capacity, especially before full prestress has been transferred. One practical solution is to place short pieces of strand or reinforcing bars at the top of the member at the picking points.

- In storage, precast members are initially very "young" and hence creep may take place on an aggravated basis. Adequate supports are needed along the member. Large hollow members may require internal struts to prevent ovalling.

Many of the earlier-cited problems and concerns have to do with forms. Form design for precasting has become an art and manufacture of steel forms for precasting is highly specialized. In general, there is a trend to the use of heavier steel plate, formed to close tolerances, with all welds and surface blemishes ground smooth, so as not to restrict movement.

Whenever possible, forms are designed so as to have no joints. For example, forms for square piles are designed with a slight draft so as to permit removal by direct lift. The flexibility of steel plays an important role in permitting lifting out of the product with minimum restraint.

End closures by necessity are separate pieces for most products. Usually, the preservation of an exact plane normal to the axis of the member is important. It will often be found that making up a double end gate, with rigid struts between the plates, of a length of 0.5 to 1.0 meter (2 to 3 feet) to set in the forms, will facilitate maintenance of a true normal plane.

In pretensioning, and in many conventionally reinforced members, the strands or bars must protrude through the end gates. This requires segmental end gates and the use of sponge rubber gaskets so as to prevent leakage of paste during vibration.

Heavy steel forms, with machined surfaces, can produce concrete members with true surfaces and dimensions, accurate to about 0.5 mm ($\frac{1}{64}$ inch), generally better than can be attained by grinding later.

Many steel forms have incorporated within them the means for hot water or hot oil curing, attachments for external vibrators, rails for screeding devices, the means for deflecting strands, or the means for fixing inserts. Although casting of concrete is heavy construction work, since precasting will usually involve many repetitious uses of the same forms, care should be exercised not to damage them in use.

Concrete forms have been used for shallow depth members but are generally of only marginal success because of their great rigidity. Especially in corners, it may be necessary to install either flexible metal devices or a removable-and-reinstallable drop-out block.

Match-casting, in which one element is precast against another, is discussed in Section 4.3.5.

Reinforcing steel for precast elements is very often prefabricated and pre-assembled on a template or stand, where the location of every piece is marked and perhaps guided by templates. Tack welding is often used to hold the members in place, but needs extreme care to prevent undercutting, with subsequent reduction of dynamic and fatigue strength.

In highly mechanized plants producing thousands of identical items, flash welding is employed, which limits the duration of contact and hence the heat.

Because of the problem of control in most cases where manual tack welding would be employed, welding is often prohibited. This is especially true in prestressed precast manufacture, since an accidental discharge into the prestressing strand will drastically lower its strength.

However welded wire or bar mesh prefabricated in the steel fabricator's plant, is increasingly used as web reinforcement, because of the ease of placement and support.

The alternatives to welding are tying and clamping. These are satisfactory means but only if they are done well, i.e., very tightly, so as to prevent slippage in later handling of the prefabricated cage.

The end blocks of prestressed girders are typically so heavily congested with reinforcement as to make it very difficult to fit the bars in place and even more difficult to place and consolidate the concrete (Fig. 4.6).

Internal vibration is normally essential, supplemented in many cases by external vibration. However, even relatively intensive external vibration is inadequate by itself to prevent internal rock pockets and segregation.

Moreover, it is difficult to insert the internal vibrator into the fresh concrete when the reinforcing steel is placed so densely in all three planes. A number of procedures have been developed to cope with this problem:

1. Use a "pencil" vibrator, of small diameter.
2. Mark rebar locations on the forms, so their position is apparent even when they are submerged in fresh concrete.
3. Shift bars and bundle them, as permitted by the design engineer.
4. Just prior to concreting, insert a tube or tubes through the steel. As concreting starts in a particular zone, insert the vibrator through the tube, then withdraw the tube over the vibrator as the concrete rises.
5. Use a concrete mix with small sized coarse aggregate in the congested zone, increasing the cement content or lowering the W/C ratio by added dosage of water-reducing agent.

When pretensioned strands are stretched and stressed and when reinforcing steel and inserts are placed, all are usually in the correct position. However, the place-

Figure 4.6. Heavily reinforced end block of large prestressed concrete girder makes concrete placement difficult.

ment of concrete and the vibration often produce systematic as well as erratic dislocations. As a heavy liquid, the concrete will produce upward forces on any hollow tubes or hole formers. Where reinforcement is very closely spaced in the horizontal plane, the concrete will similarly try to raise the cage. Similar dislocations sideways can occur due to heavy vibration.

This dislocation is not readily apparent, since the bars are covered by concrete. Their movements can only be detected by tell-tale tabs attached, by subsequent probing, or by use of a cover meter.

Most dislocations are upward.

The most extreme dislocations usually occur where the cage of bars or strand are tied to a hole former and the entire assemblage is forced upward.

Horizontal slip-forming, using mandrels, also can cause dislocations. Therefore thorough consideration must be given to holding down all the steel, including inserts, in proper position during the concreting operation.

Mechanical screeding is frequently employed in precasting. Relatively simple devices are capable of screeding any upper surface from concave to flat to convex as long as the maximum slope of the fresh concrete is less than 30°. Since on many products a slope of 45° is required, such screeding requires special design of the mix and the screeding device, in order to prevent slumping.

The use of internal sliding mandrels to form hollow cores and voids, while widely employed, is another operation requiring careful integration of concrete mix, mandrel design and operation, and rate of slipping.

The larger diameter voids, such as used with hollow-core piles, are most susceptible to slumping. The slumping may take the form of actual fall out below the strand-spiral cage or of delamination on an internal plane. This latter is of course most difficult to detect.

When starting the mandrel, it is desirable to lubricate its surface with water or a cement wash or grout, so as to prevent excessive drag on the concrete.

Vibrators in the mandrels are generally inadvisable: their energy is transmitted too far toward the tails where firm support is essential.

Heaters in the mandrels have been used, but are probably not very effective in producing a pseudo-set because of the short duration they can operate at any one location.

Concreting machines for hollow-core slabs are now highly sophisticated and integrated machinery. Similar concreting machines, which carry out all the functions of placement, forming, vibration, and screeding, have been developed for other products.

Many precasting plants are out of doors, without protection from the weather. Under many conditions—hot weather, wind, rain, cold weather—it is desirable to cover the fresh concrete immediately after screeding and finishing, so as to prevent excessive evaporation, shrinkage, or other damage.

Finishing of the screeded surface may often be unnecessary, but where there are upward sloping surfaces, e.g., round or octagonal piling or bulb girders, bleed water and entrapped air may collect under the forms, requiring finishing, either before or after curing. Water should never be added to the surface of fresh concrete to facilitate finishing, as it greatly increases the W/C ratio.

Fog sprays have been successfully used in extremely hot and drying environments to reduce surface crazing due to rapid evaporation: however, covering with a polyethylene sheet is probably sounder practice.

The subsequent curing regime may include one of several methods, depending on the product and environment.

1. Covering exposed surface with plastic, leaving forms in place.
2. Water soaking, using garden soaker hoses on burlap.
3. Membrane curing compound. Note that in hot climates, or with elements that have just emerged from steam curing, the membrane curing compound may evaporate. A second coat after four to eight hours may be needed. White reflective compound is most effective in reducing surface heat.
4. Curing by heat, e.g., hot oil or water piped under the bed, or enclosing so that heat of hydration is contained. It is important that ample moisture is available to maintain high humidity despite the higher temperatures.

5. Low-pressure (atmospheric pressure) steam curing. See Section 1.1.10.

6. High-pressure steam curing (autoclaving) is usually limited to mass production of relatively small products, although it is applied in Japan to 12-meter long segments of piles and in the former USSR to 6 × 2 meter slabs.

Forms are usually stripped, or the product removed from the molds, before full curing has taken place. At this time it is important to protect the product from sudden cooling and drying. Cold winter winds with low humidity are especially bad. Night radiation in the desert can drop the temperature by more than 20°C (36°F). Therefore supplemental protection is often required, in the form of blanketing. Similarly supplemental curing by water spray or membrane curing may be employed.

Removal of the precast element from the forms requires proper lifting points and adequate flexibility in the form. Depending on the form design, the forms may be loosened and spread, or allowed to flex. Jacks which are allowed to self-center may be able to break the element loose, after which it may be raised with relative ease.

To lift precast elements from the forms, inserts are usually embedded in the concrete. They must transmit not only the dead load, plus the dynamic force, but also that needed to overcome suction. In addition, when segmental forms are used, fins of grout may have caught in the joints. The lift on the insert may not be truly axial.

Hence inserts must be sized and embedded to develop the design force plus an adequate factor of safety. Since an insert is part of the rigging, it would be reasonable to use a factor of safety of 5 on the dead load. Of this factor, the multiple of 2.5 is for impact, suction, and nonaxial pull, leaving a factor of safety of 2.

To develop this amplified force requires carefully designed inserts well embedded in the concrete, anchored behind the reinforcement in the compression zone. Avoid anchorages in thin flanges.

Because of the high peaks of negative bending moments at picking points, a few short pieces of bar or strand are often placed at the top of the member, as cast, to resist tension.

The angle which the sling or line makes with the lifting loop, at all positions during picking and handling, should be considered and provision made therefor. Either the angle may be fixed by using a picking beam, or the loop may be designed to be effective at all angles. Obviously, the increased force due to angular lead must be used in design. The member should be checked for the axial component of load to insure against buckling. (See Fig. 4.7.)

Consideration should also be given to sway or swing, which results in bending of the picking loop sideways. This will cause sharp bending stresses in the picking loops and may cause local concrete crushing.

Bundled loops of strand may be employed, provided they are equalized so all strands will work together. Their ends can be splayed out in the concrete. Smooth, mild-steel bars may be used, provided adequate embedment against pull-out is provided. The diameter must be such that localized failure cannot occur by bearing on the shackle pin.

Figure 4.7. *Lifting prestressed lightweight concrete double-tee unit for parking structure roof.*

Fabricated lifting inserts (e.g., fabricated plates) may be used, provided the following rules are applied:

a. Their pull-out value is ensured by mechanical or positive fastening in the concrete.
b. There are no welds transverse to the principal tension.
c. Plates are thick enough in themselves to take bearing from shackle pins—no built-up washers or cheek plate reinforcement of the eye.
d. Eyes are designed for shear, moment, and tension on the minimum section.
e. Steel used is ductile (serious failures have occurred when hard-grade brittle steels were used).

For both mild-steel bars and mild-steel plates, the steel should have a Charpy impact value in the range of 20 N-m (15 foot-pounds) at the temperature at which picks will be made. (Actual tests are expensive and are not normally required.) Most US domestically produced steels (ASTM A-36) have adequate impact values at temperatures down to about minus 5°C (20°F) but, at lower temperatures, a specially rated steel should be used. Prestressing strand and bars fortunately have good low-temperature impact strength. Particular care should be employed with unknown or foreign steels since, at low temperatures, they may fail in brittle fracture at 50% or less of their nominal strength.

Fabricated inserts may be prestressed by bars to the concrete, but care should be taken to provide adequate embedment and to use the above specified safety factors in design, including design for embedment.

Deformed reinforcing bars should not be used for picking loops, as the deformations result in stress concentrations from the shackle pin. Also, reinforcing bars are often hard-grade or rerolled rail steel, with low ductility and low impact strength.

Good rigging practices must be followed for safety. Shackles should be used in preference to hooks: if hooks are employed, they should be prevented from accidental release by a mechanical "keeper" or by seizing with rope. When prestressed piles are lifted from the forms, best practice is to lift them up a few feet by slings to the inserts, then set them on blocks just above the forms, and change to the use of the slings secured by several turns of wire rope around the pile at each picking point.

Precast members are then taken to storage for curing and aging. Here they must be supported in such a manner as to minimize tensile stresses and creep. Girders will need to be supported near their ends, while uniformly prestressed elements like piles will need a number of supports designed to minimize positive and negative moments.

Large hollow members, like pipe and shafts, if being stored horizontally, will need internal strutting to prevent ovalling due to creep of the young concrete.

Discussion so far has concentrated on precasting in a yard, established for the purpose on firm ground. Where very large and heavy members are to be cast in areas which have been recently filled, special soil consolidation and densification measures may be necessary. In extreme cases, pile-supported slabs may be required.

Sometimes it is necessary to precast members on a flexible support such as a barge deck. Since the deadweight will cause the barge to deflect, the maintenance of true line is very difficult. In this case, subdividing of the unit into segments may be desirable; the joints between may then be cast when the segments have been accurately aligned.

Moderate-size precast concrete units are usually transported by truck and erected by crane. Rail may be used for long distance shipments and for over-length segments; but often requires supplemental transportation to the actual job site. Barge transportation is very economical and practicable for movement to water sites, and can be used to transport heavy and over-size units.

During transport, precast members are subjected to dead-load stresses, dynamic (impact) stresses, and lateral instability. Ideally, a member being transported will be supported at the same points as in the final structure. However, this is not always practicable, and thus special design studies must be made for the transport condition. If there is excessive overhang, tensile stresses at the point of negative moment may be countered by additional mild reinforcing steel or external steel beams bolted on.

When a deep, heavily stressed girder is tipped from its normal attitude, dead-load stresses are reduced and the unit tends to buckle. Tipping may be prevented by proper blocking and securing. Chains are preferable to wire rope because wire rope stretches.

Lateral instability is the cause of most accidents in transportation. Hog-rodding is often necessary on long, deep girders. Sometimes two or more such units can be transported together, side by side, and tied together to provide the necessary lateral strength (Fig. 4.8).

Dynamic stresses are usually just considered as an increase in stress. However, for long, thin members supported at their ends only, a dangerous condition may arise due to bounce. The dead load of the member may be temporarily relieved by

Figure 4.8. Installing "hog-rods" to give lateral stability to large prestressed concrete girder during shipment and erection.

the bounce, and the prestress, no longer countered by dead load, may cause excessive tensile stress and even failure in reverse bending.

Care must be taken to prevent surface damage to precast concrete members during transportation. Spalling of edges may be prevented by proper attention to blocking and softeners, especially where tie-down chains or lifting slings pass around the member. Surfaces should be protected against staining and discoloration during transport; in some cases this may require covering.

When a precast girder is to be supported on more than one railroad car, a bolster or other means must be provided to permit rotation of the supports as the cars go around curves.

4.3.5 Special Provisions for Manufacture of Precast Segments—"Match-Casting"

In addition to the general procedures for manufacture of precast concrete elements, special techniques are often employed for the manufacture of segments which will later be joined, usually by post-tensioning, to form a complete girder or other structural element.

When the jointing is to be done by cast-in-place concrete, then the tolerances on the faces of the segments are not excessively critical. The faces do, however, have to be properly prepared so as to bond with the joint cement. This is best done by high-pressure water jetting, although sand and shot blasting may also be used. The face should be cut back sufficiently to expose the coarse aggregate: usually a depth of 6 to 10 mm ($\frac{1}{4}$ to $\frac{3}{8}$ inch). Bush-hammering is not acceptable as it loosens the bond of the coarse aggregate particles (Figs. 4.9, 4.10).

What is critical for tolerance at the joint faces is the positioning and alignment of the ducts for the post-tensioning. The end gates in the casting bed should therefore be rigid, with close-fitting holes through which the duct can be inserted. Gaskets will prevent grout leakage.

A mandrel should be inserted in the duct: this can be a length of thin-walled pipe or conduit. The purpose is to ensure that the duct will not sag or be displaced during concreting and that it will not have any abrupt curvature near the end gate. Prior

Figure 4.9. *Precast concrete segment for 3.5-meter diameter cylinder pile. Note holes for anchorages of future post-tensioning since this is an end segment. Intermediate segment with serrated joint in background.*

Figure 4.10. *Preparation for cast-in-place joint between pile segments. Circular pipes are for subsequent steam curing.*

experience, before the use of mandrels became general, showed that the ducts often had a deflection immediately behind the end gate. When later, tendons were stressed, this sharp bend produced a localized downward force which often resulted in a crack. Such a sharp angle change, even of a few degrees, made it difficult to achieve a grout-tight joint. It also increased wobble friction for the tendon.

When precast segments are to be joined with dry or glued joints, then very tight tolerances are required. Faces must be truly normal and surfaces accurate to about 0.4 mm ($\frac{1}{64}$ inch).

Such a tolerance is possible but difficult to achieve by grinding, due to wear of the grinding wheels themselves.

Such tolerances can be achieved by casting against very heavy steel forms (for example, 40 mm ($1\frac{5}{8}$ inch) steel plates). Otherwise, under repeated use, the end forms (end gates) will spread.

If the end forms must be removed in order to strip the segment out of the forms, then some positive means of reestablishing their location for the next segment must be provided. In one successful operation, the heavy steel mating surfaces at the joints of the forms were milled to .005 inch tolerance. The end gate was controlled by large fine-threaded bolts, which could be screwed back so as to close the joint tightly. After each use, the joint face was carefully cleaned so as to prevent build up of mortar.

The most prevalent means of casting segments for dry or glued joints is the match-casting technique. The first segment is cast and cured and the forms stripped (Fig. 4.11). Then the forms for the second segment are assembled adjacent to the first segment, so that the second segment is cast against the first (Fig. 4.12). A bond breaker is sprayed onto the matching face: this can later be removed by water blast.

Pipe mandrels inserted in the ducts assure exact location and alignment. The faces will inherently match correctly, even if there are imperfections or intentional offsets due to shear keys.

A number of precautions are necessary.

The forms for the second segment must align accurately with the first segment so that the finished girder or element will have the proper profile.

Steam curing causes thermal expansion. Therefore, the steam hoods should enclose both the first, already cured, segment and the new segment.

The most efficient manufacturing system for precast match-cast segments is the short-line system.

Segment 1 is cast and cured in Position A.

Segment 2 is cast against it in Position B.

Then Segment 1 is removed to storage and Segment 2 is moved to Position A.

Segment 3 is cast against Segment 2 and so on (Fig. 4.13).

If the completed girder has a horizontal or vertical curve, this is most easily accommodated by rotating the previously cast segment. Thus the casting position, B in the above example, can be fixed as to alignment and attitude, with all variations being made in the previous segment. By this procedure, a daily cycle can be attained and forming costs minimized.

A variant for match-casting is the long line method, in which the entire length of the girder is formed, the segments then being cast in succession. When all segments of a girder have been cast, they are then disassembled for transport to the construction site (Fig. 4.14).

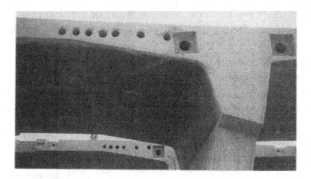

Figure 4.11. Match-casting of precast segments for prestressed concrete box girder bridge. Forms have been stripped from first segment.

Figure 4.12. *Match-casting. The second set of segments has just been cast against the first*

Match-cast segments may also be cast vertically, the second one on top of the first. While most suitable for bridge piers, hollow-core shafts, and intake and discharge towers, this method can also be used for casting the successive segments of girders which have a constant cross section.

An example of vertical casting of segments follows.

Segment 1 is cast as a hollow box lying on its open side. Then the forms are raised and Segment 2 is cast and cured.

Segment 2 is then lifted and set in a second casting position alongside. Segment 1 is removed to storage. Segment 3 is cast on Segment 2. When cured, it is lifted and set in the initial casting position.

This procedure was successfully used to match-cast the complex barrier wall of the Ninian Central Offshore Platform. By using mandrels, the exact matching of

Figure 4.13. *Match-casting. A series of match-cast segments is now ready for erection.*

Figure 4.14. Manufacture of precast match-cast segments for curved box girder bridge, Riyadh, Saudi Arabia.

Figure 4.15. Match-cast curved segments for hull of prestressed concrete vessel for floating storage of liquefied petroleum gas (LPG).

ducts for prestressing tendons could be assured, as well as the exact fit of shear keys.

When precast segments are to be jointed with dry or glued contact, means must be taken, through mix design and vibration, to minimize water and air-bubble entrapment on joint faces, particularly if these joints are thin and highly stressed (Fig. 4.15).

When precast elements are manufactured in positions normal or at an angle to their final position, they must then be rotated without damage. Prestressed elements are very sensitive to positions other than their final position, and the temporary tensile stresses must be computed for all the positions in which they will be handled, rotated (turned), or stored. Additional mild-steel reinforcement, external steel beams bolted on, handling in pairs, and temporary additional prestressing are some of the techniques used to counter temporary excessive tensile stresses.

Crushing of the concrete edges in turning may be prevented by turning in the air, using rigging designed for tilt-up wall slabs, by resting the lower edge in sand, by using wood-block softeners, or by turning in a turning frame (cradle). Very heavy units have been successfully tiled or turned in these structural-steel turning frames, which give full support to the unit during turning and prevent localized high-bearing stresses on the concrete edges.

4.3.6 Lifting and Erecting Precast Elements

Units should be picked and handled so as not to produce cracking. Buckling should be prevented by suitable stays or supports. Auxiliary trussing or support may be required for long, thin girders. Particular care should be taken at points of picking to avoid concentrated stresses and cracking; this may require careful detailing and additional mild-steel reinforcement. It should be noted that when precast units are removed from forms, there may be additional load due to friction, cement paste fins, and suction.

Picking and handling should be treated as dynamic loads.

Lifting loops for picking and handling must be designed with a safety factor of 5 or 6, that is, their ultimate capacity should be 2.5 or 3 $(DL + I)$.

Lifting at the site may involve different rigging and different inserts than the lifting carried out in the plant. Each lift should be carefully engineered and adequately transmitted to the field crew (Fig. 4.16).

For example, the lift in the plant may have been made with a picking beam and multiple pick points, resulting in low stresses in the member and low lifting loads at each point, all of which are vertical. At the site, fewer picking points will be employed and usually lifts will be by angled slings. There will be greater impact. So loads will be higher and stresses in the member will be greater and more complex.

Obviously the picking insert in the field will carry greater and much different loads than in the plant.

This change is quite dramatic in the lifting of piles, since moments in the pile will be a function not only of the location of the insert but of the angle of the sling as the pile is picked from horizontal to vertical.

Figure 4.16. Site erection of prestressed tee-girder for helical parking garage. Note transition from double-tee at one end to single-tee at other end.

Most lifting is done by cranes, either mobile cranes for moderate heights or tower cranes for high-rise buildings and great heights. When one crane is insufficient, two may be used (sometimes even more) (Fig. 4.17). Vertical lift jacking systems are widely employed in bridge girder erection. Even helicopters have been employed for high lifts of limited weight.

In some cases, due to reach or interference with the structure, it will be expedient to first set the precast element in a temporary position, then skid or roll it to its final position. Segments which have been stressed together may be jacked forward, sliding on Teflon and stainless steel supports.

The consequences of failure in the field are far greater than in the plant. In the plant, the lifts are relatively low, and there are neither adjacent structures nor public nearby. At the site, there may be critical structures nearby, as well as the structure itself that is under construction. Workers, the public, transport lines, services, and utilities will typically be in close proximity. The support conditions for the lifting equipment will vary for each lift; some may be of marginal stability. Many lifts will be high in the air; some will involve such complexities as turning or change of support.

With the height of lift come the potential problems of wind, as well as those of catching under the structure or a previously erected member as the member is lifted.

The set down in the field must be accurately located, either in the member's final position or in a temporary position.

Figure 4.17. *Erecting single-tee girder.*

When set in place, before releasing, the member must be stabilized against tipping and shifting or slipping.

All of these aspects make it essential that proper engineering and supervision be employed at all stages of the erection process.

Erection methods and techniques are determined by the span, height, and type of the structure, its location, the topography, the weight, size, and configuration of the precast elements, the method of jointing, and the erection equipment available. The demonstrated ingenuity of contractors in erecting precast members under extremely adverse conditions justifies confidence in specifying precast concrete construction. This ingenuity should be encouraged.

The success of erection of large, heavy elements to close tolerances depends on the skill and care of the construction engineer. Thorough calculations of rigging, temporary support, and stress conditions are required for each stage of erection. Additional strengthening or support may be provided internally (by additional reinforcement or temporary stressing) or externally (by means of attached trusses or hog-rodding).

Lifting loops or similar devices must be provided for all precast units for handling and erection.

Care must be taken to ensure lateral stability against twisting, buckling, and tilting, both during erection and after erection.

Consideration must be given to external forces on precast elements during erection, such as wind, current, waves, etc., as applicable. Erection should be suspended during unfavorable and unsafe conditions.

Temporary lateral support must be provided to prevent sidewise displacement, buckling, and torsion. With curved girders, guys and blocking will generally be necessary to prevent twisting and overturning. Cross-bracing to adjacent girders is often a practicable solution.

In seismic areas, it may be necessary to ensure sufficient tying and bracing of major portions of the global structure so that the structure as a whole is safe under earthquake after assembly is complete. It is generally considered unnecessary to tie each member as it is erected against this unusual loading unless its dislodgment could have "stack-of-cards" or other catastrophic consequences, or unless the structure must remain in a partially completed state for an extended period.

Steel inserts may be embedded in precast elements to facilitate their temporary or permanent fixation after erection. All the erection techniques common to structural steel erection may be utilized, e.g., erection pins, high-strength bolting, and welding.

With welding, special attention must be given to the heat generated so as not to cause spalling of the adjacent concrete nor warping of the insert.

A small reveal or groove may be formed around embedded inserts using wood or polyurethane chamfer strips. This will prevent cracks at the corners of the insert due to prestressing or welding. These may be sealed after the welding is finished, using epoxy putty.

To prevent the heat of the welding from causing edge spalling of adjacent concrete, the weld should be kept as distant as practicable from the concrete surface, intermittent and low-heat procedures should be adopted, and the steel shape should be of thick plate and well anchored into the concrete to resist local distortion.

Continuous butt welds, with concentric joints, are preferred over fillet welds. Critical welds should be inspected by radiography. Welds should comply with the specifications of the American Welding Society, using certified welders and procedures. Steels used should be selected for weldability. In particular, avoid welding of hard-grade reinforcing steel. Some reinforcing bars sold as "intermediate grade" actually are high-carbon bars and are not adaptable to good field welding.

4.3.7 Segmental Construction and Joints

Prestressing makes practicable the use of segmental construction in which precast concrete elements are joined together, so as to act in monolithic fashion. Segmental construction, in turn, permits higher quality and more economical construction in prestressed concrete.

Segmental construction has been widely employed in buildings and bridges. Some of the longest span bridges in Europe, notably the Oleron Viaduct in France, the Gladesville Bridge in Australia, the Oosterschelde Bridge in the Netherlands,

Figure 4.18. *Match-cast precast segments are post-tensioned in cantilever-segmental construction to form long-span viaduct with aesthetic contour. Riyadh, Saudi Arabia.*

and bridges across the Moscow, Dnieper, and Volga Rivers in the former USSR, employ segmental construction. Similarly, precast concrete segments have been used in such notable spans as the Columbia River I-205 Bridge, the Sunshine Skyway Bridge in Florida, the Ministry of Petroleum Viaduct in Riyadh (Fig. 4.18) and many lesser but important spans in Canada and the USA. Segmental construction can be extended to two- and three-dimensional frames, as in roof trusses and space frames. It can be used to excellent advantage in cable-suspended construction, permitting adjustments of profile prior to jointing.

Although composite construction will be dealt with separately, it should be noted that a combination of segmental construction, employing precast elements, with cast-in-place concrete, all working monolithically, may be the most efficient solution in many cases.

Similarly, segmental construction can be carried out progressively with cast-in-place methods. As the segments are cast and gain strength, they are prestressed to the previous structure (Fig. 4.19).

Segmental construction obviously is a practicable means of utilizing very high-strength concrete. Precast elements, manufactured with special techniques under close control, can be joined together for efficient structural action in the largest and most complex structures.

Precast units can be made with very thin sections, cast in the most practicable position, so as to insure high quality and close tolerance. Dimensions and quality can be verified and corrections made, if necessary, prior to assembly into the structure.

Figure 4.19. *Cast-in-place segmental construction was employed on the Alsea Bridge in Oregon.*

Prestressing ducts and anchorage assemblies can be accurately located and fixed during concreting. Concrete can be properly placed and consolidated around congested reinforcing steel, as at the anchorage zones.

Segmental construction is especially applicable to winter construction, as on-site concreting and other work is kept to a minimum.

Precast segments for bridge construction usually comprise part of or the entire cross section of the bridge. They can be erected as segments on falsework, by progressive stressing back (cantilevering), or by hanging from suspenders; the joints are then completed, and the segments stressed longitudinally.

Similarly, bridge pier shafts and columns may be constructed by erecting precast segments on top of each other, then stressing them vertically.

When a transverse cross-sectional segment would still be too large or heavy to erect, or when other structural requirements dictate, the segments may be longitudinal segments, stressed together transversely.

Segmental construction permits more rapid completion, since the segments can be manufactured while the site preparation work and foundations are being constructed. Proper techniques of erection, jointing, and stressing make it possible to attain rapid erection of the superstructure.

Segmental construction makes it possible to apply the structural and economical advantages of precasting to large and complex structures.

4.3.8 Assembly of Precast Segments

Accurate alignment of precast segments must be obtained prior to final jointing and stressing (Fig. 4.20).

With dry joints, alignments may be obtained by means of keys or dowels.

Precast segments can be jacked, wedged or shimmed to proper alignment and attitude before joint concreting. Multiple hydraulic jacks, and flat jacks have been used for this purpose.

When erecting on falsework, final alignment should be made after all units are in place, so as to compensate for dead-load deflection and anticipated creep deflections. With successive cantilever construction, continuous observation and calculations must be made of the effect of creep and elastic deformation, and corrections made in aligning succeeding segments.

Figure 4.20. Assembling precast concrete segments for pontoons of floating bridge, Washington.

Temporary vertical support for segments may be provided by blocking up from the falsework. With the cantilever method of erection, steps or seats may be provided. Bolts and steel frames, either underneath or above the new segment, may be used to support it.

Design of temporary vertical supports should take into account impact from setting, the possibility of eccentricity during initial setting, side away or tipping in erection, and localized concentrated bearing.

When setting heavy units from a barge-mounted derrick, subject to wave or swell action, the precast segment may be cushioned by landing it on rubber bumpers, timbers, or on inflated rubber tires, then jacked down to final position.

Temporary lateral support must be provided to prevent sidewise displacement, overturning, buckling, and torsion. Cross-bracing to adjacent segments is often practicable. Where segments are part of a curved girder, guys and blocking will probably be needed to prevent twisting and overturning.

Temporary stressing, using post-tensioning tendons or bridge strands, is often an effective means for providing temporary support and for holding the segments in proper position during assembly. Tendons have been used inside the boxes of box girders, outside the webs of I-girders, and on or above the deck in cantilevered construction. This latter is a transitional form of cable-stayed construction. Such temporary stressing may also be utilized to pull the segment into final position, and to counter temporary tensile stresses. Prestressing bars are well-suited to temporary installation, stressing, and later removal for reuse.

The ability to jack a tensioned or partially tensioned external tendon up or down, and thus change its length or profile, has been utilized to make adjustments in camber or deflection.

Temporary stressing may also be used to temporarily fix a hammerhead segment to a pier. The hammerhead section may be set on wedges or flat jacks, and bar tendons used to stress it vertically against the pier shaft. When the need for fixed connection is over, the tendons can be released and the wedges or flat jacks removed.

Precast deck sections and cable-suspended structures may be assembled in such a way that the main longitudinal and transverse stressing is achieved by tendons located in the joints. The precast sections are first assembled on falsework or, in some cases, on the cables. The tendons are then placed in ducts through the zone of the joint, the joint is poured, and the entire assembly stressed. With cable-suspended roofs, the precast units are hung on the stretched cables in a carefully predetermined pattern, the stress is readjusted in stages, and finally, the joints are concreted.

4.3.9 Special Provisions for Post-Tensioning of Precast Segments

ANCHORAGES. Anchorage pockets may be provided in precast segments with a high degree of accuracy and assurance of quality. By precasting under most favorable plant conditions, very high-strength concrete, accurate alignment of the ducts, and an accurate bearing surface may be attained, even with the normal congestion of reinforcing steel in the anchorage area (Fig. 4.21).

DUCTS. Duct diameters for segmental construction should be slightly larger (say, 5 mm = $\frac{1}{4}$ inch) than for other post-tensioned work to eliminate excessive friction at joints due to minor misalignment.

Duct splices may be made by using flexible metal or rigid metal sleeves, with well-taped or, even better, heat-shrink joints.

A review of experience with the jointing of precast segments in bridge girders shows an unacceptable number of cases of "cross-over" of grout during the grouting of the stressed tendons. This is, of course, a result of having all the tendon ducts spliced at the same location and the fact that the joint concrete of cast-in-place joints may not have been thoroughly consolidated due to congestion.

Studies and tests by the Federal Highway Administration show that corrosion of tendons is more prevalent at these joints.

Therefore, the detailed matter of splicing ducts takes on great importance.

Screwed joints of moderately thick-walled pipe are obviously best, but may be difficult to align. Use of heat-shrink plastic has shown excellent performance in laboratory tests. Sleeving with a relatively tight fitting sleeve and use of an epoxy sealer can be very satisfactory if the workmanship is properly carried out.

Figure 4.21. Anchorage pockets formed in precast segments.

Sleeving, with watertight taping afterwards, is the most used system and has given moderately satisfactory results. The failures, that is, the cases of subsequent grout cross-over, are believed due to:

• Loose-fitting sleeves.
• Careless taping.
• Displacement by vibrator during subsequent concreting.
• Poor concreting of joint (mix and workmanship).
• Poor alignment of ducts, preventing proper fit of sleeve.

With match-cast segments, the problem is exacerbated. There are no sleeves and no duct continuity across the joints.

Use of thin felt gaskets or rubber O-rings set in a recess have been tried; they appear to work but are time-consuming to execute and are often dislodged in the field.

Best results seem to have been obtained by smearing epoxy glue relatively thickly around each duct opening. If the joint is being shimmed to correct alignment, then a rubber or felt gasket is needed, even with the epoxy glue. A mandrel should be run in after the epoxy has set, to be sure that the hole is open.

Correction of alignment of precast match-cast segments requires very careful engineering projection as to the effects of each alignment correction. Otherwise, over-correction and exaggerated S-curves will result.

The best method for alignment protection is to insert small shims of welded wire mesh. These can be fixed in place by epoxy glue.

Where tendon splices are required, they may be located in joints between segments. The sections of the tendon are coupled at the joint and the stressing accomplished either externally, i.e., at the ends, or by jacking the segments apart by jacks in or across the joint.

The joint, or a pocket at the joint, must have sufficient length to accommodate the coupler in its extended position and, in addition, must accommodate the movement when stressing takes place. Errors in this latter have caused later collapse of entire structures.

When coupling multiple tendons across a joint, their length and initial stress should be equalized, for example by a calibrated ratchet jack.

Provision of passive reinforcing steel across joints between precast segments is often required to develop ultimate strength and to provide against unexpected reversals of stress. Several means have been developed to accomplish this.

With wide joints, of 600 mm (24 inch) width or more, the reinforcing steel may be extended, so as to overlap corresponding bars from the adjoining segments. Larger bars will require hooks, or heads to transfer tension. One system uses U-bars which are staggered to lap those from the other segment, through which a bar or bars can be run at right angles.

Welding of opposing bars may be employed. This is usually best done by welding each bar to the trough of a steel splice angle.

Direct butt welding, e.g., full penetration welding, requires that the bars be placed in each segment with great accuracy. This is hard to achieve due to the displacement effects of concreting and vibrating.

Another system applicable to wide joints has spirally corrugated bars stubbed out only a short distance, e.g., 75 mm, and set in a pocket. Then a splice connector is screwed on each bar.

With direct contact joints, one means of providing continuity is to have dowels extending from one segment which are inserted into corresponding holes of the adjoining segment. Epoxy grout or polyester resin is used to bond the dowels to the segments.

This type of splice has been extensively used when joining vertically oriented segments, such as columns, shafts, and piles. The dowel holes are formed in the lower precast segment, preferably by light-gauge metal duct which is held in alignment by a mandrel during concreting. A smooth hole, such as formed by a conventional rubber hole former, may not develop adequate bond.

After the lower segment has been installed, the upper segment, with reinforcing bar dowels extending, is brought above the lower segment. The holes in the latter are filled with epoxy, and the top section lowered into contact. It must of course be held rigidly until the epoxy or polyester has set. Alignment of the dowels may be expedited by making one of the dowels a little longer than the others, so that it can be entered first.

The setting time may be reduced by heating the joint, as with a steam hood or electric-resistance heating.

For direct contact or narrow joints of horizontal girders, the most practicable means for providing additional steel area across joints may be to run unstressed strand through extra ducts provided for the purpose. If the runs are fairly straight and short, mild-steel bars may similarly be run through ducts. The ducts are then injected with grout in the same manner as prestressing tendons.

Wide joints may require enclosing confinement, such as stirrups. This of course will congest the joint and make subsequent concreting and consolidation even more difficult.

4.3.10 Dry, Glued, and Concreted Joints

Joints are subject to the same stress patterns as exist in the body of the structural element. Therefore, their structural integrity is essential. Since joints are usually visible, consideration must also be given to their appearance.

Poorly made joints may lead to water leakage initially and subsequent corrosion of the reinforcement.

Because proper jointing of segments is essential to the security and performance of the structure, a detailed description is presented of acceptable jointing techniques.

4.3.10.1 CAST-IN-PLACE WIDE CONCRETE JOINTS. Cast-in-place concrete joints are the most common method of joining precast segments into monolithic concrete. The length of joint is set at a sufficient spacing to permit splicing of the mild reinforcing steel bars; thus, joint widths are usually in the range of 400 to 600 mm or more (18 to 24 inches). The reinforcing steel bars extend from the segments and may be lapped the required number of diameters, or be spliced by an approved welded joint. Concentric welds should be employed, or eccentric welds checked for the effect of eccentricity. Welds should be checked for fatigue loadings, where applicable. All welds should conform to applicable AWS Specifications. Looped bars, with a vertical rod linking the overlapping loops, may permit some reduction in joint length.

Joint surfaces should be cleaned and roughened, by sandblasting or high-pressure water jets, to a depth of 6 to 10 mm, so as to expose the coarse aggregate. They should be thoroughly pre-wetted or else an epoxy-bounding agent should be applied just before concreting. Epoxy-bonding materials should be applied only to clean, dry surfaces, in strict accordance with manufacturer's recommendations.

High-strength concrete with 10-mm ($\frac{3}{8}$-inch) maximum coarse aggregate is then placed. Internal vibration is essential. Curing may be by wet burlap, but it is often accelerated by low-pressure steam in a temporary hood or jacket.

Ducts for prestressing tendons will have been spliced and taped to prevent grout intrusion. However, because of the absolute importance of keeping these ducts open, it is often desirable to insert an inflatable rubber tube all the way through the ducts and joints. If this is not practicable, a small wire line may be run the full length of the duct, and then run back and forth while the concrete in the joint has still not set, in order to insure continuity of the duct.

4.3.10.2 POURED "FINE" CONCRETE JOINTS. Poured "fine" concrete joints should not be less than 75 mm (3 inches) thick. Generally, no reinforcing steel splices are provided, but duct tubes must be spliced and taped and kept open as provided in Section 4.3.10.1 above.

Joint faces are prepared as in Section 4.3.10.1 and "fine" concrete, using 8-mm ($\frac{3}{8}$-inch) maximum size coarse aggregate, is poured, vibrated internally, and properly cured. A pencil vibrator is required.

4.3.10.3 DRY-PACKED MORTAR JOINTS. Dry-packed mortar joints have been employed in the past, especially for horizontal joints between vertical segments, but may give inherent problems of local stress concentrations. If used, they should be about 12 mm ($\frac{1}{2}$ inch) in thickness, and the opposing faces should be clean, smooth, and without high spots. The mortar should be of dry consistency and placed in stages by uniformly packing with a steel tool, working simultaneously from both sides of the joint. The extreme outer surface of the joint should employ a more workable mix so as to seal the surface, leaving a small reveal to prevent edge stress concentration.

4.3.10.4 BUTTERED MORTAR JOINTS. Buttered mortar joints have been frequently used when one precast segment has been set on top of another, e.g., a hammerhead segment set on a pier. The edges may be sealed and protected against load concentrations and edge spalling by a perimeter strip of compressible material such as neoprene or asphalt-impregnated fiberboard. Then a thin coat of cement mortar is troweled onto the surface. The surface must have been properly prepared.

Buttered or mortar joints have not proven very satisfactory for vertical joints, such as joining segments of girders, due to uneven compaction and stress concentrations.

4.3.10.5 "DRY" (EXACT FIT) JOINTS. "Dry" (exact fit) joints are increasingly used for joining precast segments, as they permit rapid progress, assure uniform properties of concrete at the joint, are unaffected by weather, such as freezing conditions, and are economical.

The methods of obtaining perfectly matching surfaces are described in Section 4.3.5.

Keys or dowels must be provided to insure that the matching faces are brought together in the correct alignment and position. Temporary bolts may be used to hold the segments in position until stressed.

With dry joints, consideration must be given to the problem of continuity of ducts across the joint. If all tendons can be grouted at the same time, leakage, if any, will not plug an adjoining duct. If all ducts across a joint cannot be grouted at the same time, the continuity of the ungrouted ducts should be insured by employing an inflatable tube, or by drawing a wire rope back and forth during and after grouting of the adjacent ducts (Fig. 4.22). Alternatively, gaskets or O-rings may be used, as previously described.

The outer edges of dry joint faces should be chamfered slightly to prevent local edge stresses from causing spalling. Insofar as practicable, the joint location should be selected so that the effective stress across a dry-joint face is always compressive and as uniform as possible (including the effects of prestress and dead-load stress). It must develop sufficient compression to ensure adequate shear resistance under the most adverse loading, e.g., maximum live load on one side of joint, no live load on the other.

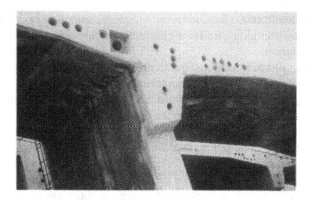

Figure 4.22. *Proper sealing of ducts across joints is a major consideration in precast segmental construction with dry or glued joints.*

4.3.10.6 EPOXY-COATED DRY JOINTS OR "GLUED" JOINTS. Epoxy-coated dry joints or "glued" joints are formed as for "dry" joints. Usually the segment is fitted to ensure that the fit is "perfect," then separated and a relatively slow-setting epoxy is applied to both faces, about 0.5 to 1.0 mm thick (0.02 to 0.04 inch). The jointing is then made and the segments stressed together wet (i.e., before the epoxy has set). Epoxy materials should be used in accordance with manufacturer's recommendation. They should be applied only to dry, clean surfaces. "Hydrophobic" epoxies have been developed which allow application to damp (but not wet) surfaces. Best results are obtained when the epoxy sets under a stress of 0.3 MPa (40 psi). Temporary prestressing bars may be used.

Rain can be especially damaging to the epoxy during this mating operation. In wet rainy climates, it may be necessary to use a tent or to build dams on the deck to prevent water running down.

4.3.10.7 SHEAR TRANSFER. Roughened surfaces, formed by sandblasting, high-pressure jetting, wire brushing, or a set-retarder plus jetting, etc., may be used to provide shear transfer from a precast joint face to a concreted joint and are usually sufficient for most cases. However, where shear forces are extremely high, shear keys or indentations may be formed in the faces of the segments; these should have dimensions two to three times the maximum size of coarse aggregate used in the joint concrete. Similar shear keys may be formed in matching faces of dry-joint segments.

Shear keys should have sloping faces, not right-angled to the joint, so as to avoid edge concentrations. They should have curved, not sharp, interior angles, and a reinforcing bar in the segment crossing this angle, to prevent cracking under local stress. With cast-in-place concrete joints, shear must be checked in the unreinforced concrete zone between the joint proper and the first shear (vertical) reinforcement. Although this zone may be relatively narrow, it may be critical for shear under some conditions of loading. This may make it necessary to provide steel across the joint in the form of diagonal bars, tendons, or mild steel dowels.

4.3.10.8 FINISH AND COLOR. Exterior faces of girders made of segmental elements should either show the joint clearly or else attempts should be made to blend the joint with the segment.

Clear showing of the joint may be accentuated by chamfers. This treatment is particularly applicable to segments assembled with dry joints.

Blending requires a uniformity of finish and color. With cast-in-place concrete joints, use of the same forming material and curing method as was used for the precast segments themselves (e.g., steam curing) will tend to minimize color differences. With "fine" concrete joints, use of some white cement in the fine concrete mix will tend to achieve uniformity of color. Prior experimentation with a mock-up will help to develop the proper method of blending.

Finish uniformity may be obtained by rubbing the entire girder face.

4.3.10.9 SHRINKAGE IN JOINTS. Shrinkage in the joints can occur with cast-in-place concrete or poured concrete joints. These shrinkage cracks at the faces of the joint may usually be prevented by keeping the joint concrete water-cured until stressing. Proper pre-saturation of surfaces will help prevent shrinkage cracks by preventing water absorption, as will use of an epoxy bonding compound.

4.4 COMPOSITE CONSTRUCTION

"Composite construction" in prestressed concrete construction generally connotes the combination of precast prestressed concrete beams or slabs with a cast-in-place concrete deck or top slab (Fig. 4.23). The two are designed to act together structurally: the prestressed member serves as the tension flange, the cast-in-place section as the compression flange.

The advantages of composite systems are the following:

- *a.* The precast tension element may be prestressed with straight tendons or minimum deflection, since, in itself, a large eccentricity of prestress is not required. Manufacture is simplified and the precast element is more easily and safely transported and erected.
- *b.* The precast element is relatively light.
- *c.* Standard precast beams may be grouped close together or spread apart as appropriate to accommodate various concentrations of loads or openings.
- *d.* The cast-in-place top slab ties the structure together and gives an unbroken surface which may be screeded for camber, super-elevation, and warped surfaces.

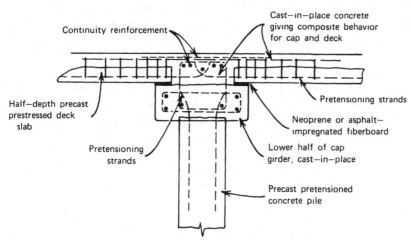

Note· As an alternative to bending up pretensioning strands
from deck slabs over cap, they may be simply lapped 30 diameters

Figure 4.23. *Composite construction: method of tying precast and cast-in-place concrete together for monolithic behavior.*

e. The precast beams, with no shoring or a minimum of shoring, support the cast-in-place deck and its forms during construction.

f. Since no prestress was induced in the compression flange, the composite section is, theoretically, more efficient structurally.

g. Full or partial continuity is more easily achieved (see Fig. 4.1).

h. Shear transfer is more easily provided.

Success with composite construction depends on insuring proper horizontal shear transfer and contact at the interface.

In the case of precast prestressed slabs which are joined in building construction with a cast-in-lace topping, suitable shear transfer may usually be obtained by bond alone, provided the upper surface of the precast member is roughened. Rough screeding, followed by a coarse broom, will generally prove most satisfactory. Then, just before pouring the top slab, the surface should be cleaned with a water jet.

Where shear loads are higher, as in bridge construction, then steel stirrups or mesh ties are necessary. As before, the top surface should be roughened.

Horizontal shear keys have been found in a number of tests to be of less benefit than formerly believed. The shear keys do not come into action until the surface bond has failed. Therefore, shear keys are generally not employed with composite slabs and then only at the ends of heavily loaded girders. When shear keys are used, they should be of greater dimension than the largest size of coarse aggregate. A roughened surface with adequate steel ties is the most effective means.

One means of providing full or partial continuity over the support of two adjoining spans is to embed mild-steel bars in the topping slab. These bars should extend well beyond the point of inflection. Staggering the ends of the bars will also help prevent a transverse crack on each side of the support.

Distributing the required negative-moment steel with many small bars is much better than a few large bars in preventing cracking at the joints between adjacent spans. Because of the rotational effects of camber and live loads, long-term shortening of the precast prestressed beams due to creep, and the possibility of positive moment at the support under unusual conditions of loading, it is usually desirable to tie the prestressed elements together at the bottom. This may usually be accomplished by letting the tendons extend through the joint, overlapping, or bending them around a transverse bar, so as to develop anchorage. Alternatively, extended reinforcing bars may be joined by welding or hooked around a transverse bar.

Since the precast element (slab or beam) is prestressed for its final service and since, in the case of a beam, it may lack an adequate compression flange, it may be necessary to shore it at the center or third points until the top slab has been poured and attained strength. Shores must be carefully placed and wedged or jacked so as not to break the beam in negative moment. Careful control is necessary.

The drying shrinkage of the top slab reduces the load carrying capacity of the composite system. In effect, it applies a small counter prestress. Thus, every effort should be taken to minimize shrinkage. Possible means are:

a. Low W/C or W/CM ratios.

b. Selection of aggregates that give low drying shrinkage.

c. Careful attention to curing: water cure is most effective.

Longitudinal shrinkage (hairline) cracks frequently form over the joint between precast slabs. This is due to the concentration of drying shrinkage. The rough surface of the precast slab prevents shrinkage cracks over the slab itself.

These can be minimized and/or prevented by taking steps to reduce shrinkage as noted above, and by proper detailing of transverse top steel. Many small bars or mesh are preferable to a few large bars. Mesh should be of sufficient steel area (gauge) to counter shrinkage stresses as well as provide the nominal transverse reinforcement required.

4.5 STRESSED CONCRETE TENSION BARS

If a bar of small cross section or a thin slab is given a high degree of prestress and then incorporated in cast-in-place concrete, it will act as a tension bar, similar to reinforcing steel but with important differences. First, there is a comparatively large surface for bond, and the bond is both adhesive and crystalline in nature (secondary crystallization can take place across the interface under proper moisture conditions). Secondly, due to creep in the highly stressed element, a degree of prestress is transferred to the adjacent cast-in-place concrete.

As a result of these and perhaps other phenomena, the tensile value of the cast-in-place concrete (the full value under the stress-strain diagram) may be able to be utilized effectively. Prestressed tension bars may thus be utilized as the transverse reinforcement in prestressed pavements and footing blocks. They have proven beneficial in crack control when used over supports to provide the negative reinforcement in continuous construction.

Since these bars normally contain only one central tendon, with that tendon encased in very dense concrete, they offer great durability.

Thin, highly prestressed slabs have been tested for use as facing panels on dams. In addition to their own inherent durability, they delay cracking in the adjoining cast-in-place concrete, thus enhancing the freeze-thaw durability and making it possible to use reduced cement contents for the cast-in-place concrete.

Similar highly stressed slabs have been used as the tensile face (and soffit form) for large prestressed girders in powerhouse construction in the former USSR.

The ability to transfer prestress from the tensile element to the cast-in-place concrete, being associated with creep, is highly time-dependent. Ideally, prestressed tensile elements should be utilized within a short specific period after casting, say, one week.

Tensile elements, being concentrically stressed and fully bonded, are not subject to the same limits of prestress that we are accustomed to work with in normal prestressed construction. The higher the prestress, the more efficient the bar, subject to the limitation that creep rates are very high at these higher prestress levels. Values up to 40% f'_c would seem acceptable upper limits, pending further research.

4.6 EPOXIES AND OTHER POLYMERS IN PRESTRESSED CONCRETE CONSTRUCTION

These exciting new materials are being adapted to prestressed construction in a number of ways:

1. For improving bond between the precast and the cast-in-place portions in composite construction.
2. For splicing piles and columns.
3. For connections between precast elements, both for primary (direct) load and for secondary forces such as seismic shear, load transfer, etc.
4. As a binder for ultra-high-strength concrete, or to coat aggregate particles to improve the cement paste–aggregate bond.
5. For coating prestressed concrete members that will be exposed to acid or other chemical attack.
6. For gluing the contact faces in segmental construction.
7. As a wearing surface. Epoxy-resin concrete may also be used in composite structural action with prestressed beams.
8. To coat tendons or to paint anchorages to protect them against corrosion.
9. As patches or injection for repair of defects, such as cracks, plucking, pockets, etc.
10. For sealing joints between precast elements.

Epoxy resins have high tensile, shear, and compressive strength, generally good impact resistance, good wearing surface, and high durability. Their limiting characteristics are short pot life, high cost, and sensitivity to temperature. In some exposed applications, ultraviolet degradation may be a consideration.

The properties of epoxies vary widely, depending on the formulation, the relative amount of catalyst added, the method of placing (e.g., whether under pressure or not), and the curing. There are so many different epoxies available that it is wise to work directly with the manufacturer's technical representative to insure selection of the one with the best properties for a particular application. Then it is extremely important to follow the directions exactly. Making one technician responsible for all the work helps to insure proper procedures. Most epoxy manufacturers offer technical service at the site.

Proper preparation of the surface is essential. Cleanliness is extremely important, for both the equipment and place of application.

Concrete must be surface dry, since water destroys the epoxy's ability to set and bond. As noted earlier in relation to epoxy-glued joints, special hydrophobic epoxies have been developed, permitting their application on damp surfaces. Similar formulations allow epoxy to be injected into water-filled cracks and even to be placed as a coating over underwater concrete.

Polyesters have not been as widely used in prestressed concrete construction applications because they lack some of the bonding and tensile strength characteristics of epoxies. However, polyesters are much less expensive and more easily applied wherever high compressive strength is required. They can, therefore, be used for applications requiring larger amounts of material, where the prime requirement is high compressive strength and accurate definition. They have been used successfully to splice pile segments.

Sealants are formed of organic materials that ideally provide bond, watertightness, flexibility, and durability. They must be able to stretch and compress to accommodate volume changes in the precast units due to temperature and moisture differentials. They must be durable in water, heat, and sunlight; they must neither ignite in fire, nor become brittle in cold weather. A number of proprietary materials are available and generally give excellent service if installed correctly. Control of joint width within allowable tolerance is important, as is cleanliness and dryness of joint at time of applying sealant.

Although most epoxy coatings tend to reflect any crack in the underlying concrete, there are epoxy coatings under development which appear able to span cracks in the concrete of widths up to at least 0.5 mm. Such a capability would give potential for preventing the action of water in the crack, even under pressure, and the absorption of corrosive substances by capillary action.

The entire subject of the use of epoxies and other organic materials in concrete construction is a complex and rapidly evolving matter. The *ACI Manual of Concrete Practice,* Part 3, contains two excellent chapters: "Guide for Use of Epoxy Compounds with Concrete" and "Guide for the Protection of Concrete Against Chemical Attack by Means of Coatings and Other Resistance Materials."

4.7 WELDING IN PRESTRESSED CONSTRUCTION

1. When welding connections between precast members, the effect of heat on the adjacent concrete must be carefully considered. Otherwise, the concrete will spall, and the anchorage of the connection plate be destroyed. Spalling is unsightly, usually requiring repairs. Burned and blackened concrete may also be unacceptable, particularly in exposed architectural applications.

Heat may be reduced by the selection of low-heat rods, of small size, and by making a series of intermittent passes, allowing the heat to dissipate. Connections and inserts should be designed so as to minimize heat transfer. Inserts should be given adequate cover and be anchored well back.

2. Never permit an accidental or intentional welding ground through a prestressing tendon. Never weld to a tendon, especially if the tendon is under stress. The point of ground or contact will burn and destroy the high-strength properties of cold-drawn wire, usually causing a rupture if the wire is under stress.

Care is especially necessary around pretensioning beds, with tendons stressed but not concreted.

3. Use of a welding torch to burn back the ends of strands at the ends of precast members can be safely done provided the tendon is not grounded, for example, as at the other end or through inserts. However, to eliminate the possibility of such trouble, a small burning torch is preferable. See also Section 4.3.6 where additional guidance is given for welding of inserts.

4.8 POWER-DRIVEN TOOLS IN PRESTRESSED CONSTRUCTION

These may be used to install fittings, studs, etc., as in all concrete construction, with the following special precautions:

1. Prestressed concrete is usually higher strength with a higher modulus of elasticity, therefore, more likely to spall.
2. Never drive a power-driven stud in the area of prestressing strands where it might shear a tendon. Tendon locations should be marked on the member by an engineer prior to use of the power-driven tool.

The general safety precautions for power-driven tools must be enforced.

4.9 CUTTING OF PRESTRESSED CONCRETE

Cutting of prestressed concrete with large jackhammers may cause extensive spalling. The concrete is under stress and, in the zone near the point of cutting, there will be transverse tensile stresses, tending to burst the concrete.

Therefore, it is best to make an initial cut with a concrete saw. A diamond blade in a hand-held power saw is often used to make a 50 mm (2 inch) deep cut. This notch protects the concrete below and results in a relatively clean cut.

In lieu of or in addition to the saw cut, a heavy clamp exerting squeezing force on the concrete will counter bursting forces and help to control the cut zone.

Use of primacord or other explosives to cut prestressed concrete has not been widely practiced. However, primacord is used to cut relatively thin-walled, non-prestressed concrete pipe. With prestressed concrete, primacord is likely to cause greater spalling due to the transverse tensile stresses.

Drilling and coring of holes in prestressed concrete members is an excellent way of fitting piping and electrical leads through floor slabs. However, extreme care must be taken not to cut prestressing tendons. The permissible zones or areas for such coring must be approved by the structural design engineer and, when in proximity to tendons, the tendon location should be marked.

Lightweight concrete cuts and drills more easily, but also tends to spall to a greater extent.

4.10 ELECTROTHERMAL STRESSING

The use of electric resistance heating for prestressing has been practiced rather extensively in the past in the former USSR and other Eastern European countries.

In one application, high-strength bars are carefully heated (usually by electrical induction), then placed in forms and anchored to the forms. Concrete is placed, the bars cool and contract, and a mild degree of prestress is introduced. The level of prestress is sufficient to overcome shrinkage cracking. However, precision is low because of the short length of tendon and the difficulty of accurately affixing the anchorages.

Electrothermal heating is also used with continuous-winding machines, in which wire is paid off under tension and laid in a predetermined pattern in the forms. The wire must be laid under stress around dowels, in a rather short radius bend. The considerable breakage of wires has been a major problem. By electric-resistance heating of the wire, the ductility of the wire is increased and breakage reduced. Also, the wire may be laid under lower stress since a portion of the prestress will be supplied by the cooling contraction. Such electrothermal heating has reportedly practically eliminated the problem of wire breakage in this application.

4.11 CHEMICAL PRESTRESSING

Several different expanding cements have been developed as a result of intensive research into gel chemistry. These have the property of first hardening, then expanding as secondary crystallization takes place. In general, the expansion is three-dimensional; thus 3-D restraint must be applied to obtain the full prestressing effect.

Chemical prestressing through expanding cement has a number of interesting possible applications. In slabs and pavements, expansion can counteract shrinkage and eliminate cracking. This, at present, is the most practicable application. If conventional prestressing is applied on one axis, the combination of weight and subgrade reaction provides the restraint on the second axis and passive steel the restraint on the transverse axis; then the expansion on the transverse axis will impart prestress.

Small-diameter pipes are produced in the former USSR, using expanding cement. The expansion reacts against the spiral reinforcement. Longitudinal prestress is imparted by wires stressed in the conventional manner.

Chemically prestressed concrete has also been successfully used for swimming pools and ice-skating rinks in the former USSR.

Expanding cement requires continuous moisture to be present during the curing period in order to promote the expansion through secondary crystallization. This moist curing may be effected by total immersion (as of small pipe segments). Lightweight concrete aggregate, batched in a saturated condition, appears to offer ideal internal curing for expanding cement mixes, and experimental work along this line has been very promising.

Control of the degree of expansion is erratic at the present stage of development. Durability appears satisfactory, but still requires additional evaluation.

Expanding cement concrete, and expanding grout may be utilized to fill connection pockets, such as the heads of hollow-core piles. Additional spiral should be installed in order to prevent excessive expansion.

Expanding grout has been extensively used to set inserts and anchor bolts. The expansion is usually accomplished by the inclusion of iron filings or aluminum powder in the mix. The iron filings oxidize and expand. Aluminum powder reacts with the cement to produce hydrogen gas bubbles.

Expanding grout has also been extensively used in the grouting of post-tensioned ducts.

However, it is now believed that the use of aluminum powder may be dangerous in that atomic hydrogen molecules may occasionally be released. These theoretically can lead to hydrogen embrittlement and sudden brittle fracture of the steel tendons. Although the probability is low, the consequences could be catastrophic; thus, there is a growing reluctance to use aluminum powder as an expanding agent in grouting post-tensioning ducts. Many recent regulations prohibit it.

Nonmetallic expanding and thixotropic admixtures have been developed. These are discussed in detail in Chapter 5.

There have also been some difficulties, particularly in Japan, with excessive expansion of grout in ducts, causing longitudinal cracks in the concrete along the duct walls.

4.12 PRESTRESSED STEEL-CONCRETE COMPOSITE BEAMS (PREFLEX)

This unique Belgian system combines structural steel and prestressed concrete in composite action to produce a very efficient beam. It makes possible a very low depth to span ratio and reduced deflections.

The manufacturing process is very simple. Two steel beams are placed side-by-side, their ends secured together. They are jacked apart at the center. The outer flanges are then encased in concrete. (These will later be the tension flanges.) After the concrete hardens, the jacks are released and the beams straighten out. What results is a steel beam, working in composite action with a lower flange of precompressed concrete. The beam is then erected and the top flange encased in cast-in-place concrete, to act as the compressive flange.

This system obviously must give careful consideration to the losses in steel due to stress relaxation and creep and shrinkage in the concrete. Such losses are acceptable only because of the large steel area in relation to concrete area. It is essential that the lower flange encasement be bonded; stud bolts or shear lugs may be necessary. Also, adequate lateral ties (stirrups) are important, in order to hold the flange concrete tightly to the flange.

Since this is a proprietary method, the procedures and techniques recommended by the parent organization should be followed in detail.

4.13 CONTINUOUS WRAPPING OF WIRE, STRAND, AND STRIP

Wrapping wire, strand, and strip under tension is widely applied to prestressed concrete tanks and containment vessels. Wire is stressed either by passing through a die or by winding around braking drums. (See Chapter 16, Prestressed Concrete Tanks.)

Similarly, in prestressed concrete pipe manufacture, wire is wrapped under tension around either a steel liner or a concrete inner pipe.

After wrapping with wire, the surface is then coated with shotcrete to provide corrosion protection. (See Chapter 19, Pipes, Penstocks, and Aqueducts.) This method of prestressing was one of the earliest applications of prestressing.

Such methods have also been applied experimentally to beams, the wire having been wrapped under tension around the sides and ends of the beam.

Continuous laying of wire under stress is carried out by two basic methods in the former USSR. In the first case, known as the turntable, the wire is wrapped around pegs or dowels that protrude up from the steel soffit. After concreting and accelerated curing, the dowels are hydraulically withdrawn.

In the second method, a machine travels down a longitudinal bed on rails. A feeding arm traverses back and forth, feeding wire through a freely rotating spindle. This wire also wraps around hydraulically actuated dowels. (See Section 4.10, Electrothermal Stressing, for the use of electric resistance heating to minimize wire breakage with this method.)

All of these methods lay out a considerable length of wire in rapid fashion. However, since most systems use only a single or double wire, the prestressing force in one pass is rather low and a large number of passes are needed.

Some newer machines, therefore, lay strand from off braked or differential-winding drums.

The strand may be spaced off of the surface of the tank by plastic or other spacers. Both black and galvanized strand are used. Shotcrete is then applied as a cover coat.

High-strength steel strip is applied by wrapping it around the tank, laying multiple layers in a groove formed by a curved steel channel. After completion of the wrapping, the channel is closed by steel plate on the exposed fourth side and the tube thus formed is injected by corrosion-inhibiting grease. This makes it possible to apply a very large circumferential force in a very limited space.

4.14 JACKING APART

Concrete may be prestressed by jacking it apart against internal or external restraint. External restraint may be in the form of abutments, similar to an arch rib's abutments. However, such abutments (and the member itself) are subject to creep, thus allowing a gradual loss of prestress. When using such external reactions, buckling must definitely be considered, as well as the probable failure mechanism, which may be explosive. For these reasons, this type of prestressing against external abutments is usually applied only to arch ribs, although it has been applied to pavements in the past. The initial need with arches is for decentering, a temporary stage not affected by creep. For its long-term behavior, any creep of abutments or concrete is compensated to a considerable degree by the arch-performance of the rib.

Jacking apart against internal restraint, on the other hand, is an entirely sound and reliable method, which can be effectively used in pavements, and in special cases where access to the ends is not available, as in tunnels (Fig. 4.24).

The tendons may be anchored at the ends, and encased in ducts, so as to be free during jacking. In this case, a post-tensioned construction results. The tendons must be joined at the central gap, equalized in length, and then the two segments jacked apart. Equalization of splices has been accomplished by using splice nuts, turned up by a torque wrench. Jacking tongs ("scissor-tongs") might similarly be employed, working against a fixed strut.

If the members are pretensioned, then the length of tendon to be stressed is only the short length at the gap. It is thus extremely difficult to ensure equalization of length of the several tendons, and to prevent excessive prestress loss.

However, this method has been utilized in special cases by actually casting the

Figure 4.24. *Prestressed lightweight concrete roadway segments are erected in the tunnel section of the San Francisco–Oakland Bay Bridge. Two segments are erected, the tendons coupled in the center, and the segments are jacked apart. Then the joint is concreted, using steam to accelerate cure, so that the roadway is put back in service two days later.*

full length (both segments) in one pretensioning line, with the gap formed; then folding one segment on top of the other and shipping the two segments, still joined together by the tendons. After erection, the two halves are jacked apart. In this case, the tendons would automatically be of the right length. (See Fig. 4.25.)

Concreting of the gap should be with concrete of low shrinkage and accelerated strength gain, attained by the use of low water/cement ratio, and careful selection of aggregates and cements. An external steam jacket or electric-resistance heating may be used to accelerate the gain of strength, along with the use of a high early-strength cement.

A variation of this method has been utilized in Northern Europe where two halves of a pretensioned concrete tee were erected as a folded plate roof.

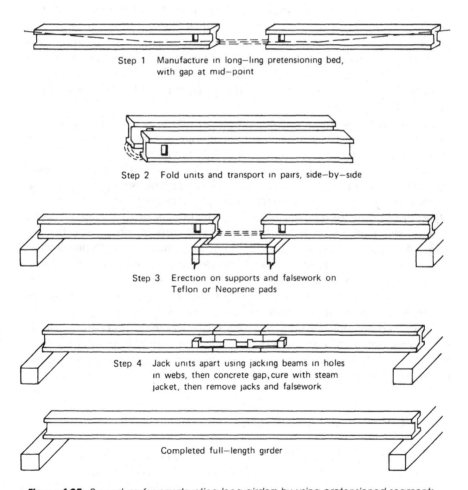

Step 1 Manufacture in long–ling pretensioning bed, with gap at mid–point

Step 2 Fold units and transport in pairs, side–by–side

Step 3 Erection on supports and falsework on Teflon or Neoprene pads

Step 4 Jack units apart using jacking beams in holes in webs, then concrete gap, cure with steam jacket, then remove jacks and falsework

Completed full–length girder

Figure 4.25. *Procedure for constructing long girders by using pretensioned segments.*

4.15 PRESTRESSING OF OTHER COMPRESSIVE MATERIALS

Prestressing may be beneficially applied to other materials to produce a state of stress that will counter service-imposed stresses. The same principles apply as for prestressed concrete. Special consideration must be given to the modulus of elasticity and creep.

Among the materials that have been prestressed with success are stone, brick, cast iron, ceramics, steel, and wood.

Marble, granite, ceramics, brick and glass facings are often embedded in a thin, prestressed concrete slab, then erected as a large panel. The concrete slab and facing material work in composite action under prestress. Therefore, it is necessary to evaluate the different moduli of elasticity of the facing materials and the concrete, and locate the tendons so as to prevent excessive warping or camber.

Native rock is sometimes prestressed to prevent its failure in tension and shear. This is an extension of the rock-bolting principle. Because of the higher stresses used, care must be taken to provide a proper bearing seat for the anchorage, such as a concrete pad (precast or cast-in-place). Grout or epoxy injection may be used in conjunction with prestressing to fill cracks or seams. Usually the tendons must be re-stressed to compensate for early creep of weaker material in fractured zones or seams of the rock.

Soil may be precompressed by prestressing anchors in soil. Creep here is a major problem, being of a much greater order of magnitude than with other materials. Therefore, it may be necessary to re-stress soil anchors one or more times.

Cast iron and cast steel offer opportunities for prestressing, using essentially segmental construction methods. Girders and trusses have been prestressed (e.g., for hangar roofs), using tendons to impart an upward camber, and making use of the efficiency of the high-strength steel tendon. Care must be taken to prevent buckling of the lower flange (chord) and to be sure that truss member connections are designed for the secondary stresses induced by axial shortening of the lower chord. A structural steel bridge with concrete composite deck, crossing the Columbia River, was post-tensioned to offset the dead-load stresses.

Prestressing has been used in the construction of large sculptures, to provide structural support and to prevent long-term deformations.

Finally, prestressing has been used very effectively in the restoration of ancient monuments, where existing segments are drilled, a tendon placed and stressed, and the hole grouted. The temple columns of Baalbek, in Lebanon, are among the ruins restored in this manner.

4.16 HORIZONTALLY CURVED BEAMS

Many structures can be most efficiently and aesthetically constructed with horizontal curvature. This especially applies to highway overpasses and bridges, and also to such structures are monorail girders. The tendency in the past was to form such horizontal curvature by a series of short chords; this applied to both precast and cast-in-place structures. However, such solutions were distinctly not elegant and presented deficiencies in aesthetics as well as in efficiency.

If the center of prestressing force is concentric with the center of gravity of the section, then the structural element will retain its curvature, since at every transverse plane, the stress will be concentric.

Many cast-in-place prestressed concrete structures have been cast with horizontal curvature, sometimes of very short radius. Provided the ducts are accurately located, the subsequent post-tensioning will cause no subsequent deflection in curvature. However, such structures are sensitive to deviations in alignment; thus, ducts should either be rigid ducts, or else flexible metal ducts with an internal mandrel. Sometimes, especially with bars, the tendon itself may be used as the mandrel. Furthermore, accuracy of concrete section must be maintained in casting. The radial local forces must of course be countered with stirrups and ties.

A number of important structures on horizontal and even three-dimensional curvature have been constructed by using precast segments. This has permitted more accurate casting of thin sections and better control of duct position.

There has been a reluctance on the part of the designers to recognize the potential for horizontal curvature of precast prestressed girders. In recent years, however, a significant use has been made of I-girders and box girders which were curved in plan. In manufacture, the side forms have been made flexible, for example, by utilizing steel forms with vertical angle stiffening only. Pulling, jacking, or wedging

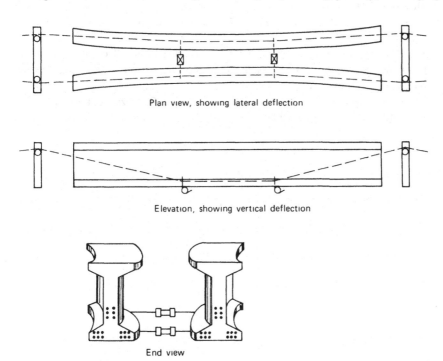

Plan view, showing lateral deflection

Elevation, showing vertical deflection

End view

Figure 4.26. *Manufacture of horizontally curved pretensioned concrete beams*

them at a few critical points automatically forces the form to achieve a smooth curvature (Fig. 4.26).

When horizontally (or vertically) curved members are stressed, the tendon will exert a radial component. While globally this is transformed into circumferential compression, locally it develops radial shear and circumferential tension, as it tries to pull the outer portion of the concrete into arch compression. Such delamination and shear failure has led to pull-through of the tendons of curved bridge girders.

Thus ties or stirrups are important in order to prevent delamination and shear failure, especially if several tendons are located close to each other. Then there may be insufficient concrete area to transfer the local tensions.

Among the notable examples of horizontally curved prestressed concrete girders are the precast monorail girders for DisneyWorld and the many cast-in-place pre-stressed concrete overcrossings in California and elsewhere.

SELECTED REFERENCES

1. *ACI Manual of Concrete Practice*, Detroit, Michigan (revised biannually). ACI 213, "Lightweight Concrete"; ACI 304, "Lightweight Concrete"; ACI 363, "High Strength Concrete."

2. ACI Publication SP-121, *High Strength Concrete*, American Concrete Institute, Detroit, Michigan, 1990.

3. Libby, J., *Modern Prestressed Concrete*, 3rd edition, Van Nostrand-Reinhold, New York 1984.

4. Freyssinet International, "Anchorages for External and Internal Tendons," Paris, 1991.

5. PCI *Recommended Practice for Erection of Precast Concrete*, Prestressed Concrete Institute, Chicago, Illinois, 1985.

6. *FIP State of Art Reports*, Structural Engineering Trading Organization, London. "Special Concretes," 1988; "Condensed Silica Fume in Concrete," 1988.

CHAPTER 5

Durability

5.1 GENERAL

Prestressed and reinforced concrete is intended to provide proper serviceability and ultimate strength over its design life with a minimum of maintenance. Indeed, durability is one of the greatest advantages of concrete as a structural system; concrete is especially appropriate for use in adverse environments.

However, durability is not automatically ensured by the production of concrete of appropriate strength. In recent years, the combination of more extensive experience in varied environments and increased laboratory research using modern instruments, such as the scanning electron microscope, has clarified many of the complex phenomena which degrade the long-term performance of the concrete.

Therefore, design for durability is now achieving equal importance to design for strength. However, the design must be implemented by proper construction technology, under strict quality control procedures, if durability is to be achieved.

Two quotations from recent papers are highly relevant.

In his paper "Concrete—50 Years of Progress," P. K. Mehta states:

> To solve durability-related problems, we need to divert more resources towards integration of our present knowledge into a "holistic" understanding of the durability issue.

Translated to the matters affecting prestressed concrete, this means that attention must be given to the system as a whole, integrating the phenomena attacking all the components of the structure: concrete and steels—and taking into account their interaction.

The second quotation is from O. Valenta in "Durability of Concrete," published in the *Proceedings of the 5th International Symposium on the Chemistry of Cements*, Tokyo, Japan, 1968:

> The durability of concrete cannot be separated from the durability of concrete structures. This (latter) must be ensured on a complex basis, starting with the choice of building site and suitable system of construction and ending with the proper design, taking into account the nature of the surrounding environment and the technology of concrete, with due recognition of the importance of the choice of suitable materials, concrete composition, and production, placing, and curing. It is imperative that proper care be accorded to the finished structures: this includes protection of the concrete and the concrete structures against the effects of aggressive media, regular checks (monitoring) and timely maintenance or repair.

This eloquent statement can be best understood by an explicit recognition that by "structure," Valenta includes the passive and prestressing steels and all appurtenances.

Lack of durability may show up as cracking and spalling of the concrete, disintegration, especially of the surface, rust staining, corrosion of the reinforcing or prestressing steel, corrosion of anchorages and embedments, excessive deflection, and in the extreme cases, by failure of the member.

The above phenomena may interact so as to intensify the disruption. For example, cracking of the concrete surface may lead to accelerated corrosion of the reinforcing steel, which in turn causes further spalling of the concrete cover, exposing more steel.

Corrosion and disintegration are not random or sport occurrences. Rather, when disintegration and disruption do take place, it is usually due to some fundamental error or neglect; the cause is real and definite and often extends to the entire structure. Thus, except for some localized spot of impact or accident, if disintegration is found, a thorough investigation should be made of the entire structure.

Among the more common environmental attacks are the following:

A. Those causing or accelerating disintegration of or change in the concrete:
1. Reactive aggregates: those containing critical percentages of reactive silica, combined with alkalis in the surrounding soil or water, or in the cement.
2. Unsound aggregates, subject to attack by sulfates.
3. Cement containing high percentages of tricalcium aluminate (C3A) combined with sulfates in the surrounding soil or water.
4. Multiple freeze-thaw cycles, especially on saturated concrete.
5. Erosion and abrasion from cavitation, moving sand and ice.
6. Aggressive marine organisms, such as mollusks.
7. Micro- and macrocracks, allowing greater penetration of water, chlorides, and alkalis and exposing more surfaces to freeze-thaw attack.
8. Salt scaling.

9. Acids, sulfates, nitrates, or organic substances in mixing water or in surrounding water, as at the discharge from chemical plants or in sewage structures or crude oil storage where anaerobic bacteria may generate H_2S and, in the presence of oxygen, weak concentrations of H_2SO_4.

10. Spalling of the concrete cover, due to impact, excessive stress, or corrosion of the steel.

B. Those causing or accelerating corrosion of steel:

1. Salt on aggregates or in mixing water or admixtures.

2. Chlorides in water used for curing.

3. Chlorides in water surrounding the concrete, e.g., seawater, and in the atmosphere as salt fog or spray.

4. Carbon dioxide in the atmosphere, leading to carbonization of the concrete cover and lower pH.

5. Inadequate thickness of concrete cover, high permeability of cover, or micro- and macrocracks, allowing penetration of chlorides and CO_2.

6. Oxygen in the air or dissolved in water.

7. Sulfides combined with moisture on stressed tendons prior to encasement or protection, in stack gasses, such as occur in railroad undercrossings and tunnels, or adjacent to refineries.

8. Stray electric currents.

9. Alkalis in surrounding soils.

10. High temperature.

11. Embedded metals other than steel, particularly copper and aluminum.

12. Cement chemistry (e.g., too low C3A), thus allowing higher chloride permeability.

13. Deicing salts, acids, or other aggressive chemicals.

14. Steels sensitive to corrosion.

15. Incipient fatigue of steel, leading to corrosion-accelerated fatigue.

16. Excessive width or number of visible cracks (macrocracks).

Fortunately, many of the steps to be adopted to overcome these forms of attack are complementary to each other. Most such precautions are normally adopted as standard good practice. Occasionally, one has been neglected and, as pointed out above, the effects have been extremely serious.

Potential problems of durability arise in both reinforced concrete and prestressed concrete. Prestressing, by itself, improves durability by preventing cracking, which in turn minimizes the penetration of water and air. Prestressing also greatly reduces cracking due to fatigue and for most real-life cases, eliminates it. Since prestressed concrete has "higher added value," greater care is normally taken in its detailing, material selection, mix design, and construction. To achieve the higher long-term and short-term strength required for prestressed concrete, lower W/C ratio and richer cement contents are normally employed. These all tend to result in greater inherent durability. However, prestressing by itself does not significantly improve the permeability of uncracked concrete.

On the other hand, prestressing introduces some special durability problems of its own, which need to be properly addressed.

In the sections that follow, the various phenomena affecting durability will be considered, along with preventive or mitigating measures. It should of course be emphasized that not all aggressive actions occur for every structure: these depend on the environment and the service.

5.2 CORROSION OF REINFORCEMENT

Corrosion of reinforcing (and prestressing) steel is the most frequent and most serious form of degradation and disruption of prestressed concrete structures. In its most extreme form, it may occur very rapidly, so that, in some cases, severe damage has occurred prior to the structure being placed in service. In the majority of cases, the attack is less sudden but perhaps more sinister, in that it arises with little warning during the most active part of the structure's life.

Actually, warning signs do occur, if there is someone alert to observe them before they reach serious levels: rust staining, cracking, and spalling of the concrete surface. Monitoring of important or sensitive structures may give earlier indications of corrosion, in time for remedial action. Monitoring methods include electrolytic half-cell (copper-copper sulfate or silver-silver chloride); linear polarization, gamma radiography, acoustic transmission, radar, and ultrasonic pulse echo. A simple and effective method is the "chain drag" on slabs and the "hand-held" hammer on walls. The human ear can readily pick up the hollow sound due to internal delamination. Increasingly, all these methods are being employed on a routine basis on critical structures such as long-span bridges and subaqueous tunnels.

For the great majority of structures, however, the sophisticated methods are impractical and, to a large extent, unnecessary in the absence of other information and warnings. Visual examination usually suffices up until the time that early corrosion is observed.

The corrosion of reinforcing steel is a complex process. When embedded in fresh concrete, with cement mortar surrounding the steel, a thin coating of ferro-ferrous hydroxide forms on the surface of the steel to act as a protective layer. The steel surface has been passivated.

The surface remains passivated until some other action ensues. The penetration of carbon dioxide from the air may cause carbonation, with lowering of the pH until the passivating layer dissolves. Fortunately this is a slow process, especially for the impermeable concrete usually used in prestressed application. It may be a problem, however, when cover of concrete over the steel is minimal as it is in some architectural applications.

The most serious aggressive element is chlorides, carried to the concrete surface by the splash of seawater, salt air, or salt fog. These latter may spread chloride molecules up to 50 miles inland from the sea. In coastal and other above-water zones, the moisture carrying the salt evaporates, leaving a growing concentration of chlorides behind.

Chloride ions penetrate by both capillary action and diffusion through water-filled pores. When they reach the steel, the protective layer dissolves. If oxygen molecules are also able to reach electrically connected reinforcement, the corrosion process is initiated. Although a concentration of chlorides of 0.2% by weight of cement is enough to cause corrosion in extreme cases, values of 0.4 to 0.5% at the surface of the steel are more usually considered the threshold level.

There is today a great deal of profound technical discussion concerning the relative roles of permeability and diffusivity, etc. From a pragmatic point of view, chloride corrosion usually occurs first and most seriously on the windward (seaward) face of exposed structural elements in and just above the splash zone.

For steel in concrete to corrode, it needs chlorides, carbonates, or sulfides, all of which form dilute acids. It also needs oxygen. It needs an electrolyte, which is generally furnished by the water in the pores of the concrete. Since corrosion is an electrochemical phenomenon, resistivity of the concrete is a measure of the rate at which corrosion can occur. Dry concrete and impermeable concretes have much higher resistivity.

For structural elements which are permanently below water, the oxygen availability is greatly restricted and the rate of corrosion will normally be extremely slow. For elements below both the permanent water level and the soil, there is usually insufficient oxygen to sustain corrosion (unless the soil is highly organic, as in an old garbage dump, or very pervious). Conversely, the most dangerous exposure for reinforced and prestressed concrete is when it is alternately wet and dry. This cyclic exposure, by leading to microcracking, increases permeation of chloride ions and oxygen. Graving docks, seawater intakes, and piling in the tidal zone are particularly subject to corrosion of the steel.

The penetration of chlorides is proportional to the permeability and diffusivity of the concrete. Aggregates typically have very low permeability, as does cement paste, but concrete, a mix of the two, has relatively high permeability. This is due to microcracks around the coarse aggregate particles and the reinforcement, caused by trapped bleed water, thermal and shrinkage strains, impact, and in exceptional cases, by fatigue. The rate of chloride ion penetration is accelerated by high temperatures.

There is evidence that carbonation from the CO_2 of the air can act synergistically with the chlorides to increase the penetration rate of the latter.

Some coarse aggregates such as limestone have significant permeability.

Structural or thermal cracks, which are generally individual, widely spaced, visible cracks, may not lead directly to corrosion. Limits determined empirically indicate that, in the splash zone, crack widths at the surface of 0.15 mm (0.006 inches) do not normally lead to corrosion. Underwater, crack widths of 0.30 mm (0.012 inches) appear to be acceptable.

Although these visible cracks of small surface width do not necessarily lead to corrosion, chlorides may be drawn into the crack by capillary attraction. Chloride concentrations along the crack will often exceed the minimum for corrosion initiation. However, if the surrounding concrete is relatively impermeable, the small anode formed at a crossing bar of reinforcing steel will generally not develop into significant pitting due to the restricted flow of oxygen to adjoining reinforcement (the cathode).

If, however, the crack is on top of and parallel to a reinforcing bar, then along the bar, cells will form and local cathodes and anodes will drive the corrosion.

Because of the dominance of corrosion of steel as a disruptive phenomenon that weakens and eventually destroys reinforced and prestressed concrete, most of the research and practice has concentrated on understanding and preventing corrosion. It has been well established that this is an electrochemical phenomenon, that chlorides are both the most prevalent and damaging initiating element, that oxygen availability controls the rate of corrosion, and that an electrolyte (water) must be present in the concrete. Chlorides come not only from the marine environment but from salt that is intentionally spread on bridge decks to prevent icing. Vehicle tires then carry this salt into parking garages, where the relatively thin decks are rapidly subjected to corrosion. Tires may also splash chlorides onto adjacent walls, bridge piers, and columns.

Particularly vulnerable areas include the undersides of wharf decks and bridge piers, which are frequently splashed by waves. There, salt is concentrated and there is ample oxygen and electrolyte to sustain a high corrosion rate.

In recent years, a number of tunnels have been built under arms of the sea. These have been both the precast submerged tunnel, which is often prestressed, and the driven tunnel, where concrete liners, both precast and cast-in-place, are installed. Saline water is available under pressure on the outside: it moves to the inside through leaks and intentional bleeding and it also permeates and diffuses through the concrete. Capillary attraction and diffusion then move the chloride ions to the inside, where they evaporate, leaving the salt on the inside surface. This then becomes increasingly concentrated.

A similar concentration occurs on over-water structures such as bridge decks where wind-blown spray, fog, and rain bring the chloride-laden moisture to the surface of the structure. On subsequent evaporation, the salt concentrates. The potential exists for such concentration on the dry inside face of concrete barges, and within the dry shafts of offshore structures.

Being an electrochemical phenomenon, both anodes and cathodes are required. Corrosion occurs at the anode where the iron ion leaves the bar in the form of rust (ferrous oxide): pitting may be its most serious form. The electromotive driving force comes from the cathode, which is the mat of reinforcing steel bars to which the oxygen molecules have access. The cathode may be an adjacent zone on the same reinforcing bar or a remote mat of reinforcing steel that is electrically connected.

Concerns have been expressed where there are very large areas of reinforcing steel exposed to potential oxygen permeation, the decks of bridges and wharves being prime examples. Will these act as giant cathodes to drive the corrosion of the steel near the water? Certainly, these large areas of steel with only minimal coverage can act as large cathodes and be a major contributor to the corrosion phenomenon.

Unfortunately, it is very difficult to seal these against oxygen permeation. Even coatings are only partial barriers. The rate of corrosion can be slowed but not stopped by impermeable concrete, adequate cover, and coatings on the steel or concrete surface. Therefore, these obvious zones of large cathodic potential are frequently isolated from the zone of potential corrosion by providing a positive break in the electric circuit. For example, the Port of Oakland, California, coats the protruding reinforcing bars of piles with epoxy so as to isolate the pile steel from the deck reinforcement.

Corrosion of reinforcement may also be generated when dissimilar metals are used in electrical contact. This has usually arisen where aluminum conduits have been embedded, or copper or its alloys used in connecting panels.

For example, special stainless steel connectors were specified for the precast panels for a saltwater cooling tower: these were found to have the potential for corrosion and a different alloy was selected instead.

Stray currents from D.C. operating equipment, subway systems, welding machines, electrically powered cranes, etc., can produce aggravated corrosion where the current discharges to ground. Separate grounding should be provided for all such equipment as well as for lightning.

The preceding discussions have all dealt with the problems of chlorides entering the mix from external sources, after the concrete has hardened, and indeed, this is the phenomenon which is the most serious and least understood in practice.

However, chlorides may be incorporated in the fresh concrete, either intentionally or unintentionally. The intentional act would be for the purpose of accelerating set. Addition of 1 to 2% of calcium chloride, by weight of cement, has in the past years been a widespread practice when placing concrete in below-freezing weather. It is still practiced in some northern building construction. If the concrete so produced is relatively impermeable and if the subsequent environment in which it serves is more or less completely dry, with low relative humidity, then it is probable that no significant corrosion will ensue. However, if an electrolyte becomes present in service, active and rapid corrosion may take place.

The incorporation of chloride in the mix prevents the formation of the protective, passivating coat on the surface of the steel. Thus the steel remains ready to corrode as soon as the electrolyte and oxygen become available.

Some of the most notorious cases have been in aqueducts, penstocks, and pipelines carrying water. The water permeates the concrete fully and corrosion proceeds at a rapid pace.

There is one mitigating process, however. Chlorides present in small amounts in the mix will combine preferentially with the C3A in the cement and hence will not be free to depassivate the steel.

Chlorides may also be incorporated unintentionally. They may be present in the mixing water, or on the surface of the aggregate. In desert regions, groundwater migrates to the surface and evaporates, leaving deposits of salt on the surface deposits of sand and gravel. In dry regions adjoining seas, for example, the shores of the Arabian Gulf or the coasts of southern Peru, salt fogs and groundwater evaporation repeatedly deposit minute amounts of chlorides on the sands, amounts which accumulate to significant concentrations. Some admixtures contain small amounts of chlorides.

Recognizing the special character of prestressing steel, especially its high strength and state of stress, along with its small diameter, special efforts are taken to eliminate all significant amounts of chloride from the concrete mix. Typical specifications limit the amounts of chloride in the mixing water to 650 ppm (0.065%) and of soluble chlorides in the aggregates to 0.02% by weight of cement.

5.3 MEASURES TO PREVENT CORROSION OF REINFORCEMENT

5.3.1 Impermeability

The single most effective countermeasure is to produce concrete of very low permeability. It is now possible to measure the diffusion and permeability coefficients: values of 10^{-12} to 10^{-13} m/s for both D and K are desirable in a corrosive environment. Impermeability not only limits the penetration of chlorides but also restricts the flow of oxygen (Fig. 5.1).

Permeability coefficients of 3 to 5×10^{-12} m/s have been attained on North Sea Offshore Platforms, and of 5×10^{-13} m/s on precast tunnel liner segments for the Channel Tunnel.

5.3.1.1 REDUCTION OF THE RATIO OF WATER TO CEMENTITIOUS MATERIALS. Lower W/CM ratios reduce permeability and increase resistivity. Use of substantial doses of conventional water-reducing agents or of HRWR agents (super-plasticizers) now makes it possible to achieve W/CM ratios of 0.37 to 0.38 in concrete for prestressed applications. Even lower values are practically obtainable in some precast products, e.g., 0.32.

Figure 5.1. Impermeable concrete has provided 30 years of durability to wharves in the harsh environment of the Arabian Gulf.

It has been found pragmatically that most cases of corrosion involve concrete with a W/CM ratio above 0.45 or 0.50.

5.3.1.2 INCLUSION OF POZZOLANIC MATERIALS IN THE MIX. Pulverized fly ash (PFA) is commonly available and can be used to replace the cement in quantities up to 20% (and even 30%) in typical prestressed concrete.

PFA can increase the impermeability by 100 to 1000 times.

5.3.1.3 INCLUSION OF MICROSILICA IN THE MIX. This material is condensed from silica fumes and consists of very fine particles, about $\frac{1}{100}$ the size of cement grains, of almost pure silica. When incorporated in the mix, they reduce permeability and diffusivity, while increasing the resistivity.

Although higher percentages are possible, this author prefers to limit the amount of silica fume to 4 to 6%, since more tends to make the mix excessively cohesive and "sticky."

5.3.1.4 USE OF A PORTLAND CEMENT CONTAINING 6 TO 10% C3A. The C3A (tricalcium aluminate) combines with the chloride ion as it penetrates and forms insoluble chloro-aluminates which block the pores. It is probably this phenomenon which accounts for the reduced rates of chloride penetration with time and the amazing longevity of some of our older concrete structures which have surprisingly defied the corrosive phenomenon.

In some areas, excessive concern over sulfate attack on the concrete has led to adoption of Type V cement, containing 0 to 4% C3A. Unfortunately, this cement is much more permeable to chloride penetration. Its use in the Middle East, in coastal zones has led to very serious corrosion of the steel at an early age.

5.3.1.5 SELECTION OF SMALL-SIZE COARSE AGGREGATE. Since the permeation takes place mostly under and around the coarse aggregate particles, use of small maximum size (typically 20 mm ($\frac{3}{4}$ inch) or less) increases the length of the pathway and provides discontinuities.

5.3.1.6 USE OF AGGREGATE WITH A COEFFICIENT OF THERMAL EXPANSION AS CLOSE AS POSSIBLE TO THAT OF CEMENT PASTE. This is usually only practicable on very large, specialized projects. Limestone is generally more compatible than siliceous rock.

5.3.1.7 USE OF NONPERMEABLE AGGREGATES OF SMALL MAXIMUM SIZE. Most coarse aggregates are highly impermeable in themselves. However, some limestone aggregates, especially in the Middle East, have both high porosity and high permeability, allowing rapid penetration of chlorides. If their use cannot be avoided, special steps are essential: use of very small coarse aggregate, i.e., less than 10 mm, and a rich mix high in PFA. Other steps to prevent corrosion, such as coating of the reinforcement, may be necessary.

It is important to differentiate between permeability and porosity. Many aggregates have discrete, nonconnected pores. Expanded lightweight aggregate is an example of a porous aggregate which is very impermeable and consequently produces very durable reinforced concrete.

Small maximum size coarse aggregates distribute microcracks and lengthen the paths of potential permeation. Thus in most prestressed applications, the nominal maximum size of coarse aggregate should be 18 to 20 mm ($\frac{3}{4}$ inch).

5.3.1.8 PROPER CURING. The surface skin of a cement paste is very effective in limiting penetration of water, chlorides, and also oxygen. Its impermeability depends on full hydration. Curing, by water, humidity, a membrane, or polyethylene, etc., is important to attaining full hydration.

It is important to provide full coverage of the exposed surfaces, despite the problems caused by wind.

5.3.1.9 MINIMIZING THERMAL STRAINS. The heat of hydration of concrete causes expansion of the fresh mix. After setting, the concrete cools and contracts. This induces strains in the concrete. Measures to reduce heat include:

- Pre-cooling of the mix by batching ice as the mixing water, by chilling the aggregates by use of water-soaker hoses and evaporation, or even by vacuum cooling, or by injecting liquid nitrogen into aggregate or the fresh concrete.
- Using cement that generates less heat: i.e., Type II ASTM with its coarser grind generates less heat than Types I or III.
- Replacing part of the cement with PFA, which reacts much more slowly. 15% replacement by PFA lowers the typical concrete's maximum temperature by 4 to 5°C (8 to 10°F).
- Using BFS cement, which generates heat more slowly. This is only effective if the BFS is coarse ground, e.g., less than 3800 cm²/g. (Note that some manufacturers, in an attempt to get high early strength, grind BFS to 5500 cm²/g, which liberates heat rapidly.)
- Insulating forms where exposed to low temperatures and wind.
- Draping blankets or tarpaulins over concrete surfaces after stripping forms to protect from very cold winds.
- Water curing and using a white reflective curing compound, both of which reduce the maximum temperature, especially near the surface.

Surface microcracking may arise from excessive shrinkage or from cycles of wetting and drying. Rich cement mixes and those containing condensed silica fume are especially subject to crazing and shrinkage cracks. These can be minimized by proper curing. Usually water curing on horizontal surfaces and membrane curing on vertical surfaces are most effective.

Membrane curing compounds degrade with the heat from the concrete and sun: in many cases two applications are necessary, one immediately following form removal and the other some eight hours later. Of course leaving forms on will also prevent moisture loss. Steam curing and high humidity curing are appropriate means for precast products.

5.3.1.10 IMPREGNATION OF SURFACE. The surface of concrete may be impregnated with a monomer which is then polymerized to form an essentially impermeable surface layer, typically 30 to 50 mm ($1\frac{1}{4}$ to 2 inches) in depth.

The process typically involves steaming so as to displace the air in the concrete pores with steam, then allowing it to condense, while excluding air. This creates a vacuum in the pores. The monomer is then introduced, usually in a water mix, and penetrates the pores. Heating is then used to polymerize the monomer.

The most common monomer for this purpose is methyl methacrylate, a highly toxic and inflammable material which must be handled with extreme care.

This process is reportedly being applied to impregnate and render impermeable the decks of the Florida Keys Bridges.

Other materials may be used to impregnate the surface and, by filling the pores, render them impermeable to a high degree. These are generally proprietary compounds containing silica which reacts with the calcium hydroxide. They are most effective with highly permeable concrete, i.e., concrete of high W/CM ratio, because they can penetrate more readily. However, it is generally accepted that they are of some value even with high-quality concrete, as they also seal shrinkage cracks.

Silane is a similar penetrant: it has the advantage of being able to allow water vapor to escape, while limiting the entry of water and chlorides. Since it is water soluble, the silane must be periodically reapplied. 100% silane is used on some European bridges to seal the decks and curbs, while in the United States compounds of 20 to 40% silane are marketed. These are cut back, either with petroleum derivatives or with water, this latter in order to reduce toxicity.

5.3.1.11 COATINGS ON THE CONCRETE SURFACE. Exterior coatings have been utilized with varied success. When the coating itself is not degraded or damaged, it protects the concrete surface from microcracking due to alternate wetting and drying.

Epoxy and polyurethane coats, properly applied to concrete whose surface has been adequately prepared, are impermeable to chlorides and, to a limited degree, to oxygen. However these coatings are subject to ultraviolet degradation; the life of the outer coat is typically 5 to 10 years.

Epoxy (and polyurethane) coats have a potential problem of "holidays" due to vapor bubbles. The water vapor migrates to the surface and pops a hole in the coating. If the concrete is thoroughly dry prior to application, and if the epoxy has good tensile strength and elasticity, this problem can be minimized or prevented.

The three coats of epoxy on the splash zone of the piling of the King Fahd Bridges between Saudi Arabia and Bahrain have prevented all chloride penetration during the so-far 10 years of exposure.

5.3.2 Incorporation of Anti-Corrosive Admixtures in the Concrete Mix

Nitrites (both sodium nitrite and calcium nitrite) have been used for many decades as an inhibitor of corrosion. Calcium nitrite has emerged as the most reliable compound and is now widely available. The available evidence definitely indicates a beneficial effect in the early years, but one which may diminish with time.

Other proprietary compounds are available which reportedly reduce the diffusivity, the migration of the chloride ion through the pores, and interstitial water in the concrete.

5.3.3 Minimization of Entrapped Air

It is important to distinguish between entrained and entrapped air.

Air entrainment provides very small, widely dispersed bubbles which may reduce permeability by acting as crack arresters to prevent the propagation of microcracks.

Entrapped air consists of relatively large bubbles, visible to the naked eye. Some research indicates that anodes preferentially form at the site of an entrapped air void alongside the reinforcing bar.

Air entrapment may be minimized by reducing the amount of extreme fines (silt fraction) in the sand, e.g., by re-washing. Internal vibration is effective in driving out entrapped air. External vibration alone may not be able to accomplish this at the depth of the steel, but is effective to at least the depth of cover.

In cases where available aggregates entrap excessive air (over 1 to 2%) then a de-aerating admixture may be employed, although this is not common. Similarly, bleed water tubes, the so-called bug holes, may be minimized by vibration and, where judged necessary, by the incorporation of an anti-bleed admixture. PFA or microsilica also reduce bleed.

5.3.4 Seal Around Embedments and Anchorages

Mortar patches often have small shrinkage cracks around the edges.

Welding onto an embedded plate will often cause it to warp at the edges. Prestressing may cause cracking at the corners and edges of embedded steel plates.

Sealing may be accomplished by epoxies, placed as a putty, or by flexible polymers, such as Thiokol. Note however that epoxy coatings are unreliable in freeze-thaw environments, since the water vapor is trapped beneath the epoxy and freezes, popping the cover off. Silane or latex coatings are preferable in such cases.

5.3.5 Treatment of Construction Joints

Construction joints are often a source of water and chloride penetration, with resultant corrosion. The usually higher concentration of reinforcing steel at such joints, due to lapping of the bars, often presents practical problems of maintaining proper cover, while leading to increased corrosion potential and disruptive effects. Construction joints frequently have honeycomb or poorly consolidated concrete.

By careful removal of laitance and exposure of the coarse aggregate, a proper surface may be prepared. This should be done by wire brushing after initial set, or high-pressure water jetting of the hardened surface, or by wet sandblasting. Bush hammering must never be used, as this loosens the coarse aggregate, creating microcracks around the coarse particles.

5.3.6 Repair of Impact Spalls

Impact and intensely concentrated loads may produce both visible cracks and micro-cracks, whose interlocking network is especially conducive to local corrosion. Cracks larger than 0.15 mm (0.006 inches) should be promptly repaired, making sure that this is done before salt can be deposited on the cracked and spalled face.

Zones where it is known that impact may occur can be specially reinforced by stirrups or wire mesh near the surface or by inclusion of fibers in the mix.

5.3.7 Provision of Adequate Cover

While low permeability lengthens the time for chlorides, carbon dioxide and oxygen to reach the steel, provision of adequate cover is also essential. The cover should normally be 35 to 50 mm (1.5 to 2.5 inches). Other empirical (but valid) rules are: "At least twice the size of the largest coarse aggregate particle," and "Twice the diameter of the reinforcing steel bar." Too small a cover allows carbonation and chloride penetration to progress to the level of the steel. Too much cover permits cracking from shrinkage and loadings.

5.3.8 Use of Small Size Reinforcing Bars

Heavy concentrations of reinforcing steel increase the current available to drive the electrochemical process. Large bars tend to have more microcracking around them. Use of small bars, closely spread, is thus preferable to the use of large bars. Well-distributed bars tend to limit the extent of damage in the event of corrosion. Macro-cracking is also reduced, with more fine cracks and fewer wide cracks.

5.3.9 Epoxy Coating of Reinforcement

Epoxy coatings have been increasingly employed to prevent corrosion in bridge decks and parking structures. These are generally applied by electrostatic fusion to straight reinforcing bars. Bending of bars, if required, takes place thereafter.

It is however becoming practicable to apply this process to relatively small compact bent bars such as stirrups and spiral.

Bending afterwards may produce cracking and sometimes delamination at the bend. In placing the bars, areas may become abraded. These areas are then touched up by hand.

Such abraded areas or scratches, if well enclosed in impermeable concrete, do not necessarily lead to corrosion, since, while the bare spot becomes the anode, there is not enough cathodic area to support the current flow.

Delamination of the epoxy coating may focus corrosion, especially if it occurs at corners where oxygen can flow in from both sides.

Another process, hot dip epoxy coating, is especially applicable to fabricated and welded cages of steel of restricted dimensions, e.g., the cages for precast tunnel liners. This coats the welded areas and the bent areas equally.

Epoxy coated bars do not have adhesive bond, but rely instead on the deformations to transfer stress in shear. Under high stress, these lugs tend to wedge the concrete and develop cracking along the bars, leading to delamination of the concrete. These problems can be successfully overcome by use of more and smaller bars and by providing adequate cover.

Epoxy coatings with fine sand embedded are now available. They develop excellent bond.

5.3.10 Zinc Coating (Galvanizing)

This has been extensively used with architectural concrete to prevent rust staining. It is important that this coating be passivated with a chromate wash after galvanizing, to prevent liberation of hydrogen ions when it contacts the alkali in the cement. This is normal practice in most developed countries but is not universal. Hot dip galvanizing and sherardizing are less conducive to hydrogen liberation, and hence are to be preferred over electro-galvanizing.

Concerns have recently been raised about the electro-fusion method of galvanization and, in the light of present knowledge, this process should be avoided.

5.3.11 Other Coatings

Dipping bars in a cement wash was extensively employed by the US Navy for construction in remote islands where fresh water was unavailable so that seawater had to be used in the mix. In India, similarly, bars have been given a wash coat of phosphate in order to provide a more durable passivating layer. Both practices are labor intensive and obviously expose the bars to abrasion and scratches during installation, so are not widely employed today.

5.3.12 Noncorroding Tendons

Reinforcing bars (and prestressing tendons) may be made of inherently noncorrosive steels, e.g., stainless steel. A Japanese steel company has recently developed a somewhat less expensive steel alloy which is highly corrosion resistant. Titanium bars have been used to repair ancient structures and monuments such as the Parthenon.

5.3.13 Cathodic Protection

Cathodic protection may be very effectively applied to prevent corrosion. It is based on the principle of providing an anodic receiver which draws electrons from the steel at a voltage which is adequate to make all the steel cathodic. The anodic receiver, whether it be metallic or simply a D.C. source, is oxidized and consumed in the process.

With cathodic protection, it is critically important that all steel elements within the concrete, reinforcing bars, mesh, prestressing steel, ducts, and all inserts be bonded: otherwise the unbonded elements will become anodes and corrode at an accelerated rate.

Sacrificial anodes are a simple and effective way of preventing further corrosion, even if it has already commenced. However, they are only effective if both the anode and the concrete to be protected are in an electrolyte, e.g., under water. Thus in a seawater intake, or conveyance, and in discharge structures, the underwater concrete can be protected by installation of zinc anodes. Electrical connection must be provided to complete the circuit. Anodes must be located so that they "see" the surface of the concrete element to be protected, since the electrons only move on line-of-sight straight paths.

Sacrificial anodes are quite cost-competitive with other special forms of protection. However, since they are confined to underwater applications, where the corrosion risk is already minimal, they are normally not applied. Exceptions are those structures which must be occasionally emptied, such as graving docks, those where the water level varies, exposing the concrete periodically, or those subject to stray currents. For these, expert advice is essential.

Recent application has been made of flame-sprayed zinc, applied to the exposed bars and adjacent concrete surface after the cover has been removed by shot blasting. This method appears to hold promise for structures which have begun to corrode and which are located in environments of warm temperature and high humidity.

The other form of cathodic protection, impressed current, is normally not cost-effective as an initial installation: hence it will be considered only in Chapter 24, Repairs. It can be applied both above and below water but requires the construction of a special surface layer to serve as the anode. An external current source is provided which is rectified to D.C. to supply the anodic current.

5.4 ATTACKS ON CONCRETE AND ITS CONSTITUENTS

5.4.1 Sulfate Attack

Sulfate attack on concrete has been a serious concern in the past, especially on irrigation and conveyance canals and on dams in arid environments. These facilities are characterized by aggregates reclaimed from the adjoining lands which have not been subject to intense weathering over the years, and which therefore may be unsound. Fresh water, of low pH, picks up sulfates in solution which react chemically with the aggregate to cause it to expand. ASTM Standard C33 contains tests to determine soundness of aggregates, which, however, are not 100% conclusive. Experience in use in the environment in which the structure will serve is the best guide. Note that water is required for the attack, fresh water being most dangerous. Prestressed concrete in cooling towers could be vulnerable to this form of attack.

Sulfates can also replace the aluminate compounds in cement, that is in the C3A which constitutes up to 12% and even 17% of the cement, and which is responsible for early generation of heat and strength. Here also, the principal concern is in fresh water. Seawater inhibits the replacement and in general, there is little or no danger from sulfate attack on the cement of saltwater marine structures where the cement content is greater than 330 Kg/m^3 (564 lbs/yd^3).

Nevertheless, it is generally agreed that an upper limit of 8 to 10% C3A is advisable for concrete in seawater service.

The action of sulfates can also be inhibited by the provision of silica in the form of pozzolans (PFA) or microsilica. 15% replacement of cement by PFA allows the use of higher percentages of C3A in the cement. Up to 17% C3A has then been permitted with no apparent adverse results.

Note that restriction of C3A below about 5% renders the structure much more susceptible to chloride corrosion of the reinforcing steel.

In Saudi Arabia, where sulfates in groundwater are very high, asphaltic membranes are placed under and around basement slabs and walls. C3A in the cement is usually limited to 1% or even 0%, but this has been partially responsible for the high incidence of chloride corrosion of reinforcing steel in structures near the Arabian Gulf and Red Sea.

5.4.2 Alkali-Aggregate Reaction

Perhaps the most insidious of all attacks on durability is that of alkali-aggregate reaction, since its symptoms may be delayed for many years. This reaction is caused by the chemical reaction between reactive silica in the aggregate and alkali in the cement or from external sources.

Not all silica in aggregate is reactive; in fact the great portion is obviously not. However, chert, opal, and flint are suspect particles. These usually are found in siliceous aggregates; however in recent years more and more cases of delayed reaction are being found in limestone, due to the particles of opal and flint which are included.

Many aggregate sources which have shown satisfactory experience in the past are now being found to have some reactivity. This is due to the production of cement with greater alkali content, and the increased use of concrete in environments, such as marine, where alkalis are present from the seawater, and higher proportions of cement are used in the mix. Since the reaction requires an electrolyte, many older structures, e.g., buildings, have not suffered due to being continually dry.

A great many structures built in the 1950's and 1960's used moderately high alkali cement. As they age, increasing cases of reactive phenomena are being reported.

There are three approaches to the problem, based on a recognition that there is a sharply peaked "pessimum" amount of reactive silica (about 4%), at which the disruption is most serious.

The first approach is to test aggregates and eliminate those with significant amounts of reactive silica. Silica contents of 3% are not uncommon, e.g., in granite, and are potentially dangerous.

The second approach is to limit the available alkali: Typical specifications limit the alkali in cement to 0.65% (Na$_2$O + 0.65 K$_2$O).

The third approach, one which is gaining favor, is to add reactive silica in an amount to push it well above the pessimum point. This usually means incorporation of 20% or more PFA in the mix (by weight of cement) or 6 to 10% condensed silica fume. See also Section 1.1.2.

5.4.3 Freeze-Thaw Attack

When concrete is exposed to freezing, the pore water will migrate to the cold face, where it will freeze. Freezing of course causes expansive forces. Most concrete can resist a number of cycles of freezing, with subsequent thawing. However, for typical structural concrete, after 200 to 300 cycles of freezing and thawing, internal fatigue sets in, the cells of the concrete rupture, and spalling of the face occurs.

Bridge girders in the northern climates may experience a sustained period of freezing each winter. The interior girders see only two or three cycles per year, so remain undamaged. The exterior girder whose face is exposed to the south, however, experiences thawing from the sun every day and freezing every night, with 50 or more cycles per year.

The problem is exacerbated when sufficient moisture is present to keep the concrete saturated beyond a critical level. This takes place on the decks of wharves and bridges, especially where drainage is inadequate.

An extreme case is presented in the marine environment where the temperature falls below freezing for long periods of time, e.g., in Eastern Canada and the Northeastern United States, and the tide rises and falls twice per day, leading to saturation plus two cycles of freeze-thaw daily.

Water retaining structures present another unique case. In northern climates, the air may be well below freezing for long periods. The inside, being full of water, saturates the concrete walls. A freeze-front develops about half-way deep in the concrete wall. As minor changes in temperature occur, this freeze front moves in and out, creating a laminar zone that alternately freezes and thaws. Delamination is the apparent end result after a number of years.

In this latter case, an impervious membrane on the inside of the tank appears to be the answer.

For all concrete exposed to freeze-thaw attack, air entrainment has been found essential. Although concrete of higher tensile strength will resist freeze-thaw attack for a long period, unless there is adequate air entrainment, damage will eventually result.

Air entraining admixtures are employed. The amount required depends to some extent on the maximum coarse aggregate size, but ranges from 5 to 8%. This is usually measured on the fresh concrete by an air meter.

However, the air meter cannot distinguish between entrapped air and entrained air, nor can it discern the character of the entrained bubbles. At present this can only be done by microscopic examination of slices cut from cores in the hardened concrete. These examinations give the specific surface and the spacing. Values of specific surface of pores of 25 mm^{-1} and calculated air void spacing factors of 0.20 mm to 0.23 mm maximum, are target values frequently prescribed.

The amount and character of air entrainment may be affected by other admixtures, especially super-plasticizers, and the sequence in which they are added to the mix. The amount and character may also be affected by the placement method: pumping to great heights can reduce both the number and size of the air bubbles.

Impermeable membranes or coatings, such as epoxies, may cause a special problem in freezing climates. The moisture in the concrete migrates to the cold face, where it collects behind the membrane or coating and freezes. This tends to push or pop off the barrier, often with rupture back inside the concrete some 30 to 50 mm. This presents a special problem in connection with post-tensioned anchorages, for which epoxy coating is often specified in order to prevent chloride penetration. Other forms of protection, such as latex mortar, should be employed in freezing environments.

5.4.4 Salt Scaling

When saline water enters the pores of concrete and then evaporates, the crystals exert a bursting force similar to those of freezing water. The result is a scaling of the surface.

Obviously, the more permeable the concrete, the greater the salinity, the hotter the temperature, and the lower the humidity, the more severe is the attack. Thus, the phenomenon is especially prevalent in the tidal and splash zones of the structures in such locations as the Arabian-Persian Gulf.

Salt scaling and freeze-thaw attack are synergistic and together can accelerate the surface deterioration. Freezing rejects the salt, which then crystallizes.

Scaling and abrasion, as by moving sands, also interact to accelerate the degradation and wear.

Therefore, making the concrete impermeable to as high a degree as possible is most important. Penetrants and coatings appear to be very effective.

BFS-cement appears more prone to salt-scaling than Portland cements.

5.4.5 Conversion of High Alumina Cement

High alumina cement has developed such a negative image that it is no longer used except in special applications. Its problem is that, under sustained warm temperatures and high humidity, it suffers a conversion of its gel structure over time, with a drastic reduction in strength. Thus its use is banned by the building codes of many European countries.

Its special qualities for particular applications are rapid setting and gain of strength, and the ability to hydrate even at below-zero temperatures. Thus, it fills a niche in grouting and concreting in arctic and subarctic applications where the service temperatures will not rise above 20°C.

5.4.6 Thermal Cracking

Although thermal cracking of massive unreinforced concrete structures such as dams and locks has been recognized since the 1930's, the similar problem with typical structural elements has not been adequately appreciated until recent years as a cause of degradation.

Initial problems arise during the curing period. The heat of hydration of the cementitious materials causes the plastic concrete to expand. After hardening, the exterior layers of concrete cool and contract while the inside still stays warm and expanded. The concrete is still young, and has little tensile strength. The result is cracking of the surface, often attributed erroneously to shrinkage. Shrinkage is primarily a drying-out phenomenon which usually requires an extended period of time, whereas the thermal cracking phenomenon occurs at an early age, measured usually in days.

Although this early cracking is especially severe in members with a thickness of 600 mm (24 inches) or more, where the internal core retains its heat for many days, the problem has been noted even in relatively thin-walled members. For example, when prestressed concrete piles are first uncovered to the atmosphere after steam curing, they may develop cracks. The cracks are typically normal to the axis of restraint. For example, a concrete pile which has not yet been prestressed may develop transverse cracks, whereas for one which has been prestressed, the cracks will be longitudinal.

If there is sufficient reinforcing steel so that after cracking the steel is still below yield, then upon cooling of the core, the steel will pull the cracks closed. Conversely, if there is no or only minimal steel across the crack, the steel will be stretched beyond the yield point and the crack will remain open.

This problem, carried to the extreme, has been mislabeled as "thermal shock." For example, concrete products moved directly from steam curing into subzero weather can develop multiple cracks.

These cracks referred to above are visible and on the surface. Internal microcracking can also develop around large aggregates if they have radically different coefficients of thermal expansion from the cement-sand matrix. Even more serious are the internal cracks that may form around heavy concentrations of embedded steel, for example, bundled reinforcing bars, especially if steam curing has been employed.

Thermal cracking may also develop in service, where radiant heat from the sun raises the temperature on one side while the other side remains cool. Alternatively, a member may be subjected to freezing temperatures on one side while the other side is warm.

Water tanks and hollow-core piles are two applications which are especially prone to cracking from temperature differentials.

Thermal differences of 15° to 20°C (30° to 40°F) are sufficient to cause such cracking. Once again, the problems become serious from a durability point of view only when there is insufficient steel area to distribute the tensile stresses and to later close the crack when the thermal differential is removed.

For hollow-core prestressed concrete piles, the critical thermal strains are circumferential, which leads to vertical cracks unless there is sufficient spiral binding.

Thermal cracks which do not close, like their structural counterparts under sustained load, are entry paths for moisture and air. Water may freeze in the crack and progressively jack the crack even more widely open. In northern climates, freeze-thaw attack may set up along the cracked surfaces. Carbonation gains access to the interior. Chloride penetration immediately adjacent to the crack will be increased.

Prevention lies primarily in providing adequate reinforcing steel on the faces of all members subject to these thermal strains. Prestressing of course will prevent transverse cracks.

The temperature rise of the hydrating concrete can be minimized by the replacement of a portion of cement by PFA (about 5°C reduction for each 10% replacement), by avoiding richer content of cement than necessary, and by pre-cooling of the mix.

Surfaces may be insulated so that they cool slowly, giving them a chance to gain strength before contracting. This may require the use of insulated forms and the re-covering of the surface by blankets or tarpaulins once the forms are removed.

Steam curing of products may require the continued covering until the member has been prestressed.

Of especial concern are cold dry winter winds, which rapidly remove heat from the surface as the forms are stripped. Blanketing may be necessary.

5.4.7 Shrinkage

Drying shrinkage occurs as a result of the loss of excess water from the concrete pores. In much prestressed concrete today, there is little more water in the mix than that needed for complete hydration of the mix (about 30%) and hence drying shrinkage of the membrane as a whole will not cause sufficient internal strains to produce cracking.

Drying shrinkage of the member as a whole requires time: the more impermeable the mix the longer the time required, which may extend to months or years depending on the relative humidity. Over such an extended period, the tensile forces will have been largely accommodated by creep.

Therefore it is desirable to keep the ratio of water to cementitious materials as low as possible. As noted elsewhere, a W/C ratio of 0.37% to 0.38% appears obtainable in high-quality cast-in-place concrete, ranging down to as low as 0.32% in some precast elements. These, however, are relatively severe limits, applicable to demanding service and more typical values for conventional service are 0.40 or even 0.42.

As concrete is vibrated and sets, bleed water accumulates on the surface to some degree. Thus the surface layer of concrete may have a higher W/C ratio and suffer surface cracking or crazing. Once again, concretes of low W/C ratio have less bleed. The admixtures selected also affect the amount of bleed.

When microsilica is added to the concrete mix, the migration of bleed water and the extremely fine particles of silica to the surface also tend to increase shrinkage and surface crazing.

Since shrinkage of all the above types is caused by drying out, i.e., loss of water due to evaporation, curing which retains or supplies moisture or high humidity to the surface will tend to prevent shrinkage until such time as the concrete has sufficient tensile strength to prevent cracking. Where practicable, water curing or curing in a high humidity enclosure appears best. The use of membranes, such as polyethylene sheets or sprayed-on membranes, may be more practicable in other applications.

Early exposure to drying winds of unprotected surfaces must be avoided.

5.4.8 Alternate Wetting and Drying

When the surface of concrete is alternately wet and dried, as for example by wave splash or tidal action, it undergoes cyclic changes of the cement paste, in volume leading to crazing in a "spiderweb" pattern. This facilitates the permeation of oxygen, carbon dioxide, and chlorides.

This effect can be minimized by external coatings, or by penetrating substances which resist water penetration, e.g., silane, but is most effectively dealt with by making the concrete impermeable in itself.

5.4.9 The "Corner Effect"

The corners of piling, girders, and beams expose two sides to the elements, hence experience twice the permeation of fluids and gases. In addition, due to the discontinuity, local thermal and structural strains are increased.

These corners are also the most subject to impact and abrasive damage.

Thermal variations, e.g., freezing and thawing, also have twice the area on which to act.

Hence to the extent practicable, sharp corners should be eliminated by rounding or chamfering.

5.5 CORROSION OF PRESTRESSING STEELS AND ITS PREVENTION

5.5.1 General

Because prestressing steels are stressed to five times the tensile stress that is seen by most conventional reinforcing steel, any loss of area due to corrosion is much more critical. Thus prestressing steels are classified as "sensitive to corrosion." Furthermore any pitting may cause stress concentrations leading to early fracture.

Manufacturing techniques and/or use of improper steel chemistry may produce microcracking, especially at anchorages, at sharp bends (as around deflectors), and in wire-drawing, making the steel more susceptible to stress corrosion.

Stress corrosion is the propagation of a microcrack which has been originally produced in the steel by very high local stresses (as around a bend), by use of steels whose chemistry is not suitable, or by fatigue under cyclic loads. See Section 5.8 for further details.

The potential for hydrogen embrittlement due to liberation of monomolecular hydrogen ions from zinc and aluminum has been previously discussed. It appears as a remote possibility with the electro-galvanizing process but is believed to be essentially a nonproblem when the galvanizing has been carried out by the hot-dip or sherardizing processes.

Both stress corrosion and hydrogen embrittlement can lead to sudden fracture of steels which are under high stress.

These two forms of corrosion, although possible, have been almost completely eliminated as a practical matter by the adoption of strict codes and quality assurance in manufacture and by good construction practices.

The much more likely forms of corrosion are those due to chlorides and carbonation, and it is these which the construction practices presented in the following paragraphs are designed to prevent.

In general, all the forms of attack on conventional reinforcement are relevant, as are the means of prevention. In particular, cracks need to be prevented from conveying water, CO_2, and/or chlorides to the level of the steel. The concrete must be rendered as impermeable as possible.

Increased cover is usually provided over prestressing steel. Increases of 12 to 25 mm ($\frac{1}{2}$ to 1 inch) are normally provided over the prestressing steel as compared to the values adopted for the conventional steel, giving total covers from 35 to 75 or even 100 mm (2 to 3 or even 4 inches).

Partially prestressed concrete is extensively applied in buildings and the philosophy is being extended to engineering structures including bridges. The concept is essentially that the member will be prestressed so as to meet its serviceability requirements with the concrete remaining uncracked. Under design load (ultimate load), the structure will be allowed to crack. Ultimate strength in tension and shear will be provided by passive steel, such as mild-steel reinforcing bars.

The durability of partially prestressed concrete requires that the number and widths of cracks be limited to values which will not lead to corrosion in the environment in which the member serves. Typically this restricts crack width to a value of 0.2 to 0.5 mm (0.01 to 0.02 inch).

Measures which give added assurance and protection of the prestressing tendons are the provision of additional cover and the installation of mild-steel bars outside of the prestressing steel.

Special attention is directed to the need to store prestressing tendons, anchorages, and ducts in a dry enclosure, above the ground. In humid climates they will also require dehumidification, or the use of vapor-phase inhibitors. Coating of the tendons to be used with post-tensioning with water-soluble oil is satisfactory as long as the tendons are then protected from water ingress, but of course such coating is not suitable for tendons to be pretensioned.

5.5.2 Pretensioned Wire and Strand

Pretensioned tendons, being bare, bright steel, are subject to rapid atmospheric corrosion of the surface. Therefore, they should not be taken out from storage for any longer period than that needed to place and tension them, then place the concrete.

In most climates, up to 24 hours open exposure will not produce more than a superficial trace of light colored rust. This does not appear to adversely affect the durability and, in fact, improves the bond by corroding the mill scale and evaporating lubricating compounds used in the cold-drawing process during manufacture. The degree of rust allowable is often defined as "that which can be removed by wiping with a soft dry cloth."

Prolonged storage in the open in a coastal region may allow salt crystals to deposit in the rust. Hence covered storage is especially important in coastal environments.

Pretensioned strands extending to the surface of the concrete, e.g., at the ends of beams and girders, present a paradox. Experience over many years exposure has shown that corrosion of the tips of the wires occurs early but subsequently does not proceed inward more than 2 to 3 mm ($\frac{1}{8}$ inch). This is especially true where the oxygen supply is limited, i.e., underground or underwater.

In air, such exposed tips should be cut back 25 mm (1 inch), using a welding rod, and plugged with epoxy.

Very few cases of corrosion of pretensioned wires or strands have been reported. However, decks of bridges to which salts have been applied are certainly subject to potential corrosion. Concrete piles damaged by thermal cracks and freeze-thaw attack have experienced moderate corrosion of exposed strands. Similarly, the ends of pretensioned strands exposed to periodic saltwater immersion in tests at Treat Island, Maine, have experienced corrosion over lengths where the concrete has also been disrupted.

Epoxy coated strand is now commercially available, for which fine sand particles have been embedded in the epoxy in order to ensure bond. Tests show greater bond than bare strand.

5.5.3 Post-Tensioning Tendons and Appendages

5.5.3.1 GENERAL. These consist of the tendons (wires, strands, or bars), the anchorages and splice fittings, and the ducts.

5.5.3.2 DUCTS. The ducts are important because they are usually of thin-walled steel. Large surfaces are exposed to corrosion and corrosion can produce holes, which in turn can allow the cement paste to run in during concreting so as to block the duct, or grout to escape outward at accidental discontinuities.

Therefore they also should be protected in storage.

They usually are installed and externally encased in concrete within a short period of 1 to 2 days. However, they may sit empty, without tendons or grout, for days or weeks. During this period they should be drained, although bleed water from the concrete may have seeped in during and after concreting. Fortunately, this will be alkaline.

The ducts should be sealed, by plastic caps, to prevent rainwater from entering.

On one major bridge, the caps were left off. Heavy rains filled the ducts. Subsequent freezing weather caused the ducts to freeze, expand, and rupture the concrete.

Steel ducting may be galvanized to prevent corrosion and to reduce friction. Concerns over hydrogen ion liberation to embrittle tendons are probably grossly exaggerated, but greater assurance can be obtained by using hot-dip galvanizing and by requiring a chromate wash after galvanizing. This is normally a standard practice.

Fully watertight ducting, whether steel or plastic, with watertight couplers at the splices, is obviously a secondary barrier against penetration of chlorides and carbonates to the prestressing steels, whereas the semi-watertight duct is only a partial barrier. Especial care should be taken at duct splices to maintain the degree of tightness that is required in order to prevent in-leakage during concreting and crossover during grouting.

Recent developments have made available polyethylene-sheathed strand which is infilled with corrosion-inhibiting grease. Epoxy-coated bars of high tensile strength steel are also available.

5.5.3.3 PROTECTION OF PRESTRESSING MATERIALS. Prestressing steels for post-tensioning, including the anchorage hardware, must be protected in transit, storage, site delivery, and installation, until grouted and encased.

The tendons and anchorages must be protected during shipping, storage, and installation. This is accomplished by one or more of the following steps.

Shipping the coils or bars wrapped and sealed in watertight "export" packaging, protected by wood or steel framing (See Fig. 1.6).

Precoating the steels with water-soluble oil or injecting VPI powder into the containers before shipping.

Storage at site in a watertight shed or building, on a floor above ground.

Dehumidifying the storage shed by heat or other means.

Adding additional water-soluble oil as the tendons are threaded into the ducts.

Sealing the ends of the ducts until grouting, except during the intervals when they are opened to place anchorages and to stress the tendons.

5.5.3.4 GROUTING. After stressing, grouting should be carried out promptly. There are several cases on record where grouting was delayed for weeks or months due to strikes or to enable re-stressing after creep losses had occurred. Industrial and refinery pollutants in fog and smog occasioned severe corrosion of the tendons, which showed up by breakage of wires and even entire tendons during stressing.

The grouting procedures adopted must be such as to expel all water and air, and to completely fill the duct and the anchorage recesses. In particular, voids due to bleed water, entrapped water, or sedimentation must be avoided.

A full description of grouting mixes and procedures is given in Section 5.10.

Alternatively, for unbonded tendons, such as those in containment structures, corrosion-inhibiting grease may be injected. In this case, the end anchorages must receive special attention to prevent loss of grease, entry of water, and damage by fire.

5.5.3.5 UNBONDED TENDONS. Unbonded posttensioning tendons may be internal or external to the concrete cross section.

External tendons are usually protected by cement grout inside of polyethylene sleeves. The materials and procedures are described in more detail in Section 2.3.6.

Unbonded internal tendons, that is, internal to the concrete element's cross section, are usually protected by anti-corrosive grease.

The tendons may be placed in a void, formed by a duct, and then be encased in the grease, which is injected through the anchor, or they may be pre-greased and pre-encased in plastic sheathing. A plastic or steel cap may be placed over the anchor plate and injected with grease.

A further description of the injection of grease after stressing is given in Section 17.9. Recently, epoxy-coated bars have become available for use as unbonded tendons.

A further description of pre-encasement, using sheathed tendons injected prior to placement and stressing, is given in Section 2.3.6 and in Chapter 10.

5.5.3.6 PROTECTION OF ANCHORAGES. The anchorages, in post-tensioned construction, must be protected from corrosion. They are vital hardware, of special material, machined to a high standard. The wires at the anchorages are under higher stress than anywhere else. In wedge-type anchorages, bearing and bending stresses are added to direct tensile stress. With buttonhead anchorages, the head of the wire is cold-worked to the point of having microcracks.

This matter is especially important when high-capacity tendons are used. In such cases little help can be provided by the bond established by grouting. The anchorage must be maintained, and the terminals of the tendons placed in a recess formed in the ends of the concrete structural element. After stressing the protruding tendons are cut off about 12 to 15 mm ($\frac{1}{2}$ inch) outside the anchor.

The anchorage hardware is now coated with epoxy. For large tendons, requiring large recesses, small reinforcing bars are bent down so as to tie the concrete fill to the structure (Figs. 5.2, 5.3). The recess is now filled with fine concrete or grout, in

Figure 5.2. *Recessed anchorage pocket. After stressing, bars will be bent in, the anchorage heads and concrete pocket sprayed with bonding epoxy, and fine concrete placed to completely fill the recess.*

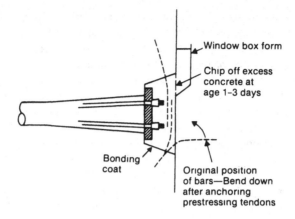

Figure 5.3. "Window box" technique for concreting anchorage pocket.

such a way as to prevent gaps due to shrinkage and bleed at the top, e.g., by the use of the "window box" technique. The exterior surface of the patch is then coated with an epoxy or other sealant, except in freeze-thaw environments, where a heavy coat of silane is preferable.

Recessed anchorage pockets are the most reliable, but external protection can work if properly tied and bonded.

Smaller anchorages are normally just filled with cement grout or epoxy mortar.

Special protective caps have been devised, which permit grout or grease to be injected around protruding strands and over the anchorage plate.

5.5.3.7 PROTECTIVE COATINGS. Special protective coatings have been devised for prestressing tendons for use in post-tensioning.

The use of high-density polyethylene sheaths (HDPE) with injected grease has been previously mentioned. Individual strands, protected in this way, are extensively used in unbonded prestressing for building floors. Groups of such individually protected strands are increasingly used in post-tensioning of bridge girders and pressure vessels.

Galvanizing of tendons was extensively employed in prior years: its use was temporarily suspended because of concern about possible hydrogen embrittlement since one case of failure (in France) was believed due to hydrogen embrittlement. However, recent in-depth investigations by the Federation Internationale de la Pre-contrainte (FIP) have established procedures believed to be fully safe, namely the use of the hot-dip process for galvanizing and the follow-up procedure of passivation by means of a chromate wash.

5.6 FATIGUE

Conventional reinforced concrete which is repeatedly cycled from higher to lower stresses may eventually be subject to fatigue. The greater the stress range, the greater the internal damage. Cycling from compression to tension is especially damaging.

Low-amplitude, high-cycle fatigue is a potential problem in concrete railroad bridges, in floating structures, and in marine structures. In these applications, the structure may see hundreds of thousands to millions of cycles in its design life. High-amplitude, low-cycle fatigue may occur in structures such as piling, which sustain repeated high impacts during driving.

Typical failures of reinforced concrete include cracking, even though the maxima tensile stresses are below the static tensile strength, leading to progressive bond slip, and eventually, brittle failure of the reinforcing steel.

Prestressing, by establishing a state of compression in the structure, minimizes or eliminates the excursions into the tensile stress range. Thus it is extensively used in structures and elements subject to cyclic loading. Its use has been largely responsible for the increasing use of concrete for marine structures, railroad bridge girders, and piling.

However, in some applications and in some environments, fatigue may cause loss of durability, even for prestressed concrete. For example, prestressed concrete shows substantially reduced fatigue endurance when immersed in water and then cycled over a large number of cycles to high stress ranges. Such conditions can occur in certain marine applications.

The buildup of high pore pressure in the concrete in the compressive portion of the cycle, followed by reduced pressure in the low compressive or tensile portion, leads to breakdown of the concrete and cracking. Subsequent cycles cause water to flush in and out of the crack, generating hydraulic ram effects and washing out the cement mortar particles. Eventually the steel is highly stressed at the crack, leading to final failure in brittle fracture of the steel.

This may be a partial explanation of the problems which occur during some installations of prestressed concrete piles, where care has not been taken to minimize the stress range in the piles. See Chapter 11.

The problem of cyclic loadings on members in service is essentially a matter for proper structural design. Insofar as construction is concerned, the endurance can be enhanced by minimizing microcracks due to bleed. Such measures as choice of aggregates with proper surface characteristics, the inclusion of PFA, the inclusion of microsilica in the mix, and low W/C ratio are all beneficial. Lightweight aggregate with microsilica shows especially good fatigue endurance.

Failure due to fatigue generally shows up internally as loss of bond between cement paste and aggregate, and cement paste and steel. Measures which enhance bond will also enhance fatigue endurance.

5.7 HIGH TEMPERATURES—FIRE

Heat can weaken steel. Prestressing steel, being under high stress, can lose significant tension at temperatures above 750°C (1400°F). It is therefore important that end anchorages be properly protected and insulated in those applications where fire or high temperatures are a potential occurrence.

When concrete, which contains excess water in its pores, is subjected to temperatures over 100°C (212°F), the resultant steam and water vapor must escape. Here is the one case in which dry, highly permeable concrete performs better than high-strength impermeable concrete. In the latter case, the steam is trapped, pressures build up, and explosive spalling may occur, thus exposing the reinforcing steel and tendons to the direct heat.

Air entrainment will provide a moderate amount of protection. The best solution, where otherwise impermeable concrete is required, is the reduction in free water through control of the W/C ratio and the use of aggregates of low water absorption.

Steels which must serve at moderately high temperatures will be more subject to corrosion processes. Hence greater care must be taken in their protection, although the techniques are those previously described. Greater cover and the use of lightweight aggregate concrete will provide thermal insulation for the steel.

Lightweight aggregates used in potential fire exposure should have low moisture content and/or be able to allow water vapor to escape through their surface (see Section 4.2).

5.8 STRESS CORROSION

This is an extremely rare phenomenon but, unfortunately, the occurrence of the word "stress" in both stress corrosion and prestress has led to concern. Stress corrosion usually is associated with minute traces of chlorides or sulfides and, possibly, other negative ions, occurring in a humid atmosphere.

Heat-treated steel wire, as used extensively in Europe, in previous years and oil-quenched-and-tempered steel wire are believed to be more susceptible to stress corrosion than the cold-drawn wire now in common use worldwide.

A few isolated cases of serious corrosion, caused by a combination of stress and sulfide ions, have occurred when heat-treated wire, shipped in closely wound coils, was stored in contaminated mud. After installation, and indeed after grouting, the wires failed progressively. Similar cases have occurred when wires have been stressed and then left exposed to moist air in the vicinity of refinery or industrial plants. It is believed that the minute concentration of H_2S in the moist air leads to corrosion and brittle fracture. Therefore tendons should preferably be protected as quickly after stressing as possible. If they must remain stressed but ungrouted, then positive steps must be taken to prevent moisture entry, and the tendons must be protected by water-soluble oil or VPI powder. Tendons should never be coiled or bent more tightly than as originally shipped by the manufacturer.

These failures above are generally classified as one form of hydrogen embrittlement. Hydrogen embrittlement occurs when monomolecular hydrogen ions are able to enter between the steel molecules. Hydrogen embrittlement is also theoretically possible when dissimilar metals, such as aluminum or zinc, are used in the vicinity of steel, for example, the use of aluminum powder in grout to cause expansion. In laboratory tests it has been very hard to produce such a reaction, but a few cases are believed to have occurred in actual practice.

A case of serious corrosion of post-tensioned tendons is reported in which the precast architectural units were dipped in dilute hydrochloric acid for surface etching but apparently ducts were not flushed with water afterwards. Subsequent grouting was also not thorough, resulting in voids containing dilute acid. Failure of some tendons occurred several months after completion, and subsequent examination revealed serious corrosion on other tendons.

5.9 ABRASION

Abrasion can occur with marine concrete due to the movement of sand, gravel, and ice. The Baltic Sea Lighthouses located in the areas of fast moving, hard, ice have shown 50 to 75 mm (2–3 inches) of abrasion near the waterline, in some cases tearing out the exposed reinforcing steel bars.

Abrasion can also occur with concrete railway ties, due to wear from the ballast. Highway pavements can be subjected to severe abrasion from studded tires and repeated use. The process is aggravated by freeze-thaw and the spreading of salt or sand.

It used to be thought that abrasion was more or less entirely a phenomenon attacking the aggregates and that by selecting very hard, strong aggregates, it could be minimized. Now we know that the process is more complex. In the Baltic Sea, for example, the ice seems to be wearing away the matrix, thus exposing the coarse aggregate, which is eventually plucked out.

Tests in the laboratory and field, as yet incomplete, indicate that matrix, bond, and aggregate type are all important.

For example, expanded clay aggregate with a hard-burned exterior shows essentially as good or better resistance to ice abrasion as selected silica crushed rock.

Silica fume can improve the bond between matrix and coarse aggregate. Low W/C ratios certainly help.

Coatings of dense epoxy (no solvent) and dense polyurethane will resist abrasion for several years, but they must then be replaced. Resurfacing may be very difficult and costly, depending on location and traffic conditions.

The only concrete exterior surface which appears able to successfully resist high abrasion is polymer-impregnated concrete.

The other alternative is to attach sacrificial steel armor plate. This is the solution currently adopted for the Baltic Lighthouses.

Note that with polymer impregnation and with steel armor, there is a potential for freeze-thaw delamination in the concrete just inside the impermeable skin. The water vapor migrates to the cold face, where it freezes, since it is unable to escape. This has reportedly occurred where steel plates have been used to armor ship locks.

Anchors for steel plates should therefore be deeply embedded.

Use of very low W/C ratio in the concrete behind as well as a mix selected for low permeability will minimize the amount of water and its rate of accumulation behind the exterior skin.

5.10 GROUTING OF POSTTENSIONED TENDONS

5.10.1 General

Grouting with cement grout is the most widely used method of protecting tendons in the ducts of posttensioning systems. The grout fulfills a number of purposes:

1. Encasement of the steel in an alkaline environment for corrosion protection.
2. Filling of the duct to prevent water from collecting and freezing.
3. Provisions of bond between tendons and structural concrete.
4. Completion of the concrete cross section.

Grouting should preferably take place within 48 hours after placing the steel and within 24 hours after stressing. If tendons must remain longer than this, then, they should be left unstressed and should have special protection against corrosion. Dusting VPI powder on the tendons as they are inserted and then sealing the duct and vent tube ends with tape is one such means.

Alternatively, and believed more practicable, is the thorough coating of the tendons with water-soluble oil as they are inserted. Later, as the grouting process is initiated, the grout is used to remove this oil. The initial discharge will be contaminated with the oil and discolored: it must be discarded.

The technique of grouting is one of the most important aspects of the entire prestressing construction sequence. Since it is somewhat complex, and since it depends on proper and skilled performance by workers, a number of rules and procedures have been developed to ensure proper performance (Fig. 5.4).

The grout, ideally, will fill the entire cross section, without voids, such as those due to entrapped water or bleed water. It should also be free from high expansion pressures that might produce longitudinal cracks in the structural concrete. It should be dense and homogeneous so as not to leave voids along the steel. It must flow easily so as to fill the interstices between wires, and between wires and duct where the wires bear. The grout should not be susceptible to freezing and thawing disruption. It must have adequate compressive strength and high bond strength. It should attain reasonably high strength as soon as possible. It must have a high cement content.

Materials, mixtures, and injection procedures should be selected to minimize bleeding of grout after injection. Bleeding is due to the fact that the water has a specific gravity only one-half that of cement; thus, sedimentation tends to take place and lenses of bleed water may be left in the ducts, tending to migrate and collect at high points in the profile. This entrapped bleed water may freeze and rupture the member, or it may be reabsorbed, leaving air voids along the tendons, thus permitting an oxygen gradient to develop, with subsequent corrosion. Some specifications limit bleeding to 2% at 65°C in 3 hours, with a maximum of 4%, and require that the separated water must be reabsorbed within 24 hours. Recently developed admixtures have thixotropic properties which effectively prevent bleed almost entirely. Their use is recommended.

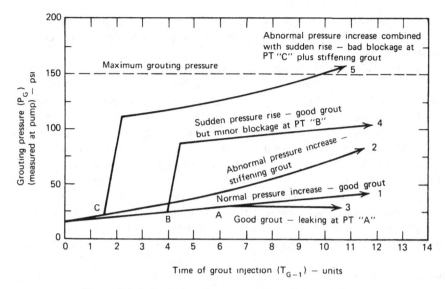

Figure 5.4. Indicated performance of grouting operations.

Expansion agents have been used in the past to try to achieve complete filling. They must not have too much expansive force or the duct may be ruptured. Total expansion should not exceed 10% (measured unconfined). During the test, the testing tube should be covered to prevent evaporation. Use of aluminum powder, which evolves hydrogen gas when in contact with cement, is definitely not recommended because of the remote yet real possibility of the release of monomolecular hydrogen and consequent hydrogen embrittlement of the steel. Admixtures are available commercially which liberate a relatively inert gas such as nitrogen.

Furthermore, the expansion is worse than useless if it takes place during mixing rather than after injection.

The best assurance of obtaining satisfactory grout and, therefore, the most important control is the W/C ratio. This should not exceed 0.45. The grout must also be thoroughly mixed by machine to ensure complete mixing and to obtain a workable grout for injection.

Neat-cement-and-water grout has proven satisfactory for ducts up to 75 to 125 mm (3 to 5 inches) in diameter.

Previously, in grouting large ducts, a sand-and-cement grout was sometimes used. In these cases neat-cement grout was injected first, followed by the sand-cement grout. After initial set (usually $1\frac{1}{2}$ to 2 hours), a second injection of neat-cement grout was made. With the current availability of anti-bleed and thixotropic admixtures, it is recommended that only neat-cement and water grout be used even for large ducts.

The duct size should be selected to give a ratio of steel tendon area to duct cross-sectional area of approximately 0.5. If too large a percentage of the duct is filled with steel, more voids will be formed. Tests have shown that fewer voids are found when strands are used as tendons than when wires are used, presumably because air bubbles can escape more easily around strand, and grout can more easily penetrate the spiral paths.

Proper grouting can be obtained only when the entire system is free of holes or joints which would allow grout to leak out of the ducts. This requires that a positive means of sealing be adopted for joints between segments and between precast members. The duct system must also be free of blockages or restrictions. These can be caused by collapsed or dented ducts, debris and foreign matter in the duct, and ice. Sudden changes in alignment and cross section of ducts should be avoided.

Leaks from one duct to another may occur at joints in segments and between precast members, unless properly sealed, or when two or more ducts are bundled next to each other and the ducts (or the joints in the ducts) are not grout-tight.

As noted earlier, ducts, particularly those of thin metal, are often rendered nontight by corrosion in transit or storage, by tearing and ripping in handling, or by tearing when placing adjoining reinforcing steel. Duct joints may be accidentally pulled apart when inserting a tendon prior to concreting. Ducts may be inadvertently holed by carpenters drilling holes for form ties or inserts. Ducts may also be holed by rough use of an internal vibrator.

Ducts may be sealed or repaired by several wraps of waterproof tape, or even more positively, by heat-shrink tape. When holes or gaps are larger than $\frac{1}{4}$ inch, they should be sealed by a metal strip taped in place over the hole.

Grout ports and vents are required for the injection of grout and for the escape of air and water at all high points, changes in duct section, and at the terminus of any grouting length. These must also be sealed to the duct with tape. Vents or drains are also desirable at the low points to prevent water from collecting.

After concreting, but prior to insertion of the tendons, the ducts may be checked for blockages by a rubber ball of smaller diameter passing through on the end of a messenger wire or blown through by compressed air. After tendons are inserted, see-sawing back and forth will give an indication of any blockage (and will help remove minor restrictions).

The cement used for grouting should be ASTM Type I, II, or III. The cement must be free from any lumps, else serious difficulties are inevitable. PFA, up to 20%, may be used to replace part of the cement.

Water/cement ratio should be limited to a W/C ratio of 0.45 or less, using a water-reducing admixture as necessary.

Admixtures must not contain chlorides as Cl⁻ in excess of 0.25% by weight of the admixture.

There are a number of admixtures commercially available which promote fluidity and have thixotropic and anti-bleed properties. Since recent investigations have shown the inadequacies of past practices, the use of these advanced admixtures is strongly recommended.

Air entrainment in amounts of 4 to 8% should be provided whenever there is danger of freezing and thawing.

If sand is used (as in grouting very large ducts), it should be fine sand and the mix should contain about 600 Kg/m³ (940 lb/yd³). Use of sand is not generally recommended.

Grouting equipment should be capable of producing grout of a uniform and, if possible, colloidal consistency. Mixing should be done mechanically. Water should be added first to the mixer, then cement. Additives should be added during the latter part of the mixing time. Mixing time should normally be 2 to 4 minutes. After mixing, the grout should be kept continuously agitated or moving.

The grouting pump should be a positive displacement pump, capable of producing at least 1 MPa (150 psi) discharge pressure and having adequate seals to prevent introduction of oil, air, or other foreign substances into the grout and to prevent loss of grout or water. The pumping pressure should be limited to a maximum of 1.7 MPa (250 psi) so as to prevent damage. Use of special grouting admixtures will reduce the need for high pressure and hence minimize the incidence of crossovers. A pressure gauge should be installed at the discharge of the pump.

The grout should be screened prior to introduction into the pump. Screens should have clear openings of about 4 mm (0.17 inch or 14 mesh). The grouting pump should be fed by gravity. A suction feed tends to suck air into the grout mix. Grout hoses, valves, and fittings should be watertight. Standby water-flushing equipment, powered by a separate power source, should be available, with a pressure of 1.5 to 2.0 MPa (200 to 300 psi).

A correct balance of materials, proportions of materials, method of adding to the mix, and mixing time will produce a grout which:

a. Is easily pumped initially.

b. Will not take a false set during the planned grouting time.

c. Has minimal bleed.

d. Has good final strength.

The grout temperature should not exceed 30°C (90°F). If necessary, the mixing water should be cooled. Blockages are more common in hot weather.

Injection by compressed air is usually not satisfactory and should not be used. There is danger of introducing a slug of air and of erratic, uneven grouting.

If the ducts have been properly sealed against entry of water, then the grout can be used to expel any water-soluble oil or in-leakage of water during concreting.

However, if for any reason the duct has become water-filled, then the ducts should first be blown free with oil-free compressed air.

Grout should then be injected, initially with a low pressure (about 300 KPa, 40 psi), increasing it until grout runs out the vents. Then pressure can be slowly brought to the designed pressure. The optimum flow rate for grout in the normal-sized duct (75 to 120 mm, 3 to 5 inches diameter) is about 5 to 12 m/minute (15 to 40 ft/minute). The slower rates are preferable in that they reduce the number of voids. Grouting must be continuous until the consistency of the grout emerging from the vents is the same as that being injected, without visible slugs of water or air. If water-soluble oil has been swabbed onto the tendons during their insertion, then the initial grout will be oil-contaminated and will emerge discolored with black streaks. Enough should be discharged to completely expel the oil. A rule-of-thumb is to waste 100% of the duct volume. As grout flows out the nearest vent, it can be closed off; the grout then flows to the next vent, and so on. Grouting should be continuous as long as one-way flow of grout is maintained.

If the grout pressure exceeds 1 MPa (150 psi) and the one-way flow cannot be maintained, it indicates a blockage and the duct should immediately be flushed out with water.

All vents and the injection pipe must be closed off under pressure. Use of plastic or rubber nipples with squeeze clips is one way to accomplish this. These nipples can be cut off after the grout has hardened. If the pressure is allowed to fall, serious voids due to entrapped air and bleed water will result. Injection pipes are preferably fitted with mechanical shut-off valves. Full injection pressure should be maintained for 30 seconds to a minute before closing off the injection pipe.

Tests have shown that holding the pressure on the grout tends to cause voids to disappear and forces grout into the anchorages.

5.10.2 Grouting of Vertical Ducts

Grouting pressure will have to be sufficient to force grout to the top or to an intermediate vent, but should not exceed 2 MPa (290 psi).

Grouting of vertical ducts or those with substantial vertical projection has not been satisfactory in past practice. In the shafts of offshore platforms and the walls of nuclear reactor containment vessels, bleed water has tended to accumulate at the top, subsequently being reabsorbed but leaving a void. Such voids have been as long as 2 meters in a 70-meter high wall. Bleed volumes up to 10% have occurred in extreme cases.

Strand appears to have greater bleed than wires, due to the wick effect.

The phenomenon is caused by the high fluid pressure which forces the bleed water into the interstices of the strands, whence it flows upward to be trapped under the upper anchorage.

Several methods to remedy this problem have been used.

- Using a thixotropic admixture which causes the grout to gel as soon as pumping ceases.
- Providing a spare hole in the upper anchorage and affixing a standpipe several meters high. Meanwhile the strands are left long so as to allow the bleed water to rise above the anchorage and overflow. The standpipe is removed after final set. The recent development of anti-bleed admixtures with thixotropic properties, combined with the standpipe, is believed to be the best current practice.

Concern has also been voiced about the same phenomenon with long-span bridge girders in which the tendons have a substantial vertical rise. Fortunately the tendons at the top of the negative-moment curve are tightly stressed against the bottom of the duct, so that the bleed water is generally above them.

A vent provided at the top of the curve and, in appropriate cases, the use of a standpipe, will partially solve the problem, but the use of anti-bleed thixotropic admixtures is felt to be also important.

5.10.3 Grouting in Cold Weather

The freezing of fresh grout presents the danger of cracking of the structural walls and permanent damage to the structure. Therefore, if possible, postpone grouting until warmer weather (no danger of frost within 48 hours), keeping ducts sealed against accidental entry of water.

Before grouting after a period of cold or frosty weather, the ducts should be flushed with warm water (but not steam) to remove any ice. In temperatures below freezing the water must be blown clear with heated compressed air to prevent refreezing. At least 100% extra grout should be injected and discharged so as to clear out lenses of trapped water.

It should be recognized that, in prolonged periods of freezing weather, the entire structure may be cooled below freezing. Therefore, merely clearing the ducts of ice or warming them will not suffice: the adjacent structure will extract heat from the grout and cause it to freeze.

Some admixtures are available which lower the freezing point of the grout. Admixtures containing chlorides must never be used.

If it is forecast that the temperature of the structure will not fall below 5°C (40°F) with the next 48 hours, grouting may be undertaken, using a grout containing 6 to 8% entrained air. If grouting must be carried out in cold weather, then means must be taken to keep grout (and adjoining structure) above 5°C (40°F) for 48 hours.

As noted earlier, do not use any admixtures containing calcium chloride.

After final injection, seal all openings and vents with cement grout or epoxy.

Records of grouting mixes, times of injection, quantity of grout used, and pressures should be maintained so as to be able to verify compliance with the specifications for that particular project. These records may show up radical problems such as leaks and blockages.

Gamma radiography, using Co^{60} source, has been used in previous years in England as a means of verifying the completeness of grouting of main tendons. When voids are shown to exist, the duct is drilled into from the girder side and a secondary injection of grout is made. Two holes are required—one for injection, one for venting. A vacuum process enables injection with only one hole. This procedure is believed largely unnecessary now that appropriate admixtures and procedures have been developed, except for inspection of existing structures.

Grouting of posttensioning tendons in nuclear reactors has been the subject of special studies and tests, and certain additional procedures have been developed. These are discussed in Chapter 17.

5.11 CORROSION PROTECTION FOR UNBONDED TENDONS

Because all steel is subject to corrosion, prestressed tendons should be protected during storage, transit, construction, and after installation.

Unbonded tendons and their protective coating may be installed in one of the following ways:

a. Tendons are coated with grease against corrosion and for lubrication, then sheathed, usually with polyethylene, so as to permit slippage during tensioning and for protection of the coating where required, then cast in the concrete. After hardening of the concrete, tendons are stressed.

b. Tendons are coated against corrosion and placed externally to already hardened concrete. This is typical of galvanized or epoxy-coated tendons for repairs.

c. Tendons are inserted in plastic and steel ducts which are external to already hardened concrete. They are then injected with cement grout or grease. This method is typically employed for long-span bridges.

d. Tendons are pulled into ducts. The ducts are then filled with a suitable material, such as grease. This is typically employed for nuclear reactor containment structures.

The coating material should:

a. Remain free from cracks and not become brittle or fluid over the entire anticipated range of temperatures, at least −20°C (0°F) to 70°C (160°F).

b. Be chemically stable for the life of the structure.

c. Be nonreactive with surrounding materials such as concrete, tendons, wrapping, or ducts.

d. Be noncorrosive or corrosion-inhibiting.

e. Be impervious to moisture.

The coating material should be continuous over the entire length of tendon to be protected, since a bare spot on a coated tendon is potentially susceptible to electrolytic corrosion.

The minimum coating thickness depends on the particular coating material selected, but it should be adequate to ensure full continuity and effectiveness with a sufficient allowance for variations in application. Where the coating material is applied before tendons are pulled into the ducts or casings, the coating material should be sufficiently tough to resist abrasion.

Coating materials used in the past have run the gamut of bitumastics, asphaltic mastics, greases, wax, epoxies, and plastics. The current trend is to use the HDPE sheaths and grease. This system is believed to provide superior protection. Galvanizing, if used, should be by the hot-dip process and should have the following minimum thickness:

For strand: Class A, as specified in Table 4 of ASTM Designation A-475.

For bars and wires: current commercial specifications call for 0.2 to 0.3 Kg/m² (0.6 to 0.9 ounce/foot²) of surface.

Where tendons are exposed to a corrosive atmosphere, salt air, or high humidity, additional protection may be required.

For systems using friction grip anchorages, the coating may remain if it will not lead to slippage or excessive initial set. Otherwise it must be removed and cleared from that portion of the tendon to be gripped. If the coating is removed, this portion of the tendon must be protected from corrosion after the anchorage has been installed. Refer also to Section 24.3.5.

Where a filling material such as grease is injected after the tendons are in place, the filling material should have these additional properties:

a. No appreciable shrinkage or excessive volume increase.

b. Suitable viscosity at ambient temperature or with moderate preheating to permit injection by pumping.

Sheathing must be continuous over the entire zone to be unbonded, and should prevent the intrusion of cement paste and the loss of coating material. It should have sufficient tensile strength and water resistance to resist damage and deterioration during transit, storage on the job site, and installation. Impregnated and reinforced paper, plastic, and fiberglass have been used as wrapping material. However, experience has shown that the paper wrappings are subject to excessive damage during installation and subsequent concreting. Plastic ducts are now believed to be the most reliable.

If the wrapping is damaged, the damaged section should be removed, and the tendon inspected. If the tendon is undamaged, it should be re-coated with corrosion-protection material and the damaged portion of the wrapping replaced and sealed.

Several of the limited number of cases of extensive corrosion of prestressing steels have been with unbonded tendons which were greased and wrapped with paper. External exposure prior to concreting led to salt deposits from the salt-laden fog permeating the wrapping. Subsequent corrosion and sudden fracture of the tendons has occurred intermittently over the years as the corrosion reached failure proportions. The use of paper sheaths has now been superseded by extruded polyethylene sheaths.

Steel ducts for unbonded tendons are similar to those for post-tensioned grouted tendons. They should be mortar-tight and nonreactive with concrete, tendons, or the filler material.

The anchorages of unbonded tendons must be adequately protected from corrosion and fire. Except in special cases, anchorage zones should preferably be encased in concrete or grout free from any chlorides. Epoxy mortars have been widely used for this purpose.

Detailing of the concrete or grout encasement, design of the mix, and details of application are most important. Shrinkage cracks in the concrete may permit moisture penetration.

Where concrete or grout encasement cannot be used, the tendon anchorage should be completely coated with a corrosion-resistant paint or grease equivalent to that applied to the tendons. A suitable enclosure should be placed where necessary to prevent the entrance of moisture or the deterioration or removal of this coating.

The anchorage encasement should provide fire resistance at least equal to that required for the structure, recognizing that a failure of an anchorage could cause explosive disruption of the structure.

Wire-wrapped tanks, pipes, and other externally stressed concrete elements require special protective steps to ensure against corrosion. Because the tendons are not placed within ducts encased in hardened concrete, they have proven more susceptible to corrosion. The cover is usually of shotcrete although poured concrete is sometimes employed. Neither of these covers are prestressed, except as a minor amount of prestress may be transmitted through creep.

Shotcrete (gunite), if carelessly applied, will have zones of high porosity: any lapse from the best workmanship may entrap some rebound. The wet process is believed preferable to the dry process.

The optimum procedure for protecting exposed tendons is to coat each layer of wire with cement slurry, applied as a thin flash-coat of shotcrete. Finally, the external coating is placed by shotcrete or cast-in-place concrete. (See Chapter 16, Prestressed Concrete Tanks.)

Galvanized strand is used in one process to promote long-term durability.

It is very important to seal the top of the tank walls against water infiltration into any laminar microcracks that may have formed between the wrapped layers. Failure to do this has led to catastrophic collapse of sewage tanks.

With external tendons placed on the sides of girders, three steps may be taken to improve corrosion resistance. First, thorough coating of the tendons with an epoxy paint after installation but before concreting. Second, encasing the tendon in shotcrete or concrete. Third, painting or sealing the concrete cover with epoxy or bitumastic paint, especially at the joints between the cover concrete and girder concrete.

5.12 ELECTRICAL BONDING

Stray currents may cause corrosion, as noted previously, and are a special concern for reinforced and prestressed concrete girders and other elements on which electrified railways will operate. AC current is believed to impose only 5 to 10% of the risk that DC current imposes. Electrified railroads typically use one rail as an earthing or grounding rail. If the reinforcing steel and all other embedded elements, including prestressing steel, are completely bonded, then the potential discharge can be led to the earthing rail.

Full bonding will also permit the future application of cathodic protection. However, full bonding is obviously very difficult to achieve with pretensioned tendons and even with posttensioning tendons, although in the latter case, the anchors provide a good location for attachment of the bonding cable.

Bonding, with appropriate discharge to an earthing rail, gives good protection against lightning, although reinforced and prestressed concrete are normally shielded from it by the concrete cover.

Bonding must be thorough and complete. Hence for A.C. railroads, conventional forms of corrosion protection may be selected.

SELECTED REFERENCES

1. *FIP Guides to Good Practice,* "Grouting of Tendons in Prestressed Concrete," Structural Engineers Trading Organization Ltd., London, 1990.

2. *FIP Guides to Good Practice,* "Practical Construction," Federation Internationale de la Precontrainte, London, 1975.

3. Gerwick, B. C. Jr., "International Experience in the Performance of Marine Concrete," *Concrete International,* May 1990 American Concrete Institute, Detroit, Michigan, and F.I.P Notes 1991/1, Federation Internationale de la Precontrainte.

4. ACI Publication SP-47, "Durability of Concrete," American Concrete Institute, Detroit, Michigan, 1975.

5. ACI Publication SP-109, "Concrete in the Marine Environment," American Concrete Institute, Detroit, Michigan, 1988.

6. Schupack, M., "Corrosion Protection of Unbonded Tendons," *Concrete International,* American Concrete Institute, Detroit, Michigan, February, 1991.

7. Mehta, P. Kumar "Concrete-50 Years of Progress" ACI-SP126, Proceedings, Second International Conference on Durability of Concrete, American Concrete Institute, Detroit, Michigan, 1991.

TABLE 5.1 Techniques for Obtaining Maximum Durability of Prestressed Concrete—An Abbreviated Summary

Aggregates	Sound, nonreactive, abrasion resistant, impermeable
Cement	Type I or II cement
	Low alkali content (less than 0.6% Na_2O + K_2O)
	Moderate C3A (for marine environments)
	High cement factor
Water	Fresh water
	Free from chlorides and sulfates
Concrete mix	Low w/c ratio
	Clean aggregate
	Small size coarse aggregate
	Dense grading
	High-bond aggregate (where exposed to abrasion or cavitation)
	Limitation on chlorides from any source
	Cool as appropriate
Admixtures	Water-reducing
	Anti-corrosive
	Retarding, anti-bleed and thixotropic
Forms	Rigid
	Re-entrant angles, projections, etc., minimized
	Avoidance of sharp corners, edges
	Grout-tight
	Smooth surface
	Insulated in extreme cold weather
Placing	Thoroughly consolidated and compacted
	Bleed holes minimized
	Well-vibrated
Cover	Adequate cover over mild steel
	Adequate cover over prestressing steel
Finish	Troweled
Construction joints	Well prepared
	Well bonded (by pre-soaking and grout or by bonding epoxy)
Curing	Adequate water cure or steam cure
	Moisture available or sealed in during cooling period
	Drying after curing (for marine and freeze-thaw environments)
	Water free from chlorides or sulfates
	Protected from sudden thermal strains, insulated
Mild-steel reinforcement	Free from pitting
	Well-distributed bars
	Avoidance of very large bars
	Epoxy coating in aggressive environments
Embedded metals	Avoidance of galvanic action, especially Cu and Al
	Ducts to be tightly sealed
	Epoxy coating in aggressive environments
Prestressing steel	Free from pitting and extensive surface rust
	Clean and dry, no salt
	Galvanized or plastic or epoxy-coated (for special cases only) (coating must not be abraded)
	Stored in covered, dry storage
	Kept free from corrosion until finally grouted or protected by (a) soluble oil, (b) VPI powder, (c) sealing, (d) limited time of exposure

TABLE 5.1 *(Continued)*

Anchorages	"Flush-type," with the anchors themselves in pockets, are more thoroughly protected
	Epoxy concrete is best material (except where subject to freeze-thaw attack)
Grouting	Specified grouting practices followed.
	Special precaution for vertical ducts
Coatings on con-	Bitumastic
crete surface	Epoxy
	Metallic sheathing
	Wood lagging
	Silanes
Pozzolons (PFA)	To replace 15 to 30% of cement
microsilica	4 to 6% where appropriate
BFS-cement	Alternate to Portland plus PFA
Unbonded tendons	Properly sheathed in high-density polyethylene (HDPE) with corrosion-inhibiting grease
	Anchorages protected
Hot weather	Cooled concrete mix
	Prevention of moisture loss during initial curing
Cold weather	Protection against freezing for fresh concrete
	Insulation of young concrete from cold winds
Earthing	Bond all steel elements within concrete, including all prestressing
(grounding)	tendons, and connect to electric ground. (Only necessary where stray currents are possible.)
Cathodic	As with earthing but provide connection to external anode and
protection	provide impressed current. (Note: in conductive medium, e.g., salt water, may use sacrificial anodes.)

CHAPTER 6

Posttensioning Technology

6.1 GENERAL

Many of the most spectacular and important structures in prestressed concrete have been constructed in posttensioned cast-in-place concrete. This includes many tall towers, long-span bridges, and offshore platforms in the ocean. The use of cast-in-place concrete makes possible complicated shapes and curves and transitions. Where construction joints occur, only one face needs to be prepared and joined, as compared to the two faces of precast construction.

Cast-in-place construction also minimizes the weights to be handled. The concrete may be transported to the forms by small bucket or pump and the forms may be handled by jacking or small hoists, as well as by cranes. Climbing forms (slip forms) may be used as well as jump forms. Thus, the process is adaptable to construction in lengths or heights beyond the practicable reach of large construction equipment.

Although the basic techniques of cast-in-place concrete construction are similar, whether reinforced by conventional mild steel or prestressed, there are a number of special considerations applicable to prestressed structures.

6.2 SHORING

Horizontal cast-in-place concrete members must be supported during construction. The loads include the weight of concrete, with its embedded steel, plus the forms and the concreting equipment. After concreting, curing, and stressing, the member will be post-tensioned. It will then deflect in the vertical plane, usually cambering upward (Fig. 6.1). When it does, it transfers the full load to the end supports. The shoring at those locations must be able to support the load.

Figure 6.1. Long flat arches of Moscone Convention Center in San Francisco are supported on high falsework showing until tie rods are posttensioned.

The prestressing also shortens the member. The shoring must not unduly restrain this, as the member must shorten in order to be stressed. This shortening, through friction or binding, will force longitudinal loads into the shoring.

Failure to provide for these deformations and loads has resulted in unfortunate catastrophic accidents.

Shoring must also provide adequate support against sidesway and lateral loads due to the construction operations and wind. When constructing post-tensioned concrete over an active roadway, protection must be provided to prevent vehicle collision with the shoring.

Shoring must be adequately supported so as to prevent settlement as the loads of the concrete are applied; in extreme cases, this requires piled foundations (Fig. 6.2).

6.3 FORMS

It is quite important in long-span construction to minimize the deflections that occur during concreting. Thus, the forms must be rigid and of sufficient section to reduce elastic deformations to a minimum.

Since today it is common to use HRWR agents in the concrete mix, so as to obtain "flowing" concrete, form pressures will be high. Further, internal vibration tends to cause local deformation of the forms, such as bulging, unless the forms have sufficient rigidity.

Figure 6.2. *When bridge girder in posttensioned, the span lifts off the falsework, permitting removal and relocation of the shoring.*

Form vibration is frequently used to obtain more complete consolidation of thin sections. This generates extremely high stresses, both hydrostatic and dynamic, in the forms, requiring better form construction and securing.

The technique of prestressed concrete requires accurate location of the prestressing force relative to the center of gravity of the cross section. This means that the forms must be constructed with greater-than-normal accuracy.

After the concrete has hardened, forms are usually partially stripped; for example, the side forms are removed but the soffit forms are left in place. Shrinkage takes place. The remaining forms must be designed to permit this volume change to take place. Then at stressing, the member will shorten and may rotate (camber). The remaining forms must be designed and constructed so as to permit the shortening without undue restraint, and to accept the camber and also the shift in dead-load distribution of the member.

Finally, forms must be chosen to provide a dense, impermeable surface to the concrete, as it is this "skin" of concrete at the surface that plays the most important role in protecting the prestressing tendons and mild steel from corrosion.

6.4 MILD-STEEL REINFORCEMENT

Prestressing creates secondary (transverse) stresses, which require mild steel for confinement. In addition, the thin sections, particularly webs of girders, are subject to high shearing forces, which are usually countered by mild steel, although occasionally posttensioning also is used here.

When tensile forces change direction, they produce high transverse stresses. This is especially demanding when tendons are curved out from a slab or wall to an anchorage, thus departing from the center of resistance of the section. When tendons are installed on a curved alignment in a plane normal to the element, for example, in a web of a box girder that has horizontal curvature, there are significant radial forces.

These above are design conditions for which the designer will or should have installed proper reinforcing steel to tie the tendon and its encasement to the wall.

To the above considerations must be added unintentional curvatures. Prestressing ducts are usually installed to a tolerance of 6 to 10 mm ($\frac{1}{4}$ to $\frac{3}{8}$ inch). Within the ducts, the tendons usually can move up to 6 mm when stressed. Tendon ducts may have to be slightly deviated because of conflict with inserts, although this should be referred to the designer.

At construction joints or stops, the ducts will usually be accurately fixed in position. During concreting, they may sag in between supports. The result is a relatively sharp change in angle, although small in amount, producing a radial stress. Such deviations as those noted above have caused internal delamination and cracking. In some cases they have led to failure. Mild steel is necessary to restrain these forces.

Similarly, where several ducts are concentrated in one area, such as the negative moment area of a girder, there may be insufficient concrete cross-sectional area before grouting to withstand radial crushing or to transfer tension to the full section (Figs. 6.3 and 6.4). Reinforcing steel in the form of "hairpins" or stirrups is valuable here.

When anchorages of heavy tendons are installed, there are transverse tension stresses developed. Locally, a transverse tension field will exist around the anchor-

Figure 6.3. Interior anchorages are formed by bolsters which are floored out from the slab, so as to permit anchorage of tendons at that location.

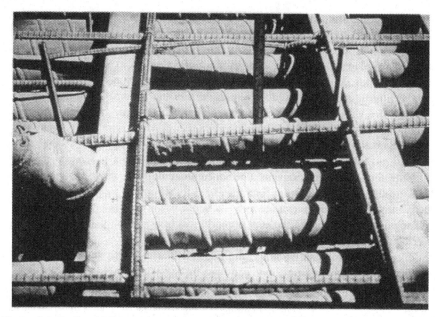

Figure 6.4. Posttensioning ducts are placed very closely together in negative moment area near pier. Stirrups are required to resist transverse tension through the slab.

age. It must be confined by spiral steel. Typically, this is an integral component of the anchorage.

Between two or more such anchorages, a tension zone may develop, tending to split the member in two. Mild-steel reinforcement is used to "stitch" the member together and prevent cracking.

Tendons are frequently anchored within the member instead of at its ends. Examples are the continuity tendons of box girders. As they are stressed, they tend to pull the compressed zone away from the remaining ends, thus developing shear along the sides of the anchorage and tension in back of it. Mild-steel reinforcement should be provided to ensure distribution and transfer back into the member.

Imposed deformations in service and heavy concentrated loads are other zones where confining reinforcement must be added, in the form of spirals, stirrups, or mechanically headed bars.

It is remarkable how a relatively few bars of small diameter, properly placed, can assure proper performance, whereas their absence can result in disastrous cracking or splitting.

All these considerations tend to require substantial amounts of mild-steel reinforcement on all three axes in critical zones. Thus, the steel may be very concentrated and hence difficult to place and hold to proper tolerances.

Prefabrication of such critical assemblies, and placement as cages is often the best solution.

The mild steel must be accurately placed and held during concreting. Since accidental impact from the vibrators may dislodge the mild steel, it is often advisable to mark the location of bars on the forms prior to concreting, so they will be readily apparent to the vibrator operator.

Current design practice emphasizes the use of more small bars rather than a lesser number of large bars. This again requires greater care in placing and restricts the space for placement and vibration of concrete. This, in turn, may require the use of smaller coarse aggregate and smaller vibrators, and more frequent tying of the steel in position.

Because of the traditional separation of contractual relationships in the United States between the general contractor (performing the concrete construction) and the reinforcing steel-placing subcontractor (placing the steel), there has been a tendency for the general contractor to shrug off responsibility for shop drawings and the accuracy of cutting, bending, and placement of the reinforcing steel. In posttensioned cast-in-place construction, it becomes of great importance that the steel be in full accordance with the design drawings. Therefore, the general contractor must check and verify the accuracy of the reinforcing steel, whether placed by a subcontractor or by his own forces.

6.5 INSTALLATION OF TENDON DUCTS AND ANCHORAGES

Anchorages must be installed normal to the tensioning force and the concrete must be thoroughly consolidated under the bearing plates. The trumpets must be properly fitted and taped to both the anchorage and the duct.

Frequently, anchorage assemblies are correctly set in the forms, only to be displaced during concreting. Therefore the template or form holding them must be rigid and sufficiently massive.

It is essential that the ducts not be ruptured or cracked by accidental contact of the vibrators. This consideration may lead to the selection of heavier gauge metal for the ducts, or rigid ducts, or the insertion of a mandrel in the ducts during concreting. This latter solves the multiple problems of rupture, wobble, and sag. Plastic grout entries and vents must be properly supported so they will not be dislodged during concreting. As with mild steel, duct locations may be marked on the forms as a guidance for the vibrator operator.

When a posttensioning tendon is anchored at some intermediate location, within the length of the member, as opposed to being anchored at the ends, there will be internal strains and stresses introduced at this intermediate point. As is common with all posttensioning anchorages, there will be lateral bursting forces around the anchorage zone and transverse tensile forces between adjacent anchors. In addition, there will be longitudinal forces behind the anchor (away from the lead of the tendon), which result in shearing stresses along the sides and tensile stresses behind. In some cases, these have resulted in sudden shearing of the entire anchorage zone; in other, even more serious cases, this phenomenon had led to a failure of the member in vertical shear.

Thus consideration of the stress conditions around the anchorage is essential.

Although principally a responsibility of the designers, in practice it is of great consequence to the constructor, and he must therefore be aware of the needs for adequate reinforcing in this zone. A similar problem arises at the blisters or bolsters which typically form the intermediate anchorages of bridge girders. At such locations the tendons curve away from the longitudinal member, producing high transverse tensile forces. The designer should calculate these and provide the stirrups needed to resist these transverse forces. However, the constructor must realize that these forces have a zipper-like performance. Therefore, the initial crack must not be allowed to progress. This means that the constructor must ensure that the stirrups are placed at the locations designed, especially at the end next to the anchorage. While this may appear self-evident, in practice it is physically extremely difficult to fit.

Splicing of tendons requires special consideration. The splice sleeve is larger than the tendon and hence requires a larger diameter duct at its location. Further, the splices must be free to move as the tendon is stressed and elongated. Finally, the transition of the duct must be configured to facilitate the grout flow.

Several posttensioning tendons may be concentrated high in the web over an intermediate pier or support in order to provide continuity. Similarly, others may be concentrated low in the web near midspan. Sufficient concrete must be left between ducts to prevent "pull-through" as the tendons are stressed. Assuming the sharpest curvature is over the pier, the lower tendons may be stressed and grouted before the upper tendons are stressed (Fig. 6.5).

With bar tendons, splices must have some means to prevent accidental reduction in the number of threads which are engaged. The first bar may have been accurately screwed on exactly half-way. When the next bar is screwed in, it typically turns the

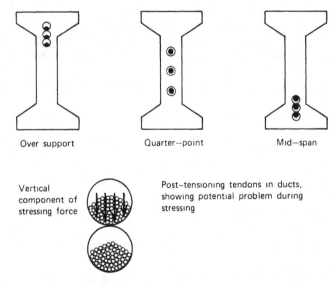

Over support   Quarter–point   Mid–span

Vertical
component of
stressing force

Post-tensioning tendons in ducts,
showing potential problem during
stressing

Figure 6.5. Adjoining ducts require special consideration.

splice as well, due to minor but inescapable misalignment. Being of opposite thread, it thus disengages the splice from the first bar. When stressed, it may strip the threads. Being embedded in the concrete structure, it is usually impracticable to repair. This sequential phenomenon may be prevented by epoxy coating the threads of the first splice as it is made, so they cannot unscrew.

For many posttensioned applications, especially in building slabs, single strand tendons, wrapped-and-greased, are installed. These must be accurately draped on chairs to follow the design vertical profile and to maintain it while concrete is being placed and vibrated.

The wrapped tendons must be handled with care so as not to rupture the sheathing.

6.6 CONCRETING

To obviate or minimize many of the problems referred to above, a mix using smaller size (e.g., 15 mm, $\frac{1}{2}$ inch, or even 10 mm, $\frac{3}{8}$ inch) nominal maximum size coarse aggregate is preferable, especially in the region of the end blocks, which will typically be congested with reinforcing steel and post-tensioning anchorages. The mix should be workable; this requires adequate cement content and sand. Use of a water-reducing, retarding, and shrinkage-reducing admixture is beneficial and recommended.

Retarders are particularly effective in slow pours such as those frequently required in prestressed cast-in-place structures. The rate of set can be controlled by varying the amount of retarder during the various stages of pour.

Internal vibration is always required to consolidate the inner portion of the concrete. External vibration is useful in eliminating surficial defects but generally is not powerful enough to consolidate the inner portion. Generally, it cannot extend more than 150 to 200 mm (6 to 8 inches) in depth. With deep end blocks, "windows" may be needed in the forms to permit placement of the concrete without segregation and to facilitate vibration.

High early-strength cements and finely ground cements are frequently employed to achieve early strength so as to enable prestressing. This is usually satisfactory, provided the cements do not contain chlorides and provided the sections are typical thin-walled structural members. With thick walls of large masses, their use may produce excessive shrinkage and excessive heat of hydration, resulting in cracking.

Steam curing is a very effective way of accelerating strength gain. The cast-in-place section is enclosed in a hood of rubberized nylon or polyethylene and steam is applied at atmospheric pressure. For smaller sections, metal or timber hoods have been employed. In some complex sections containing well-distributed ducts, hot water has been circulated; however, this should be treated with a corrosion inhibitor to prevent corrosion of the duct.

In any event, cast-in-place concrete for prestressed construction requires more thorough curing than conventional cast-in-place concrete, because sections are thinner and more highly stressed. Therefore, adequate means of curing must be specified and enforced.

6.7 INSTALLING AND STRESSING THE TENDONS

In almost all cases of cast-in-place construction, the tendons are inserted after concreting. Current practice calls for water-soluble oil to be brushed on as the tendon is inserted. Later, during grouting, the initial flow of the grout will be used to flush out the oil (Fig. 6.6).

Flushing with water, which formerly was the practice, is no longer favored, because of the difficulty in later displacing the water, even when oil-free compressed air is used. Tests and experience show that the grout is more effective in displacing the oil. Alternatively, the temporary protection against corrosion prior to grouting is obtained by the dusting on of vapor-phase-inhibiting (VPI) powder. When powder is used, it is even more important to seal the ends, since VPI powder only protects for a short distance and the dehumidified zone must be enclosed.

When inserting the tendons, care must be taken not to abrade them at the entry: a guide funnel is often used with large tendons (see Fig. 1.8).

Figure 6.6. *Pulling strand off reel for threading into ducts*

High-strength prestressing bars which are inserted in the duct prior to concreting are, of course, not nearly as sensitive to corrosion; nevertheless, they should be protected as described above. This consideration is especially relevant to "blind" anchors and "dead-end" anchors of prestressing tendons.

Strand tendons are generally inserted by pushing in, one at a time. Alternatively, several strands may be bound together, with a common flexible nosepiece to prevent catching on the duct corrugations, and pushed in, or a messenger line may be sent through by attaching it to a rubber ball that is pushed through by compressed air, or to a wire leader similar to a "plumber's snake"; then the bundled tendon may be pulled in.

If the messenger line is inserted prior to concreting, this may be used immediately after concreting to run back and forth to insure that the duct is free and unblocked. Then, later, the tendon may be attached and pulled through.

After installation, the duct openings (at each end) should be sealed with polyethylene to prevent the entry of rainwater.

Stressing to design transfer stress should be performed only after the concrete has reached the specified strength for this prestress. Otherwise, excessive creep may result.

However, subject to the approval of the design engineer and provided it is compatible with the prestressing system adopted and the forms, it may often be found advantageous to impart a small degree of prestress shortly after concreting (e.g., 12 to 24 hours). This offsets shrinkage and thermal strains and helps to prevent cracks from developing. Later, when the concrete reaches the specified strength for prestressing, the tendons may be stressed to their full transfer value.

Such early stress values are usually very low (i.e., perhaps 10% of the final tendon stress) so their influence on corrosion-susceptibility is slight. However, whenever tendons are to be left in construction in the ungrouted condition for more than 24 hours, sealing of the ducts is necessary. (See Section 5.5.3.2.)

Generally speaking, tendons should never be installed prior to steam curing, since the steam accelerates corrosion.

Jacks should be calibrated before use on each project and at 60-day intervals during a project or when discrepancies are noted.

Frictional losses should be calculated. As actual field data becomes available, the constants used in the calculations can be corrected.

Jacking at one end, followed by lift-off at the other end (by jack) will give a measure of the total losses.

In production stressing, the jacking pressure will be read while at the same time, the elongation is measured. These should agree within 3 to 7%. Elongation is generally used as the control.

Discrepancies and inconsistencies should be investigated before proceeding.

Tendons which are curved should normally be jacked from both ends in order to more or less equalize tension throughout the length. Where this is impracticable, the designer must approve single-end stressing.

After seating all the strands within the anchorage, the elongation should be

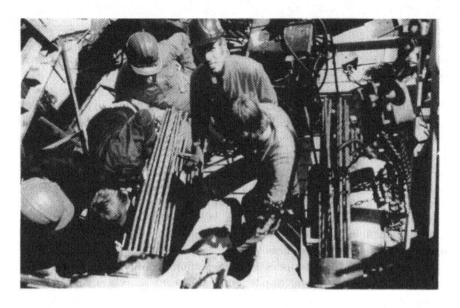

Figure 6.7. *Measuring seating loss (set) in tendons after all strands have been seated in anchorage.*

measured to verify that the seating loss or set does not exceed allowable values, usually about 6 mm ($\frac{1}{4}$ inch) (Fig. 6.7).

6.8 CONSTRUCTION JOINTS

Much of the difficulty with post-tensioned cast-in-place concrete has arisen at construction joints. The concrete faces of construction joints must be properly prepared so as to present a roughened face with adequate shear transfer. The preferred means is water jetting, using a high-pressure jet on the hardened concrete, or at least after final set, to expose the coarse aggregate. Generally speaking, a depth of about 6 to 10 mm ($\frac{1}{4}$ to $\frac{3}{8}$ inch) is appropriate.

Another means is the painting or spraying of a set retardant on the forms for the construction joint, then washing it off after stripping, so as to expose the aggregate.

For vertical construction joints, a fine mesh (about 4 mm openings) placed inside the forms will strip off the laitance when the mesh is removed. However, supplemental water jetting is still recommended.

Bush-hammering is an ancient technique that must no longer be allowed, since it has been found to loosen the coarse aggregate and to produce extensive microcracks, resulting in reduced shear transfer and potential corrosion.

Ducts must be made continuous across construction joints. With rigid or flexible metal ducts, the joints may be sleeved with a short section of duct and these secondary joints taped with waterproof tape. This is satisfactory, provided the ducts are somehow tied so they can't be accidentally pulled apart during subsequent operations, such as vibrating of the concrete. Extensive testing has, however, thrown some doubt on the reliability of such sleeved-and-taped joints. Duct joints in which the sleeve is sealed by epoxy and those sealed by heat-shrink tape have proven far more reliable in accelerated testing and use. Rubber and plastic sleeves have also been used, as have screwed connections.

Here again, the temporary insertion of inflated duct-formers or of flexible rigid conduit (such as electrical conduit) in the ducts will serve to preserve continuity across the joints.

6.9 STRIPPING

In all post-tensioned cast-in-place construction, after stressing, accurate measurements should be taken of camber, shortening of tendons, and stressing loads. These should then be correlated and checked to be sure the structure is performing in accordance with the design calculations. Current specifications allow 3 to 7% deviation between measured elongation and that calculated from the stressing load. Only after behavior has been fully verified it is safe to strip the supporting forms and falsework.

This is especially critical in thin members such as floor slabs where, if errors of placement, etc., have been made, the slab may not camber upward under prestress, but may actually be unstable. Thus a standard procedure should be instituted to verify performance prior to removal of supports.

When members such as bridge girders on long-span beams are post-tensioned, they camber upward from their temporary shoring. This then changes the support conditions, typically placing the entire reaction supports at the ends.

Similarly, posttensioning of a member inherently causes it to shorten. If shortening is restrained by the forms or supports, the member itself will not be prestressed. In turn, there may be high forces introduced into the forms or supports, leading to damage to these, or even collapse.

6.10 GROUTING AND GREASING: CORROSION PROTECTION

The permanent protection for post-tensioned tendons and of their anchorages is essential to their continued performance. Because of the importance of these operations, they are treated separately in Chapter 5, Sections 5.10 and 5.11.

Grouting bonds the tendon to the duct, which in turn is bonded to the concrete. The bonding is obviously more effective with small tendons then large ones due to the lesser force which has to be transferred. However, even with large tendons, bonding by cement grout provides force transfer over a relatively short distance due to mechanical jamming as well as adhesion. Thus it modifies the behavior under high tension so as to reduce the spacing between cracks and the crack widths.

SELECTED REFERENCES

1. VSL International, "Post Tensioning Systems," Beme, Switzerland, and Los Gatos, California, undated.
2. DSI, "Dywidag Post-Tensioning Systems," Dywidag International, USA, Long Beach, California, undated.
3. Freyssinet International, "Prestressing Technology," Paris, 1991.
4. *Post-Tensioning Manual*, 5th edition, Post-Tensioning Institute, Tucson, Arizona, 1990.
5. *FIP State of Art Report*, "Tensioning of Tendons: Force-Elongation Relationship," Structural Engineers Trading Organization, London, 1986.
6. BBR "Activities and Services, Bureau BBR Ltd. Zurich, Switzerland (undated).
7. Mocalloy Bar Systems "Design Data" McCalls Special Products, Sheffield, England, undated.
8. FIP Guide to Good Practice, "Preparation of Specifications for Post-Tensioning Work," SETO Ltd., London, 1992.

CHAPTER 7

Manufacture of Precast Pretensioned Concrete

Precast pretensioned concrete elements are manufactured either in a permanent manufacturing plant or in a temporary job-site plant. The principles are the same: the permanent plant can economically incorporate more sophisticated machinery and equipment along with the flexibility to adjust to a variety of products, whereas the job-site plant is tailored to the specific needs of that particular project.

The manufacturing plant comprises these minimum elements:

1. Storage facilities for aggregates and cementitious materials.
2. Concrete batching and mixing facilities.
3. Concrete delivery systems.
4. Prestressing strand (or wire) storage on reels, with means for laying out tendons down the beds.
5. Reinforcing-steel storage and fabricating facility.
6. Beds upon which the tendons are stressed, forms placed, and concrete poured (Fig. 7.1). Stressing stands at each end must be capable of taking the high compressive forces, as well as moment introduced by the height of tendons above the bed, and by deflection of tendons (Fig. 7.2). As an alternative, self-stressing forms, that is, forms which can withstand the high compressive forces imposed by the stressing, are often used, especially on small projects.
7. Stressing equipment (usually hydraulic jacking equipment and strand vises or grips) (Fig. 7.3).
8. Deflection mechanisms (if tendons are to be deflected) (see Fig. 2.2).
9. Forms (Fig. 7.4).
10. Concrete placing and consolidation equipment.

Figure 7.1. Pretensioning bed.

11. Means for accelerated curing (usually a low-pressure steam system). Insulated tarpaulins are used to confine the steam.
12. Lifting and handling equipment.
13. Storage area, with dunnage, i.e., blocking, on which to set the finished product.
14. Transportation equipment: trucks, barges, etc.

Figure 7.2. Stressing stand takes reaction of stressed tendons.

Figure 7.3. Long-stroke hydraulic jacks for stressing tendons.

15. Equipment for testing and inspection.
16. Facilities for maintenance and repair.
17. Utilities (water, power, fuel supply, compressed air).
18. Storage and fabrication of inserts, voids, picking loops, cushion blocks, etc.
19. Burning and welding equipment.

Figure 7.4. Steel forms for precast pretensioned elements.

20. Shop engineering for shop and working drawings and computations.
21. Yard management and administration, cost records, accounting, purchasing, expediting, and estimating, sales engineering, and service engineering.

Obviously in special cases some of the above may be performed off-site, or by outside suppliers.

While the individual items listed above are discussed in detail elsewhere, or are well-known standard practice, the coordination of all of these items into a manufacturing system requires extremely careful planning and management.

Some guidelines are the following:

Communications must be provided, particularly from point of concrete placement in forms, to batch and mixing plant. Use of two-way voice radio has been found useful. Communications are also needed, of course, to the vehicles or tugs, etc., involved in transport of the finished product.

Materials handling has been identified as the largest single component of labor. Therefore, materials flow must be laid out so as to involve the least movement and least number of handling operations.

Beds must be designed to remain level and true despite repeated loading, and frequent wetting of ground, e.g., by water used in curing and clean-up. Use of short piles, or cast-in-place concrete posts in drilled holes, as a support for the beds, will often be found advantageous in this regard. The height of the bed should be set at the optimum working level, particularly where considerable hand work is required as, for example, in the manufacture of pretensioned railroad ties (sleepers).

Most beds operate on a one-day cycle. Thus, the member must have a substantial strength at 14 to 16 hours after casting. This required strength is a function of the prestress level but typically is two to four times the initial precompression under prestress.

Such rapid gain in strength is attained by rich cement mixes, fine grind of the cement, low W/C ratios (using water-reducing agents) and heat.

Proper storage must be provided for prestressing steel, mild reinforcing steel, and inserts, to keep them clean and dry. Prestressing steel, in particular, and the associated fittings must be stored up off the ground in a weather-tight warehouse. German specifications require that this storage space be dehumidified, so as to reduce the relative humidity to 20% or less: this is usually done by heat within an insulated warehouse. The tendons should be protected from extended exposure to salt air (fog) and from refinery or chemical airborne effluents. However, it is generally accepted that one day's or overnight exposure, resulting in slight surface rusting, is not detrimental to long-term durability. One measure of such "acceptable" rusting is that it can be wiped off with a dry cloth.

Roads and drainage must be provided in all work areas and in storage areas. It is important to eliminate soft spots, ruts, or holes, any of which might cause a crane or forklift truck to tip, possibly injuring a worker or damaging product or equipment.

Utilities should be brought to the work area at convenient outlets. Boxes or guards should be installed to keep outlets and receptacles dry and clean, and to protect them from accidental impact. Adequate lighting should be provided for night work.

Hydraulic jacks and strand vises must be properly maintained, clean, and properly lubricated in accordance with manufacturer's recommendations.

Shop drawings and complete special instructions on inserts, strand tension, deflecting, etc., must be provided, in a readily readable form, at an accessible place at the work site, e.g., the casting bed.

Strands are usually pulled off the reels in groups by temporarily affixing a pulling nose, and pulling down the bed by a hoist (Fig. 7.5). Once all tendons are laid out, and anchored in their correct position, it is necessary to stretch them to assure equal length. If all strands are to be jacked (stressed) as a group, then each strand should first be stressed to a nominal figure, say, 500 to 1000 lb, and anchored.

If tendons are to be fully stressed one at a time, then each can be initially anchored at the required stress. Equalization is then not a problem. If strands are deflected, the consequent change in longitudinal tendon stress should be verified to ensure its accord with design requirements.

As noted above, strands which are bright when initially installed may get a thin coating of atmospheric rust within a few hours. This is generally considered not to detract from the durability and may actually enhance bond. However, tensioned strands should not be so exposed for more than about 24 hours, depending on the atmospheric environment and its proximity to salt-laden or contaminated air.

Concrete is placed in the forms, working from one end to the other, and vibrating it to ensure full consolidation (Fig. 7.6). External vibrators may be mounted on the forms of deep members such as bridge girders. These external vibrators may be moved along the bed progressively as the concrete advances.

Figure 7.5. *Pulling prestressing strands down the bed through pre-placed spirals.*

Figure 7.6. *Placing and vibrating concrete.*

Behind the concrete placement comes the screeding and then the finishing of the surface (Fig. 7.7). If because of high slump or cool weather, the finishing has to be delayed more than 15 minutes or so, the surface should be temporarily covered to prevent excessive evaporation.

After finishing, the insulating tarpaulins are spread, being kept clear of the concrete surface and forms by temporary frames (Fig. 7.8).

Figure 7.7. *Screeding and finishing surface of pretensioned concrete piles*

Figure 7.8. Spreading tarpaulins over frames to permit steam curing.

When the concreting is completed and the tarpaulins completely cover the bed, then the steam is turned on at a very low rate, that is, the valve is "cracked," so as to introduce moisture into the enclosure. After about four hours, the steam may be turned on in earnest, with thermostats controlling the rise in temperature to 20°C to 30°C per hour (40°F to 60°F). Maximum temperature should be 70°C (about 150°F). It is held for about six hours, then the steam is turned off to allow the members to cool to not more than 20°C (40°F) above ambient.

Test cylinders will have been stored alongside the forms so as to undergo the same thermal and humidity regimes as the elements themselves. These are now tested and, if the strength is satisfactory, the tarpaulins are stripped and the stress of the strands released into the elements.

Extra lengths of strand between elements are now burnt and the elements are ready for lifting.

After release of the stress into the concrete elements, the friction due to the deadweight of heavy girders, binding of holddown devices, etc., and binding in the forms may prevent them from moving. Until a member actually shortens, it is not prestressed.

Therefore, removal of members from the bed should start at the free end (where the strands are slack) and move progressively to the other end.

Similarly, in picking members from fixed forms, it may be necessary to overcome friction, suction, and deadweight (Fig. 7.9). The concrete may even have a slight outward expansion against the forms due to the Poisson's ratio effect of the prestress. Once the member is broken loose, the load drops to its deadweight value only. This produces a dynamic "bounce," which may cause cracking. For this reason, when picking members from fixed forms (especially thin elements), it may be better to first break them loose with self-centering hydraulic jacks, reacting against the bed or forms.

Because of these high initial picking forces, in which the dead load is augmented

Figure 7.9. Picking pretensioned element from forms.

by friction and binding in the forms and by suction, the inserts used for picking must be designed accordingly. This refers not only to the hardware itself but to its anchorage within the member. While special inserts designed for this purpose are available, picking loops of strand are also frequently used. These consist of several short pieces of waste strand, bundled together and bent in the form of a loop. Their tails are anchored well down in the concrete so as to be in the compressed zone at the lifting stage. Bent reinforcing bars should not be used since the deformations may cause stress concentrations leading to brittle failure.

Forms are generally designed for multiple reuse and should, therefore, be of steel, concrete, FRG, or heavy wood framing of equivalent strength. Attention should be paid to the laps and joints in the forms, as offsets and irregularities may produce cracking in the concrete element. Jointing surfaces must be rigidly held in true alignment and ground to remove the weld bead. Forms must be heavy and rugged enough to withstand repeated vibration without suffering fatigue (Fig. 7.10).

Forms must be cleaned immediately after removal of product. Particular attention must be paid to removal of grout from joints and holes for affixing inserts. Careful taping of such joints and holes prior to pouring concrete will eliminate much cleaning.

Sliding forms are used on some products, to form the external upper surfaces at the top of some members, such as round or octagonal piles, and to form internal voids or cores in some piles and in floor slabs.

The sliding forms must have low friction so as not to drag the concrete surface.

Figure 7.10. *Steel forms for multiple use in pretensioned concrete manufacture must be accurately made.*

Stainless steel and Teflon-coated surfaces are sometimes used, although clean, bright carbon steel will often prove sufficient if lubricated by water or fluid cement grout.

Sliding forms must be accurately guided. This is relatively simple for external forms but is complicated for internal mandrels by the transverse and spiral reinforcing steel. Guiding on the stressed strands is one means but then the strands themselves must be restrained from excessive deflection due to the mandrel's weight.

Internal void mandrels tend to raise up, due to the buoyant force of the fresh concrete mix as it is vibrated. If the mandrel is sufficiently long and stiff and its tail is in stiffened or setting concrete, then it is only necessary to develop the means to hold the leading end in place and prevent its displacement upward.

Similar principles apply to inflated mandrels, and waxed cardboard tubes, where flotation upward between supports is difficult to prevent. Both of these tend to cause excessive "wobble" of the internal void. To overcome this, ties may be installed at close intervals. The ties must have a pad under them to prevent cutting the rubber or cardboard tube due to the uplift force during vibration.

Because of the generally thinner sections and higher concentrations of reinforcement, typical of prestressed concrete products, reinforcing steel must be secured by closely spaced ties and carefully positioned in the forms (Fig. 7.11). Prefabrication of cages on a template will often permit better access for assembly. Such cages may be tack welded, but care should be taken not to underburn or notch a bar, especially in a zone of high tension or repetitive loads. If prestressing strands are to be inserted

Figure 7.11. *Reinforcing steel must be accurately positioned in forms.*

in the cage while on the template, all tack welding should be completed before the strands are threaded into place. Prestressing strands should never be welded as this causes a radical loss of strength.

Concreting has been greatly facilitated in recent years by the development of HRWR agents. These give "flowing" concrete and thus reduce the need for the screw extruding devices formerly used with low-slump concrete.

Vibration is required to consolidate the concrete, even "flowing" concrete. Except in lightly reinforced thin members, internal vibration is essential. Otherwise internal lamination and honeycomb may result.

External vibration is highly useful. Attached permanently to the forms, the vibrators can be turned on successively as the concreting proceeds. They will result in a dense surface, relatively free from rock pockets and honeycomb.

The pretensioned strands must not be released so as to transfer stress into the member until the concrete has attained adequate strength. This is normally ascertained by breaking companion test cylinders which have been cured in the same manner, for example, alongside the member, so as to be subject to the same heat cycle. Premature release may cause cracking, since the low tensile strength of the concrete may not be able to resist the transverse tensile stresses imposed. Premature release may also cause excessive stress loss due to high creep.

Curing has typically been by covering with insulated rubberized nylon tarpaulins, and injecting steam at atmospheric pressure. This insures both moisture and heat during the early critical hours.

It has been found desirable to delay the full injection of steam for a period of three to six hours. During this pre-steaming period, moisture should be available. One method is to crack the steam valves slightly. Another method is to seal in the moisture in the concrete by coating with a membrane curing compound which later degrades as full steam is introduced.

Steam temperature should be raised slowly. Limits of $\frac{1}{2}$°C (1°F) per minute are often specified. The objective is to prevent excessive thermal gradients, which obviously depend on the thickness of the member and whether normal or light-weight aggregates are being used.

Maximum temperatures must be limited. Best long-term results are achieved with an upper limit of about 70°C (150°F) but for products of low sophistication, where ultimate strength is subordinate to early strength, higher maximum temperatures, up to 80°C, are sometimes used.

After soaking the concrete for several hours at maximum temperature, the steam can be turned off and the enclosure and product allowed to cool down gradually. Steam hoods (tarpaulins) should not be removed until the internal temperature is within 20°C (36 to 40°F) of the ambient.

It is customary in plants to cure test cylinders under the hoods and to remove these for testing sometime prior to the release of prestress into the member. Uncovering to the cold air, especially on dry, windy days in winter, will result in thermal cracking. Therefore, after removal of test cylinders, the covers should be promptly replaced and not removed again until just before releasing stress.

At this latter stage, the prestress will be released into the member. This produces axial compression, but the member will still be hot. Further, even though the strands have been released, friction may prevent heavy members such as girders from shortening and hence these may be in a reduced state of prestress. Application of a supplemental membrane curing or water curing, e.g., wet burlap, will often be found necessary to prevent surface crazing.

Most cracking problems with pretensioned members occur during the cooling off and release period. It is important therefore to follow through the processes occurring in these stages for each product and each environment. For example, in Arabia, the hot temperatures in the day will evaporate curing compound, and the sudden drop of temperature at night will produce thermal strains. The combination of low humidity and low temperature will often lead to cracks, especially cracks parallel to the prestressing tendons.

Returning to curing, in a number of plants, alternative methods of attaining high early strength are used. Hot oil circulated in pipes produces radiant heat, replacing steam. Moisture is supplied by an intermittent fog spray. Electric "blankets" can be attached to the forms for heat.

In some cases, especially at job site facilities, the early strength required for release of prestress is obtained not by external heat but by insulating covers to keep in the internal heat of hydration. Cement may be ground especially fine (greater than 4000 cm²/g) to generate heat and accelerate early strength.

Deflected (curved) profiles of prestressing are efficient in matching the moment curve, so as to offset tensile stress and shear.

Pretensioned strands may be deflected so as to give favorable profiles of prestress force along the length of the member.

The strands are first stretched straight, then pulled down at intermediate points while being held up at the ends, or they may be pushed down from the top with the reaction taken against a frame straddling the forms (see Fig. 2.2). Alternatively they may be jacked up at the ends while being held down at points within the girder length. The initial "straight" stressing force is calculated so that, after deflection, the final stress will be as designed. This can be checked by "lifting off" an end anchorage.

While concreting, the deflected tendons constitute a bowstring and the hold-down inserts a potential projectile, so personnel must be trained to not expose themselves over these critical points, for example when vibrating concrete. A steel or timber crosspiece may be installed as a protection device, and the points of deflection marked on the forms.

Release of these two-dimensional systems presents a problem. If the longitudinal anchors are released first, then the hold-down may bind and prevent the shortening of the member, whereas if the hold-downs are released first, and if these forces exceed the deadweight of the member, the top flange will crack in tension.

This dilemma has been solved in several ways, for example, arranging external flexible ties so as to hold the cured member down at these same points while the deflecting device is released, then releasing the longitudinal prestress. The hold-down devices may be designed to permit longitudinal movement, since the shortening of the concrete member is usually only a very short distance, less than 25 mm (1 inch).

In any event, it is important to recognize that the member is not prestressed until it has physically shortened.

Friction of the tendon at the deflector devices is a concern. The bend radius must not be too sharp, else the combined stresses in the wires of the tendon may cause breakage.

Small rollers are often used to minimize friction and the sharpness of the bend. Other devices allow the hold-down point to translate longitudinally a small distance to accommodate the elongation.

When forms are stripped from precast concrete members, surface blemishes may be found on vertical and overhang surfaces, despite use of a proper mix and thorough vibration. These blemishes are due to trapped air bubbles (entrapped air) and bleed water.

Entrapped air may in general be minimized by proper gradation of the aggregate and if necessary by an admixture. Bleed is more difficult but here again, recently developed admixtures eliminate most of the bleed.

Less costly measures are low W/C ratio, selection of optimum form oil, and control of the internal vibration so as to drive the water to the top surface before finishing.

Elements should be stored on firm supports placed as nearly as practicable at points of final support so as to prevent adverse stresses and excessive creep. Protect thin slabs from excessive heat from the sun and from rain. Strut inside of large voids to prevent plastic deformation. Thin slabs must be lifted with special devices or equipment, such as vacuum-lifting devices (Fig. 7.12).

One of the main justifications for establishing a precast plant for the production of prestressed concrete is to get more efficient use of manpower. To achieve this requires a carefully developed work program. One effective way of achieving this is to prepare charts.

The first chart is a plan layout of the work area, showing each station of work and listing each operation at each station. Then, for each such operation, the number of workers of a particular craft are listed along with the time required in the schedule. This chart is adjusted until the schedule fits the specified production cycle (usually 24 hours) and the desired work day. From this, the number of hours per worker in each craft can be taken off. A second readjustment of the chart may then be necessary to properly utilize the full workday of each worker.

Similarly, an evaluation can be made at this point of the use of staggered shifts— also of selective overtime—in order to minimize the total labor cost.

A second chart then lists for each worker his assigned operation by location and function for each hour of the workday.

Charts such as the above require readjustments as experience is gained in actual production runs.

For some mass-produced members for buildings, more than one complete cycle per day can be achieved by a mechanized operation, in which strand and wire mesh are laid and the concrete "extruded." This process is ideally suited to flat slabs, such as floor slabs. Successive slabs may be cast, each run on top of the previous run,

Figure 7.12. *Using vacuum-lifting device to lift thin pretensioned members.*

using a bond breaker. Prestress is not released until the entire stack has attained the required strength.

The previous text has described the "long-line production" system, which is dominant throughout North America and has spread to many other parts of the world where mass production of standard building elements takes place. Although one early reason for its development was the availability of large manufacturing sites and generally benign weather, it has since been adapted to indoor production in enclosed plants as well. The long-line process can be characterized as one in which each process, e.g., concreting, is moved to the product.

An alternative concept emerged in Northern and Eastern Europe, as an extension of previous practice with precast conventionally reinforced concrete. This essentially is the flow-line pretensioning method, in which a discrete product, say a roof slab or beam for a building, is moved successively from one stage of manufacture to the next, an adaption of the conventional assembly line, in which the product is moved to the process.

As implemented in a precast concrete plant, the element is produced on an individual pallet or soffit. As this pallet moves from one station to the next, a particular operation is performed at each station.

If these several processes proceed in sequence, but are not interlocked, each will be independent of the other as far as the production cycle is concerned, but the final production rate will be determined by the slowest operation. The under-carriage pallets may be connected (as by flexible connections or articulations) and the movement interlocked and continuous. A typical sequence, by successive stations, is as follows:

Station

1. Cleaning.
2. Oiling of forms.
3. Placement of pre-assembled cage of tendons, reinforcement, and details (inserts).
4. Assembly, locking, and sealing of the forms.
5. Stressing of tendons.
6. Placement of concrete mix.
7. Vibration and compaction of the mix.
8. Screeding and finishing surface.
9. Inspection.
10. Loosening or stripping of forms.
11. Steam curing.
12. Removal of product for storage.
13. Storage.
14. Shipment.

Similar methods, but without Step 5, have been extensively developed for precast elements such as tunnel liners and building wall panels, the latter including in many cases the pre-glazed windows and door frames.

The author has proposed a system whereby a continuous ribbon of concrete moves through the several processes, not in intermittent steps but continuously. Tendons are held at the upper end on constant tension drums. At the other end, a complete cycle distant, the completed concrete is tensioned by means of tracked tensioners with polyurethane friction pads, similar to those used in the offshore pipe-laying industry.

Returning to the upper end, mesh is/rolled out continuously onto the strands. A void-former may be installed at the concreting point, using stainless steel or Teflon-coated mandrels, with internal vibration and, perhaps, fore-and-aft jiggling vibrators to prevent bond. A reinforced neoprene conveyor belt forms the soffit.

The concrete is fed on at a single point, vibrated, screeded, and finished. The ribbon of concrete, containing mesh and strands, and riding the conveyor belt, enters a low-pressure steam curing tunnel. When it emerges from the other end, tensioning rollers, as described above, provide the tension. After moving off the conveyor belt, a traveling saw cuts the precast slab (or other unit) to exact length. It then moves by roller conveyor to stacking pallets. The conveyor belt returns underneath to the concreting station.

Despite the appeal of sophisticated production-line systems like that described above, which obviously require long periods of continuous production, currently the lowest labor consumption ratios (man-hours per unit of product) in the world are being achieved by the relatively simple "long-line" pretensioning system. This system has achieved its efficiency through the adoption of "advanced forms" incorporating the following improvements:

1. Use of multiple forms side-by-side, up to a workable width of 2 meters (7 feet).
2. Multiple-handling (lifting equipment and gear capable of removing, storing, and loading several units simultaneously).
3. Use of fixed, flexible side forms, so that product may be raised from forms without moving them. Form vibrators are permanently attached, or designed to move along the form with concrete placement.
4. Where forms must be moved, use of hydraulically actuated rams to move them sideways for stripping and subsequently to close them to exact position.
5. Use of mechanized form cleaning equipment, such as rotating brushes.
6. Synchronized deflection and stressing of strands by remotely controlled hydraulic system.
7. Combination of automatically controlled form vibrators with supplemental internal vibration.

In-plant quality control is important to the achievement of economy, schedules, erection, and performance. Inspection, checking, and testing procedures must be provided to insure that the product is correctly produced, not only for strength, but for dimensional accuracy (within prescribed tolerances), and accuracy of placement of inserts.

REFERENCE

1. PCI *Manual for Quality Control for Plants and Production of Precast and Prestressed Concrete Products*, Third Edition, 1985, Prestressed Concrete Institute, Chicago, Illinois.

CHAPTER 8

Architectural Prestressed Concrete

8.1 GENERAL

The use of concrete for architectural effects, with a wide variety of color, texture, and sculpture, has experienced a rapid growth in both application and technology. It has emerged as a challenging medium for expression by artist, sculptor, and architect. Its moldability, permanence, and the flexibility with which it may be subjected to an endless variety of techniques, have been combined with concrete's performance as a structural and cladding material possessing strength, durability, weather and fire resistance, and economy.

The techniques for creating an architectural or artistic effect in concrete include those of integral color, sculptured or textured pattern, exposed aggregate, and embedded ornament (Fig. 8.1). It is beyond the scope of this book to discuss these in detail, with their multitude of processes for obtaining a wide range of artistic effects. However, there is a strong trend to integrate this artistic function with the structural function, and architectural concrete techniques are being increasingly applied to prestressed concrete. This chapter addresses the application and the effects of prestressing.

Prestressing enables the member to perform its structural function more efficiently, thus permitting thinner sections, longer spans, and overhangs. It can be used to control behavior, that is, to prevent long-term sag that might destroy the architectural effect. With sculptured members it gives increased freedom to the artist: prestressing can be used to counter adverse weight distributions.

8.2 PRESTRESSED ARCHITECTURAL BUILDING PANELS

The growing use of precast prestressed concrete building elements as architecturally exposed members means that great care must be taken to prevent surface discolora-

Figure 8.1. *White cement and white aggregates used to create attractive exterior of pre-stressed concrete double-tee slabs.*

tion and imperfections. Discoloration may be caused by variations in W/C ratio, by variations in vibration duration, by impingement of steam during curing, and by variations in thickness or composition of form oil. Even more inexcusable is staining from grease or oil during storage or erection, etc. Steel forms seem to give slightly more discoloration than FRG or wood forms, probably due to the thermal conductivity of steel during steam cure.

Exposed architectural members may require extreme care, such as wrapping in polyethylene or heavy kraft paper during shipping.

The formation of "bug holes," that is, air and water "pin holes," during manufacture has been discussed earlier. For architectural building members, where lower strengths may be permitted by specification, use of a slightly higher W/C ratio may minimize these. Form vibration is extremely effective, even though the same pin holes may be buried below the surface. Absorbent forms, such as plywood, or special absorbent liners also tend to keep the holes "submerged" so as not to present a visual problem.

In finishing, large bug holes should be patched first, using a mixture of white cement and regular cement. Then the entire surface can be rubbed. This will give a uniform surface appearance that will be quite lasting. Use of regular cement patches or, worse yet, epoxy-cement patches, will give a spotted appearance that worsens with time.

Architectural concrete can be ruined by rust. In one of the author's earlier experiences, stainless steel inserts were used to prevent rust staining, but projecting reinforcing bars, for later cast-in-place jointing, allowed rust stains to run down the panels. So any projecting steel or inserts, even if they are to be later encased, should be covered with plastic tubes and taped prior to shipment. In other cases, galvanized steel reinforcement has been employed in very thin sections to prevent later rust staining, only to have the ends of the bars burnt off, and thus ungalvanized, allowing unsightly rust stains to run down the wall panels. Exposed ends of prestressing tendons can lead to similar discoloration. They should be cut back and touched with epoxy, and then patched with white cement. A trial series of patches should be made to determine the best mix of cement to produce even coloration.

One of the great advantages of bush-hammering and sandblasting is that they erase and blend discolorations. When the structure permits sandblasting in place, this can frequently prove the best and cheapest solution. However, sandblasting will emphasize any shrinkage cracks by spalling the edges of the crack.

Coating of architectural units with silicones or pigmented linseed oil may prevent water staining. Because such coatings affect the coloration, they should be selected only after tests on samples are approved by the architect.

With open-web architectural units, and units with large openings, timber strutting or girting may facilitate handling, shipping, and erection and prevent damage.

With colored concretes, nonuniformity presents a serious problem. The controls of mix and curing become of extreme importance. Judicious sandblasting after manufacture may correct minor imperfections. Absolute cleanliness of mixers and transporting containers is essential as, of course, is absolute accuracy of batching, including the color addition. This should, wherever practicable, be added to a large quantity of cement or water, so as to attain a uniformity of mix. When colored panels are shipped and erected, it is frequently necessary to shift them at the site, so as to match color variations as far as possible. Much erection time and cost can be saved by checking all panels in the plant with a color chart, and actually working out an erection sequence and pattern. It is an added effort, but much easier to perform in a drafting room than on the side of the building, and the architect will rightly insist on its performance. His criterion is uniformity of appearance, both texture and color.

With cast-in-place concrete construction, all of the previous suggestions are valid, but the techniques differ. With very thin sections, such as lift slabs, floor slabs, and shells, extreme care should be taken to ensure uniformity of thickness and accuracy of tendon positioning. The author has observed a thin shell in Europe in which, after one crew positioned the steel and tendons with care and accuracy, the concreting crew then walked on top of them!

There is a peculiar phenomenon frequently observed with very thin concrete sections with small cover. The pattern of the reinforcing steel is shown by dark lines on the concrete surface. This is due to water content variations, usually from bleed water. This can be largely overcome by denser, less permeable concrete and smaller bars and, where practicable, greater cover. Lower W/C ratio will reduce bleed. Anti-bleed admixtures are available.

Some beautiful patterns have been specified and obtained, as at the Sydney Opera House, by casting against carefully patterned wood surfaces, and then forbidding patching. Since any patching spoils the entire effect, it is essential that placement and vibration techniques eliminate honeycomb and rock pockets. The use of a workable mix is essential for this. Wood surfaces should be field-checked to ensure reasonably uniform absorption characteristics. No form oil should be used, as it tends to absorb differentially and produce discoloration. After completion, such surfaces should be covered with paper or plywood to prevent disfiguration by workers (or engineers!) writing on them with marking crayon or other sources of discoloration and damage.

The type and location of picking inserts for exposed elements must be selected with care to prevent disfiguration of the element. In particular they should be located and, where necessary, reinforced, to resist edge spalling adjacent to the insert. Where they will be exposed in service, inserts should be galvanized, stainless, or otherwise treated to prevent rust staining.

For many architectural panels, the prestressing tendons may be placed at the center of the section, giving essentially uniform prestress. This increases the available cover over the steel, with reduced likelihood of rust-staining and spalling, and is of special value in salt-air exposure.

Prestressing, properly applied, will prevent cracks. Not only are cracks unsightly in themselves, but they often lead to rust-staining.

Because of the low degree of prestress typically used in architectural concrete, creep is not normally a problem. Rather the prestress tends to minimize out-of-plane and flexural creep. However, in heavily prestressed applications, such as sculptures, creep must be considered.

Prestressing may be one- or two-dimensional. Techniques have been developed for two-dimensional pretensioning. One of these is the wire-winding system of the former USSR, in which wire is wound in a pattern around dowels. The dowels are retracted after the concrete hardens, and the holes are then filled. The architect must be cognizant of these holes and, if possible, incorporate or hide them in the pattern, as it is very difficult to accurately match color and texture in the patching.

Two-dimensional pretensioning has also been utilized in the United States, with the transverse tendons being anchored to a side frame along the bed. After curing, the transverse tendons must be released first, before the longitudinal strands are detensioned. A cork, styrofoam, or rubber plug can be fitted at the edge, so that the strands may be cut back 25 mm (1 inch) or so from the edge, and the plug hole patched. Being on the edge, this patch is usually hidden from view.

Patching tends to always turn out darker than the adjoining concrete, particularly when damp. Trial mixes, using varying proportions of white cement, and various curing techniques should be conducted well ahead of actual production patching in order to permit determination of that mix which most nearly matches. Such patches must also prevent rust staining. A tiny, carefully applied dab of epoxy on the end of exposed strand may prevent local rust stain.

Precast plugs may be used, with a white epoxy or latex bonding cast carefully applied to the hole (not to the plug) and the plug driven in. Care must be taken that excess material does not drip or spread over the adjoining concrete.

Post-tensioned anchorages may be recessed and patched in the same manner as indicated above.

Tolerances are of great importance for architectural concrete; they must be realistic and practicable of manufacture, yet provide the required architectural appearance (Fig. 8.2). *The Manual for Quality Control for Architectural Precast Concrete Products,* published by the Prestressed Concrete Institute, lists recommended tolerances for warpage, thickness, squareness, location of anchors and inserts, block-outs and reinforcements, and joint widths. These may be summarized, in general, as follows:

1. Warpage: 3 mm ($\frac{1}{8}$ inch) per 2 meters (6 feet) length.
2. Thickness: minus 3 mm ($\frac{1}{8}$ inch), plus 6 mm ($\frac{1}{4}$ inch).
3. Squareness: 3 mm ($\frac{1}{8}$ inch) in 2 meters (6 feet) out of square as measured on the diagonal.
4. Anchors and inserts: 9 mm ($\frac{3}{8}$ inch).
5. Block-outs and reinforcements: plus or minus 6 or 12 mm ($\frac{1}{4}$ or $\frac{1}{2}$ inch).
6. Joint widths: specified widths are normally 9 mm ($\frac{3}{8}$ inch) to 15 mm ($\frac{5}{8}$ inch).

Accurate tolerances are especially required for fenestration, since typical jointing sealants, such as Thiokol, will provide a watertight joint only over a small dimensional range.

When special tolerance requirements are deemed desirable by the architect, they should be clearly designated.

If prestressed architectural panels are to be acid etched, then the exposed ends of the tendons, etc., should be thoroughly flushed with fresh water after immersion (Fig. 8.3). Ducts for post-tensioning tendons must be thoroughly flushed to remove all trace of acid. Serious corrosion has occurred when the remnants of hydrochloric acid were left in the ducts.

Figure 8.2. *Pretensioned hollow-core slabs are accurately fabricated to exacting tolerances for use as architectural wall panels of cold storage warehouses; Tokyo, Japan.*

Figure 8.3. *Pretensioned spandrel units utilize white cement and acid-etching.*

Prior to acid etching, all exposed metal surfaces should be protected with acid-resistant coatings. In working with acid, safety precautions should be set, including protective clothing, breathing masks, and eye protection as required.

Considering the many restrictions on this process, the author does not favor the acid etching process for prestressed concrete.

One of the early processes for exposed aggregate consisted of casting face downward, with the soffit painted with a retarder. After stripping, a water-jet was used to wash off the laitance and expose the aggregate. When a retarder is so applied, a positive means must be taken to prevent accidental spillage on the tendons. If the retarder is placed first, the tendons must be held up so they do not accidentally drag through the retarder.

There is a growing trend to cast architectural panels face-up, and to expose the aggregate after initial set by a water-and-air jet. In some cases, the concrete is cured and sandblasting or a hydraulic ram-jet is used to expose the aggregate. From a prestressing point of view, these processes are to be preferred as running less risk and giving less interference to normal production methods.

Flame-spalling of the concrete surface, followed by water-jetting, has been used to produce exposed aggregate surfaces on pretensioned panels (Fig. 8.4).

Honing and polishing are used to produce smooth, polished aggregate surfaces. Honed surfaces are produced by grinding with carborundum particles bonded in resin, or by diamonds set in the cutting surface. Air voids must be filled before each of the first few grinding operations, using cement or a cement-sand mixture and allowing it to harden prior to the next grinding operation.

Figure 8.4. *Flame-spalling applied to create special surface texture for airport terminal at Frankfurt, Germany.*

Sometimes a surface layer of different material is embedded in the concrete. This material may be glass, ceramic, marble or granite sheets, half bricks, or just different stone aggregates. Prestressing serves an important function of locking these into the panel. These materials have a different modulus of elasticity from that of concrete, and thus an engineering analysis must be made as to the proper location for the tendons in order to prevent warping of the panel. In addition, it is very wise to run a test panel to verify the tendon pattern.

Steam curing may adversely affect certain matrix colors, particularly blues and greens. The dripping of condensate may also spoil a surface. For this reason, particular care should be taken in selecting and controlling the curing process. In some cases it may be preferable to employ conventional water cure, or hot water or hot oil circulating through pipes in the bed or forms if accelerated cure is desired.

Grease, mud, and oil must be kept off panels during storage, transport, and erection. Many panels are wrapped in heavy kraft paper or polyethylene during delivery in order to minimize staining. After erection, they still must be protected until the joints are fully completed.

Another source of difficulty with architectural panels is efflorescence, in which lime leaches out over a period of time to form irregular white patterns on the surface. This is particularly objectionable on a colored panel. The means of minimizing this are to replace a substantial portion (15 to 30%) of the cement with PFA, to add 3 to 6% microsilica, and to seal the surface immediately after casting.

Use of a very dry mix (low W/C ratio) and heavy vibration will produce a more impermeable product, less subject to cracking and efflorescence. Sealing the surface is usually done with silanes; however, these coatings disappear through time and must be renewed every few years.

Lightweight aggregate concrete is generally adaptable to a wide range of architectural finishes, paralleling those applied to normal concrete.

Light sandblasting will give a brush-hammered effect to the surface.

White cement is often used with lightweight concrete wall panels, etc., although the basic brown or grey color of the lightweight aggregate particles may show through after sandblasting. Use of white natural sand, such as quartz or dolomite, will generally give a very satisfactory appearance.

When colors are being applied to the matrix, use of white cement and white sand, at least in part, may help keep the color pure.

Exposed aggregate finishes are commonly employed. Where ceramics, marble, quartz, or other facing materials are employed in conjunction with lightweight aggregate concrete, and the composite selection is prestressed, careful attention must be given to the difference in moduli of elasticity of the two materials. Then the strands can be positioned so as to minimize camber or sweep. Frequently this may require that the tendons be close to, or even in contact with, the facing material.

As noted in Section 4.2.5, prestressing of lightweight concrete wall panels permits thinner sections, of reduced weight, but presents greater problems of creep, which may affect joints and connections. Prestressing does prevent cracking normal to the prestress.

8.3 QUALITY CONTROL

8.3.1 Quality Control of Architectural Precast Concrete

Quality control for this type of work comprises the following:

1. Qualified personnel responsible for all stages of design, production, inspection, and installation.
2. Adequate testing and inspection of the various materials selected for use.
3. Clear and complete shop drawings.
4. Control of dimensions and tolerances, including adequate form work.
5. Inspection of all embedded hardware.
6. Mix design, proportioning, and mixing of concrete.
7. Handling, placing, and consolidation of concrete.
8. Finishing (Fig. 8.5).
9. Curing.
10. Handling, storing, transporting, and erection of elements.
11. Color uniformity.

The intensity of color of concrete will of course vary depending on whether the concrete surface is wet, saturated-but-surface-dry, or dry. Color also varies with moisture. The steam curing may cause the color to vary.

Figure 8.5. *Accurate finishing to a tolerance of 1.0 mm gives superlative finish to these architectural wall panels, which will be given a coating of translucent polyurethane after erection; Tokyo, Japan.*

Drips of water during curing will spot the product. Prior covering with a loose sheet of polyethylene will prevent this discoloration. Color charts should be used as a means of verifying the color of each panel as it is produced.

8.3.2 Defects

Defects in architectural precast products which require correction in manufacturing and/or design procedures include these:

1. Ragged or irregular edges.
2. Excessive air pockets on exposed surfaces.
3. Adjacent surfaces with different color or texture.
4. Construction joints or accidental cold joints readily visible.
5. Form joints visible, fins, honeycomb.
6. Rust or acid stains on surfaces.
7. Concentrations of aggregate, honeycomb, or gaps in aggregate proportions, or variations in aggregate appearance.
8. Areas of back-up concrete showing through facing concrete.
9. Foreign material showing on face.
10. Cracks visible after wetting.

11. Visible repairs and patches.

12. Reinforcement shadow lines.

Many architectural panels have an acceptable appearance when dry but show defects badly after a rain. This is due to differential absorption and evaporation, which in turn is caused by variations in permeability. This problem can be greatly minimized by adoption of a low W/C ratio and the incorporation of PFA.

In tropical and subtropical climates, beautiful white panels soon turn black due to fungi. Early remedies were the incorporation of toxic chemicals in the concrete mix. A much more acceptable practice today is to render the surface impermeable to water, either through the periodic application of silanes and/or the fabrication of the concrete with a high degree of impermeability through the use of a low W/C ratio and incorporation of PFA and possibly microsilica.

8.4 ERECTION

Erection of architectural panels follows the general rules for erection of precast elements except that restrictions are placed on inserts, and obviously greater care must be used to prevent accidental impact and spalling. Panels may be landed on timber softeners, and nylon slings may be used.

Architectural panels are required to have a true face, despite the fact that the structure behind them may have considerable allowable drift. For example, 25 mm (one inch) drift per story is allowed by the American Institute of Steel Construction (AISC) standard specifications. Therefore, erection and permanent fixing details must be designed to accommodate these tolerances.

The width of joints between panels may be controlled not only by aesthetical considerations but also by fire codes. Pre-marking of exact setting positions on the steel or concrete frame will help to prevent cumulative errors.

We have in concrete an ideal material for the incorporation of artistic effects into structures. Prestressing enhances this property in many ways, but care and forethought are required in application if the maximum benefits are to be achieved.

REFERENCES

1. PCI, *Architectural Precast Concrete*, Prestressed Concrete Institute, Chicago, Illinois, 1973.

2. ACI 303, "Guide to Cast-in-Place Architectural Concrete," American Concrete Institute, Detroit, Michigan.

3. PCI *The Manual for Quality Control for Architectural Precast Concrete Products*, Prestressed Concrete Institute, 1977, Chicago, Illinois.

CHAPTER 9

Safety

9.1 GENERAL

Prestressing inherently involves the use of very high forces, with steel and concrete stressed to a high percentage of their ultimate load. Prestressed concrete structures frequently involve heavy masses and high lifts. During construction and erection, lateral sway, vibration, imbalance, dynamic (acceleration) and impact forces, wind, and, in some cases, hydrodynamic forces will be significant.

Safety of personnel and equipment can be achieved only by positive efforts, including planning, temporary additional bracing, the installation of safety guards and warnings, and the indoctrination of workers through a continuing safety program.

People who are constantly exposed to dangerous situations are prone to lose their conscious fear unless they are constantly reminded of it.

Safety should never be subordinated to production expediency.

Tendons may be under tension to values of as great as 1,200 MPa (180,000 psi), which represents a tremendous stored energy. This can convert an anchorage into a deadly missile.

Guidance for safe operations has been set forth in the FIP Manual *Prestressed Concrete, Safety Precautions in Post-Tensioning* and in *Safety Precautions for Prestressing Operations,* published by the Concrete Society.

Without in any way attempting to diminish the value of a thorough study and implementation of these references, the following abbreviated list of rules is set forth. Most of these have been extracted in principle from the references, with amplification or modification based on this author's experience.

9.2 SAFETY MEASURES FOR ALL TENSIONING OPERATIONS

The operation of tensioning has more potential for serious accidents than all other phases of prestressed production combined. The following basic rules applicable to tensioning should be included in the safety requirements of all sites.

1. Before tensioning, a visible and audible signal should be given and all personnel not required to perform the tensioning should leave the immediate area.
2. Jacks should be held by such means that the jack will be prevented from flying longitudinally or laterally in case of tendon failure.
3. Personnel should never be permitted to stand at either end of the member (or bed) directly in line with the tendon being tensioned.
4. Personnel should not stand over tendons being tensioned to make elongation measurements. Such measurements should preferably be made by jigs or templates from the side or from behind shields.
5. Eye protection should be provided for personnel engaged in wedging and anchoring operations as a protection from flying pieces of steel.
6. Do not permit any welding near high-tensile prestressing steel: prestressed steel must not be used for grounding electrical equipment of any kind.
7. Keep all equipment thoroughly clean and in a workmanlike condition. Badly maintained equipment always gives rise to trouble and, consequently, is dangerous.
8. See that the wedges and the inside of barrels or cones of grips and anchorages are clean so that wedges are free to move inside the taper.
9. Arrange for stressing to take place as soon as possible after the grips have been positioned.
10. When assembling bundles of wire or strand, check each individual wire or strand for visible flaws.

9.3 ADDITIONAL SAFETY MEASURES FOR PRETENSIONING

1. Avoid kinks or nicks in strand. Use care in handling strand to avoid damage. Do not tension strand that has been nicked.
2. Screw up nuts on jacking rods as jack cylinders are elongated, in case of jack failure (Fig. 9.1).
3. Prevent accidental heating of a tensioned strand. Keep all torches and welding equipment away from tensioned strand.
4. Do not allow personnel to expose any part of their bodies above hold-down points of deflection strands. Block off directly above such hold-downs to prevent accidental dislodgement by vibrators during vibration of concrete. Hold-down inserts are potential crossbow projectiles.

Figure 9.1. Nuts or jacking rods are screwed up as jack cylinders elongate. In case of jack failure, there will be no sudden erratic release.

5. Protective guards for personnel engaged in tensioning should be provided at both ends of the bed and should be of structural steel, concrete, or heavy timbers.

9.4 SAFETY DURING STRESSING

1. Do not become casual because you have stressed hundreds of tendons before. The forces you are handling are enormous, and carelessness may lead to loss of life.
2. Regular examination of hydraulic hoses is essential, and oil in the pump reservoir must be regularly drained and filtered.
3. Use only self-sealing couplings for hydraulic pressure piping and take particular care that no bending stresses are applied to end connections.
4. It is preferable to use only hydraulic equipment supplied with a bypass valve which is pre-set to a maximum safe load before stressing. The maximum safe load should not be more than 90% of the minimum specified ultimate strength of the tendons.
5. Never stand behind a jack during stressing operations.
6. Do not strike the equipment with a hammer to adjust the alignment of the jack when the load is on.

9.5 SAFETY DURING GROUTING

1. A clear eye-shield should be worn by operator during grouting operations.
2. Before grouting, check all ducts with compressed air to make sure that they are not blocked.
3. It is preferable to use only threaded connectors between grout nozzles and grouting points. A sudden spurt of grout under pressure can cause severe injury, especially to the eyes.
4. Do not peer into duct bleeders to see if grout is coming through. Grout may jam temporarily and, as pressure is applied, it may suddenly spurt from the bleeders, or the far end of the duct, causing serious injury.

9.6 SAFETY DURING TRANSPORT

Prestressed girders typically depend on the deadweight to offset the prestressing forces. If the girder is tipped off vertical, it may buckle or "explode" as the compressive stress due to prestress exceeds the ultimate compressive strength. Girders must therefore be stayed during transport to prevent tipping. Chains should be used since wire rope stretches.

Support points during transport should be as close as possible to the permanent support points for which the member was designed. When the temporary supports are at different locations, the stress conditions should be analyzed and if necessary, mild steel added to counter the tensile stresses.

9.7 SAFETY DURING ERECTION

The normal practices for safety during erection of prestressed members apply, with these additional warnings.

1. Picking points may not be the same as the permanent supports. The stresses should be determined and mild steel added as necessary.
2. Picking with a bridle or with inclined slings may result in axial compression which, when added to the existing stress conditions, causes lateral buckling.
3. The girder may be purposely set out of vertical to produce the designed superelevation, or may be accidentally set on a slight angle. Even if set vertically, it is vulnerable to wind gusts or especially to impact from the next girder being set, or lateral pull from a line, which may tip it. Once it tips substantially, it may explode in compressive failure. Therefore, bracing and ties should be erected simultaneously with the setting.
4. The effect of wind on panels must not be underestimated. Groups of panels may generate greater unit wind forces than a single panel.

5. Concrete members being set in waves, especially surf, or in currents, will experience strong lateral surge despite their heavy weight in air, due not only to the lateral inertial and drag forces but to their reduced net weight when submerged.

6. Lifting eyes are subject to dynamic loads, and oblique loads in three dimensions. These tend to break bond and rip out the insert. The insert should therefore be designed with the same safety factor on direct load as is required for the rigging, i.e., 5 or 6; this includes bond since bond is especially reduced under impact. Anchor lifting inserts in the confined core or compressive zone, never in a tensile or unreinforced zone.

7. Never use deformed bars for picking loops: the deformations lead to stress concentrations and brittle feature.

9.8 REPORTS AND SAFETY MEETINGS

1. Proper reports should be made of all accidents and injuries, even those that are not of a serious nature. Frequently, a serious accident is preceded by several close calls. Proper attention to these may permit preventive steps to be instituted.

2. A brief but rigorous safety meeting should be held once a week, and prior to starting any new operation or use of new equipment. Safety requires a team effort!

SELECTED REFERENCE

1. FIP, *Prestressed Concrete, Safety Precautions in Post-Tensioning,* Thomas Telford Ltd., London, 1989.

2. *Safety Precautions for Prestressing Operations,* Concrete Society, London, undated.

USE OF PRESTRESSED CONCRETE

Prestressed Concrete in Buildings

10.1 GENERAL

The use of prestressed concrete in buildings has taken two paths. By far the largest use has been that of precast pretensioned elements, fabricated in standard sections in a manufacturing plant, then transported, erected, and joined in the building structure. The most widely used units have been the hollow-core slab, primarily as a floor slab (Fig. 10.1) and the double-tee girder (Fig. 10.2), for floors and roof slabs, and more recently, as walls. To support the floor and roof slabs, specially shaped girders, such as the inverted tee and "L" girder have been developed.

Some building systems also employ prestressed precast columns.

In long-span structures, such as parking garages, industrial buildings and auditoriums, the single tee may replace the double tee (Fig. 10.3).

The other trend has been that of post-tensioned cast-in-place concrete, primarily floor slabs. This system has been especially effective in producing thin-plate floor slabs which remain flat under service conditions.

Post-tensioning has also been extensively employed in special buildings such as the roofs of sports stadia, and in industrial buildings for heavily loaded floors and girders. Post-tensioning has been used for foundation slabs over weak soils.

Obviously there are many variations that have been devised to meet the needs of different building projects, as well as special applications; such as the peripheral edge beams of shell roofs.

Because of the different approaches required by the differing trends, they will be dealt with in separate sections.

At the end of this chapter will be a brief discussion of how these trends can be utilized to augment each other, as well as the need to integrate the building systems with the structural components.

Figure 10.1 *Pretensioned hollow-core floor slab.*

10.2 PRECAST BUILDING TECHNOLOGY

10.2.1 General

The general principles for manufacture are presented in Chapter 7, Manufacture of Precast Pretensioned Concrete. In this chapter, these principles will be augmented and the special considerations relevant to building will be discussed.

Figure 10.2. *Pretensioned double-tee girder.*

Figure 10.3. Pretensioned single-tee girder being loaded for shipment to site of new school auditorium.

Experience over the years has led to the standardization of those building components which have proven most efficient and satisfactory, principally the hollow-core floor slab and the double-tee girder. Within regions and countries, the dimensions have tended to become standardized as well.

Design tables have been prepared for the various span lengths for which the elements of a particular depth are suitable, enabling their ready selection by architects and engineers.

With all precast systems, the connections and joints are critical elements for which effective yet practicable details have been developed. These will obviously differ according to the specified wind and seismic loadings of the region.

Architectural finishes are also often imposed on these units and will affect the tolerances as well as the finish requirements.

10.2.2 Hollow-Core Slabs

A number of proprietary schemes for mechanized manufacture have been developed. They are generally based on the extrusion process, in which mandrels are drawn through the fresh mix which is being placed and compacted.

To prevent slumping of the hollow cores after the mandrel has passed requires very careful control of mix, especially slump, temperature, compaction effort, and rate of progress.

In most systems, the strands (or wires) are laid out initially on the bed, then stressed. Any mild-steel reinforcement bars are placed. Then a concreting machine moves down the bed on rails, extruding the concrete from a receiving hopper, compacting it, and striking it off. Final surface finish may be performed by an oscillating finishing tool attached at the rear of the same machine or on a separate machine.

Some advanced machines feed out wire mesh for the top reinforcement just in advance of the concrete extrusion.

Accelerated curing is then applied, by steam or heat. When adequate strength for transfer has been achieved, the slabs are sawn to length by a high speed corundum or diamond bladed saw.

In some cases, the saw blades stop short of the strands, which are then burnt in two by an electrode.

Slabs are then lifted off, often by insertion of a C-hook into the hollow ends, but this requires a finite length space at the end in which to insert the C-hook. More advanced plants use a vacuum lift (see Fig. 7.12) or separate the slabs as each is lifted.

In one system, several slabs are cast on top of one another, with a bond breaker, and then the side forms are stripped and slabs removed.

Both standard-weight and lightweight aggregates are used. With lightweight aggregates, fire resistance requirements will require that lightweight aggregates either have very low absorption of water, or else be fractured aggregate products, as opposed to the sealed surface types, in order to prevent explosive spalling due to generation of steam inside the aggregate particles.

When a special architectural finish is required, as for slabs which are to be used as exposed wall panels, then a trailing hopper, moving behind the concreting machine, may deposit a cement mortar, enabling the finishing machine to achieve the dense fine surface required (see Fig. 8.5).

Before leaving the plant, plastic cups are inserted in the hollow cores a specified distance, to act as end stops for field placed concrete in the joints.

Hollow-core slabs generally do not have any web reinforcement; hence their shear capacity depends on the tensile capacity of the webs. Tolerances on webs must therefore be controlled to ensure adequate thickness. The mix and process must be such as to maximize the tensile strength of the concrete while preventing shrinkage cracks and slumping.

When the members have been cured and the stress is released, the strands will take a finite set or slip into the ends of the slab. One important control on the quality of the slab is the measure of this set. Excessive slip will result in reduced camber as well as reduced strength.

Hollow-core slabs are usually erected by boom crane or tower crane, and set on top of walls or on beams. The walls may be precast wall slabs or concrete block. A cement mortar pad is spread on the seat to provide a uniform bearing (Fig. 10.4).

Good design practice requires that these wall-to-slab joints have reinforcing ties to prevent dislodgement in case of accident, such as the gas explosion in England which led to a stack-of-cards progressive collapse of the entire building. Of course in seismic regions, such reinforcement is definitely required.

This is readily achieved with hollow-core slabs by concreting the reinforcing bars in the hollow core ends.

Lateral shear transfer may be attained by grouting the longitudinal joint between

Figure 10.4. *Hollow-core floor slab being erected for industrial building.*

units. To prevent the grout from running out, a strip form may be held up by wires. However, this requires labor to place and to strip. Satisfactory joints may usually be achieved by profiling the joints and forcing down a plastic rope that wedges in place at the bottom.

While the great bulk of hollow-core slabs are utilized in floors for which the covering will be directly applied, in the case of industrial floors and in seismic-resistant design, toppings may be required. Since any topping mix will bond to the slabs, when it shrinks, cracks will appear over the joints. Use of wire mesh or closely spaced transverse bars in the topping slab will eliminate these, especially when the topping concrete mix has a low W/C ratio.

10.2.3 Double-Tee Slabs

This highly efficient structural form has gone through many variations, each adapt-ed to the needs of the dominant applications.

Modular dimensions dictated the early and still most prevalent configuration, so as to be architecturally compatible with other components.

Among the special forms of the double tee which have emerged have been the "giant tee," which is twice the width of the modular unit (Fig. 10.5), and the "M" section, which blends two double tees into one.

The dominant form remains the modular double-tee unit. This shape has been optimized and standardized for various span lengths, enabling ready selection and specification by architects and engineers.

For all but the shortest spans, the strands in the legs are depressed at the center so as to give a favorable stress profile. In some plants, the strands are initially stressed

Figure 10.5. *"Giant" double-tee 97 feet long by 8 feet wide, for roof slab of paper mill.*

at the high level, then pulled down at the center. In others, they are held down at the center, and lifted up at the ends. In all cases, rollers and other devices are used to minimize friction loss.

As noted in Chapter 7, release of strands which have been deflected requires careful planning. Obviously the hold-downs or pull-up deflectors cannot be released until the longitudinal stress is released into the member: otherwise it is not prestressed. Yet release of longitudinal prestress is accompanied by shortening of the member, hence physical movement of the hold-down. In the usual case, where relatively low stress, hence low axial shortening, is involved, proper sequencing of release and cutting of the exposed strands in succession will suffice to prevent cracking.

The stems of double tees must carry shear. At the ends, this often requires web reinforcement. For the levels of shear stress involved, welded wire mesh is usually sufficient. Use of mesh has the advantage of positive anchorage.

Even though the shear capacity may be adequate when initially installed, restraint of longitudinal creep over a period of time often leads to cracking near the supports, and has resulted in delayed shear failure. Hence the use of shear reinforcement is strongly advised.

10.2.4 Picking, Storage, Handling, and Transportation

Precast elements must be picked, stored, loaded, transported, and erected. Picking involves a very complex behavior of the member. If it is in fixed forms, with a side draft, as is common in the manufacturing practice of double tees, there is the frictional resistance on the sides and the suction on the bottom to be overcome. Experience has shown a considerable reduction in picking forces when the side forms are flexible.

When rigid forms are used, for example, steel-lined concrete molds or polished concrete, suction may present a serious problem. Attempts have been made to overcome the side friction by the use of grease; this only insures that the suction effect is greater. In some instances, injection of compressed air through the soffit has helped to free the unit.

Even with flexible forms, if these are multiple forms, the fact that adjacent forms are filled with the product at the time of pick may make them act as rigid forms.

Corner binding of closed channels is usually the most serious problem. Use of a loose corner insert in the forms, that comes out as the unit is picked, will prevent corner binding.

Perhaps the best method for overcoming side binding in rigid forms is to use self-centering jacks, which raise the member a few inches vertically, hence breaking it loose.

A heavy member, such as a girder, will probably not move lengthwise when the prestress is released into it, because of friction. Thus at the moment of initial pick it is not prestressed. For this reason the picking slings may have to be slightly offset so as to pick one end slightly ahead of the other end.

Another cause of binding is the presence of fins of grout, due to leakage between side forms and soffit or between soffit and insert holes, etc. This can be prevented by forms manufactured to a tight fit or by sealing the joints with tape.

The effect of such picking loads, in addition to the deadweight of the member, combined with the dynamic effect of breaking loose, is to increase stresses by 50 to 100%. This picking is, of course, done at a time when the concrete is relatively green. Although the concrete has attained a specified minimum compressive strength, this is probably only 75% of its 28-day strength. The tensile strength at picking may be an even lesser proportion. Damage to the element can be prevented by using more picking points, properly equalized, and by adding mild-steel bars over the picking points (points of maximum negative moment).

During storage, the unit should be supported at essentially the same points as in its final condition. However, because of the unit's early age, creep may be proceeding and provision should be made with long, heavily stressed members to permit longitudinal shortening.

During transport, the members are again subjected to dynamic loading and the addition of temporary mild steel may be necessary to prevent overstress and cracking, particularly over the points of support. The up and down whip of a cantilever during transport may be a particularly serious condition, requiring addition of mild steel to prevent both cracking and bond failure.

10.2.5 Erection

In erecting buildings, the temporary stresses depend on the means of erection and the number of support changes that take place. Because the members do not have the design live load and perhaps not even all the dead load, the chance for primary failure in bending is slight. However, end-support conditions are not as positive as in the final support; thus concentration may occur in the bearing area, as well as negative moments and shear. Provision of adequate stirrups throughout the end zone, wherever temporary support may occur, will minimize possible damage and failure (Fig. 10.6).

Where a cast-in-place floor slab is to be placed over precast prestressed beams to develop composite action, temporary props or supports may be required to support the dead load of the fresh concrete without imparting excessive stress or deflection. Means must be provided for later removal of these shores, as by knock-out double wedges, or, better yet, by screw jacks, so as to avoid upward deflection of the slab during removal.

The manufacture of precast prestressed concrete elements for buildings and the construction of prestressed building structures is characterized by the relative thinness and sometimes fragility of sections used, and by the incorporation of architectural treatments.

Figure 10.6. Erecting pretensioned channel for floor slab of industrial building.

Control of camber and warping becomes of great importance in long spans with slender beams and slabs. The main reasons for variation in camber are: (a) variations in concrete thickness; (b) variations in strand position, especially deflection; (c) variation in W/C ratio and curing (affecting the modulus of elasticity); (d) variation in storage; and (e) variation in strength at time of stressing.

All of these are susceptible to control to ensure uniformity, but only if the problems are recognized beforehand. Forms and screeds can be designed so as to keep the tolerances in cross section to a minimum. Strand positioning is largely a matter of care and supervision. W/C ratios can be carefully controlled with proper allowance for moisture content of the aggregate. With lightweight aggregate in particular, more or less continuous adjustment may be needed to ensure uniform W/C ratio at point of delivery. The curing cycle can be kept uniform by using automatic controls and recording charts and, particularly, by ensuring that the steam covers are tight enclosures to prevent loss of heat in a wind, etc. The temperature of the concrete mix should be kept approximately uniform, using ice or water-spraying in hot weather, and warm water or heated aggregates in extremely cold weather. Storage should be uniform, with the top slab in any pile covered to prevent earlier drying and excessive thermal response to the sun.

Many sections for buildings are so thin as to require special care in handling. See Fig. 10.7. Lightweight aggregate concrete is particularly sensitive to chipping and spalling on corners and edges. Rough, careless handling can cause damage which is difficult to repair. In one factory, technically and architecturally perfect hollow-core floor slabs were observed being crudely handled by picking tongs and inserts gouged into the concrete at the ends.

When double or single tees are erected, they often have slightly varying cambers. Before fixing them, therefore, each slab must be brought into conformity with its neighbor, Fig. 10.8. Simple tools can pull up and secure adjoining flanges. Shear transfer is usually by welding adjoining inserts which must be anchored back into the slab.

Double tees typically have measurable shortening due to creep. Storage prior to erection of 60 to 90 days will eliminate creep as a problem, but this is usually not practicable. The alternative is to seat the units on neoprene pads which will accommodate rotation as well as shortening of the legs of the tees.

Thus creep, as well as drying shrinkage, which adds to the problem, can be overcome as a practicable problem in most cases by:

a. Setting on neoprene pads.
b. Delaying fixing as long as practicable.
c. Low W/C ratio in fabrication.
d. Balanced design of tees, so as to avoid excessive prestress in bottom flange.

Double tees are typically incorporated into the supporting beam by embedding protruding reinforcing steel and/or strand from the top flange into the closure pour.

Some problems have arisen with double tees used for long spans. Most prevalent

Figure 10.7. *Careful handling is required for long-span roof slab.*

have been those due to shortening under creep, which creates tension near the support and reduces the resistance to shear.

A second problem has been spalling of the bearing seat due to rotation under creep and the daily and yearly thermal movements. This concentrates the load on a small area.

In cold climates, heated moist air inside may cause expansion of the stems of the tees while the top flange is cooled, leading to excessive deflection and rotation.

Long-span tee slabs, with shallow depth, will undergo substantial daily and seasonal camber variations due to the sun's heat, especially if they have been topped by black roofing. There may even be a growth in camber, as the tee rotates upward in the day and is then restrained by friction at the end supports from rotating fully back at night.

When highly stressed slab elements are erected and rigidly tied into the building frame by a topping slab or by welded end connections, prior to the creep having stabilized, severe tensile forces are imparted. These may result in cracks in the tees, near the ends of the stem, as previously noted, or in spalling at the seats or in cracks

Figure 10.8. Prestressed double-tee units form floor slabs of administration building. Particular care was exercised at all stages of fabrication and erection to control camber and deflection.

in the beams or columns. Thus creep stabilization is important or else provision for movement.

Seating the slab on neoprene pads and leaving the stems free from connection, while tying in the top slab to the supporting beam, appears to be one satisfactory solution.

Excessive camber variations due to thermal gradients, moisture gradients, creep, and shrinkage, all can be successfully minimized by proper design and construction.

The erection methods and sequence will vary widely with the building to be erected. Some buildings permit easy crane erection and the elements are stable in themselves; others are difficult to reach, may require rehandling, and are not stable until a series of elements are erected and tied together.

For all structures an erection drawing should be prepared, showing the location of the crane, radius and weight of each pick, and point of delivery. Even in the simplest building, such a drawing will minimize crane moves, and guard against overextending the crane reach. When delivery is by truck, the turning radius for the trucks should be laid out, with tolerances for backing, etc. In congested sites, this may be a major factor.

Such drawings may have to consider the three-dimensional effect, to insure that the crane boom will not encounter an obstruction. This can usually best be done by drawing sections in the vertical plane for each critical angle and position.

The need for insuring stability of the crane has been previously emphasized. It may require mats or gravel pads.

During erection, wind may cause difficulties in controlling the precast element, and may induce lateral and torsional stresses in the crane boom, to the point of endangering it. Therefore, guy lines are often necessary, particularly with panels, and may have to be supplemented by hand-jacking cable devices or air hoists. The lead and length of these guys varies during the lift, and the drawings must lay out the several positions.

For final positioning some form of close, positive control is necessary. Here ratchet or hand-jacking tools or "come-alongs" may be used effectively or, in some cases, screw jacks may be used to lower into place.

Stability must always be considered during erection of girders. If the girder tips sidewise, it may break, with serious danger to workers and structure below. In general, all members should be picked so that the lifting eyes lie above the center of gravity of the member. The horizontal component due to the sling angle may be so great as to induce buckling; this must be particularly checked for thin I-girders.

When a girder is to be moved by support at its base (e.g., by rolling), lateral support must be provided at each vertical support point to prevent tipping.

After erection, each element must be stable against wind, accidental impact from a subsequent element being erected, and from other construction operations. The degree of securing must be carefully thought out with regard to consequences to the total structure and to personnel, and to the probability of occurrence. Certainly normal maximum wind forces for the season must always be considered. Seismic forces are usually only considered when the time of unsupported erection condition will be unduly prolonged (e.g., winter shut down) or when the results will be catastrophic. This, unfortunately, was the case in a number of partially completed prestressed concrete buildings in Anchorage, Alaska, during the 1964 earthquake and led to several total building failures.

Accidental impact from subsequent erection is hard to appraise as far as force, but is, nevertheless, a very real hazard. Similarly, subsequent construction operations, e.g., raising structural steel or concreting, may produce lateral impact forces.

Of particular concern is the "stack-of-cards" type of failure, which has so often occurred with steel roof trusses. A pair of prestressed girders should have temporary trussing installed in the horizontal plane; then subsequent girders can be tied to the first pair.

During the erection and construction of very large shells, creep and shrinkage must be considered. Unbalanced snow and wind loads may be extremely troublesome, particularly the combined suction and the direct wind force. Aerodynamic stability during construction must be investigated for large cable-supported roofs and some convex shells as well (Fig. 10.9). Additional stability can sometimes be provided by properly located weights, e.g., concrete blocks or slabs, by strutting and tie cables or, in some cases, by shields or "spoilers" to minimize the suction effect. A snow-removal program can be instituted if necessary.

Temporary additional prestressing can sometimes be used effectively to ensure stability against wind. This stressing, for example, can be used to impose a downward load as well as a lateral tie.

Thermal effects during erection may assume serious proportions when large surfaces are involved. Diurnal movements of the order of 1 to 2 inches are not

Figure 10.9. Cable-supported roof for airplane hangar, Frankfurt, Germany.

uncommon. Any temporary restraints, such as guys or scaffolding, must permit these thermal movements.

Scaffolding to support precast concrete buildings during erection is subject to the usual precautions against buckling and sidesway. The only difference when precast elements are used is that these units are usually longer, and are usually supported only at the ends. Thus, the scaffolding quantity is reduced, and some of the inherent space frame action of the scaffolding, whether so designed or not, is lost. Particular attention must therefore be paid to column strength with allowance for eccentricities in setting, etc., and lateral stability of the system against wind, etc.

Decentering of long-span shells is always a "moment of truth." The shell must deflect, and uniform release is very hard to achieve. The possibility of collapse, while rare, must be considered. Therefore, screw jacks should be provided so that small retractions can be made uniformly, or in carefully determined sequence, without providing more than a very small gap in case uneven deflection or unusual sag occurs. The sequence of decentering must also be carefully engineered insofar as the stresses in the shell differ from the final condition (Fig. 10.10; see also Fig. 6.1).

Decentering by raising, whether by flat jacks or circumferential stressing, is attractive because the desired state of stress can be induced, the behavior checked progressively, and the shell raised slowly from the scaffolding.

Specific erection techniques for the erection of precast prestressed bridge girders are set forth in Chapter 12. Many of the methods and principles are applicable to the erection of buildings. In addition, there are several methods specifically applicable to building erection.

10.2.5.1 SLIDING AND ROLLING. Large structures in crowded urban areas frequently present a serious restriction of access for erection of interior members. One method

Figure 10.10 *Large-span prestressed shell roof covers International Center in Colorado.*

of overcoming this is to erect members at one side of the structure, then roll or slide them laterally to final position. Extreme precautions must be taken to:

1. Prevent tipping.
2. Insure that both ends move together, under full control.
3. Prevent lengthwise movement that would drop one end off its support.

These conditions may usually best be met by providing a structural steel cradle at each end, into which the girder end is set. This cradle, in turn, runs on a track, such as a steel channel, bolted to the cap-girder so that it is held in position. Movement is usually accomplished by air tugger or hydraulic winches, or long-stroke jacks may be used. Movements should be incremental, step-by-step. Final positioning from the cradle to its permanent seat is by jacking down.

10.2.5.2 TOWER CRANES. Tower cranes are extremely versatile for erection of tall buildings. However they usually have limited capacity at extreme reach, so precast units may have to be fed into a special bay or near side, from which they are hoisted and swung into final position.

10.2.5.3 STEEL ERECTION DERRICKS. These are frequently employed, especially where precast units are hung or combined with a structural steel frame. Since structural steel tolerances are often much greater than precast concrete tolerances and since the precast concrete is so rigid, provision should be made for accommodating dimensional variations. Double-slotted bolt holes and shims are most effective.

10.2.5.4 HELICOPTERS. Helicopters can be used very efficiently and economically for erection of precast units of small and moderate size, especially when the point of placement is difficult to reach, e.g., a tower. The helicopter lowers its hoist, slings are attached, the pilot raises the hoist. If trailing lines are attached to the unit, they must be individually handled by a competent man to prevent any possibility of snagging. Only those workers directly connected with the operation should be within the proximity of the rotor blades.

The helicopter then lifts off to the placement position, where the unit is lowered and affixed in position. Erection sequence may have to be altered, depending on the wind direction. Each unit should be immediately secured, as, for example, with erection bolts. The safety precautions for helicopter operation must be rigidly enforced, both for helicopter safety and safety of workers. Communication between pilot, ground foreman, and top foreman is essential, both voice radio and visual signal.

10.2.6 Connections of Precast Building Members

Whenever precast concrete elements are assembled into a structure, the connections and joints must be designed and executed so as to continuously perform over the life of the structure in the manner contemplated by the design. In other words, if the connection is an expansion joint, then freedom of movement must be assured regardless of the external environment, load history, and the effects of age. Hinged connections must continue to perform as hinges; fully continuous connections must maintain their integrity (Fig. 10.11).

With prefabricated structures especially, the effects of variances in manufacture and in erection, of unequal foundation settlement, and of differential thermal movements are all concentrated at the joints.

Prestressing amplifies this problem due to creep and the generally thinner sections used, which permit greater rotation. Longer spans are more common with prestressing, thus again increasing the magnitude of the movements at the connections.

By far the greatest portion of difficulties and problems that occur with precast prestressed concrete building assemblies are at the connections. Yet this subject is treated very lightly, if at all, in most reference books on prestressed concrete. The reason is that the variety of connections and combinations and details is so great that classification and generalization are very difficult (Figs. 10.12 to 10.18 inclusive; see also Fig. 4.23).

Certain principles do emerge, however. The connections must be so designed and executed that they:

1. Transmit bearing, shear, moment, axial tension, and axial compression as required by the design.

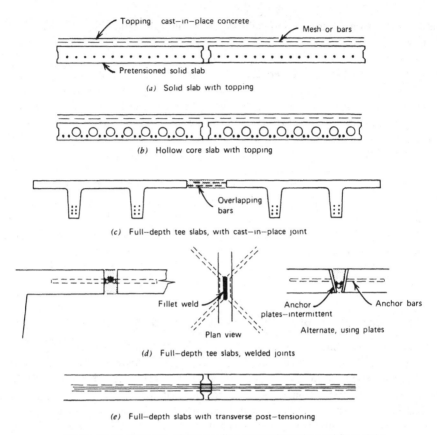

Topping cast—in—place concrete

Mesh or bars

Pretensioned solid slab

(a) Solid slab with topping

(b) Hollow core slab with topping

Overlapping bars

(c) Full—depth tee slabs, with cast—in—place joint

Fillet weld

Anchor plates—intermittent

Anchor bars

Alternate, using plates

Plan view

(d) Full—depth tee slabs, welded joints

(e) Full—depth slabs with transverse post—tensioning

Figure 10.11. *Methods of connecting precast prestressed floor slabs.*

2. Accommodate volume changes due to creep, shrinkage, and temperature without exceeding allowable stresses and permissible strains in the member, its support, and the connection assembly itself.

3. Accommodate all design loading combinations including superimposed live load, wind, and seismic loads within allowable stresses and permissible strains in the member, its support, and the connection assembly.

4. Accept overloads, that is, ultimate load, with ductility, so that failure does not occur at the joints or connections before primary failure in the member. An exception is where a joint is specifically designed to break prior to primary failure of the member.

5. Perform their connecting function as designed, e.g., expansion, continuity, or hinging, continuously without marked change in performance despite the effects of age and anticipated environment.

6. Have adequate corrosion and fire protection. A connection should accom-

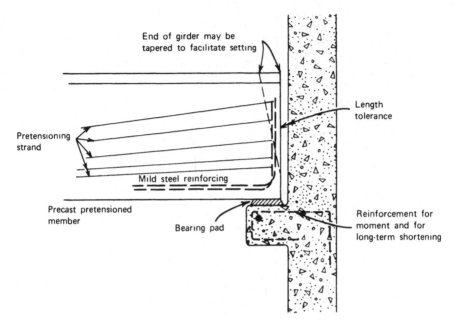

Figure 10.12. Typical bearing detail: slab or girder on wall.

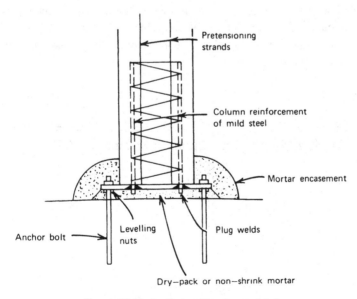

Figure 10.13. Typical column base detail.

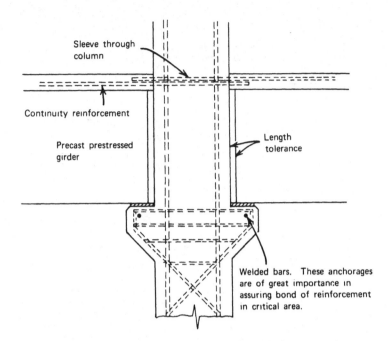

Figure 10.14. *Girder to column connection.*

modate rotations and expansion due to high temperature on the member while under fire exposure itself, for the prescribed fire duration.

7. Ensure adequate seating and performance despite the maximum permissible cumulative deviations in tolerance in manufacture and erection.

8. Insure watertightness, where appropriate, under maximum wind and volume-change conditions.

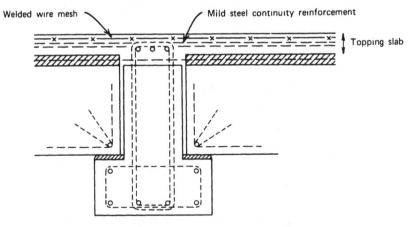

Figure 10.15. *Double-tee to inverted-tee connection.*

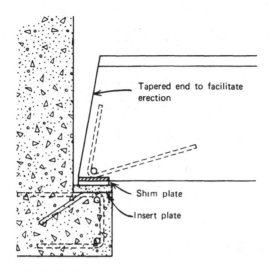

Figure 10.16. Welded connection: beam to girder or column.

9. Provide mechanical means (stops) as necessary to prevent a member falling off its seat when the design limit is exceeded, as in a severe earthquake.
10. Be practicable and economical in attachment to the members and in erection.

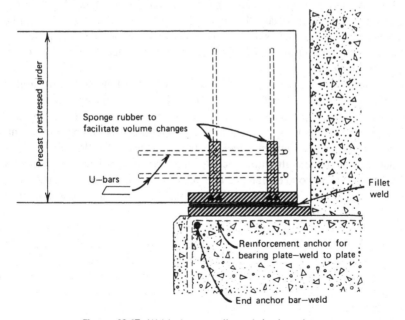

Figure 10.17. Welded connection: girder to column.

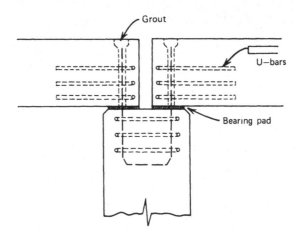

Figure 10.18. *Doweled simple spans: top of column.*

The design and detailing of joints and connections is normally the responsibility of the design engineer. However, the constructor is often thrown into a position of partial or entire responsibility when the specifications require the constructor to "propose and submit details for approval" or where the constructor has proposed a prefabricated alternative to a cast-in-place structure.

The constructor is always responsible for the careful and accurate execution of the jointing and connecting details.

When distress arises, such as cracking and spalling, the tendency is to charge the constructor with failure to make the connections properly; it then may be up to him to prove faulty design. Thus it behooves the constructor to examine joint and connection details carefully and call attention to any inadequacies he discovers in the design details.

The materials used in connections generally consist of fabricated structural steel assemblies (plates, angles, and bars), reinforcing steel (of varying grades and yield strengths), prestressing tendons (wire, strands, and especially bars), epoxies (and similar organic compounds), cement grout, concrete, and neoprene. Occasionally, other materials, such as stainless steel and bronze plates and Teflon, are employed. All materials and their structural details must be designed in accordance with applicable codes and good design practice. The stress distribution within connection details is usually very complex, requiring careful analysis under the whole range of criteria.

Entirely too frequently, errors are made in the embedment or securing of the connection into the concrete. This is usually a zone of high shear and also of transverse tensile bursting stress, due to prestress. Minor eccentricity of misalignment may produce stress concentrations. Under dynamic or shock loading, bond may fail. Finally, this is usually a congested area where it is difficult to place and consolidate the concrete.

Most of these difficulties can be overcome by increasing the depth of embedment and by liberal use of binding ties of reinforcement (stirrups or spiral).

In the member itself, shear and diagonal tension are usually at their maximum at the ends adjacent to the joints or connections. Adequate stirrups are essential to bind the concrete. It must be recognized that even with binding, there is always a zone (i.e., the cover) outside and beyond the last stirrup, which is unreinforced.

Welding of connections may cause spalling of concrete due to heat. This may be minimized by an intermittent welding procedure, careful selection of electrodes, and provision of adequate length of connection steel, both that extending beyond the face of the concrete and that embedded in the concrete.

The many connection schemes that have been developed and successfully used fall generally into the following categories:

1. Steel plates and angles, welded to reinforcing bars or anchors.
2. Bolts through sleeves in the concrete or through embedded steel plates. Double nuts may be used to facilitate adjustment.
3. Dowels entered into corresponding holes in the adjoining member and secured with low-shrink grout or epoxy (see Fig. 11.18). Polyesters have been used, with their set accelerated with an embedded wire through which a current is passed.
4. Socket connection, with the precast member set in a socket and fixed with cement grout or concrete.
5. Cast-in-place concrete joints, enclosing reinforcing bars projecting from the precast members, which are lapped, welded, or mechanically connected.
6. Posttensioned bars or other tendons.
7. Fabricated steel hangers.
8. "Glued" epoxy joints.

Earlier in this section, the importance of permitting movement and rotation was emphasized. Neoprene bearing pads have proven widely applicable for this purpose (Fig. 10.12). Some excellent connection details utilize neoprene or sponge rubber inserts within the concrete member itself to relieve stress concentrations and permit a restricted degree of translation and rotation. Seats for the support of precast members must be adequate to ensure full bearing without excessive edge concentration, even if members are cumulatively out of tolerance by their maximum permissible deviation. Edges should preferably be chamfered or rounded to prevent spalling.

When posttensioning is extended by means of couplers at the joints, the sleeve for each coupler must permit the required longitudinal movement of the tendon without jamming. It must permit grout to flow around it without blockage. Since sleeves for couplers take out so much of the gross concrete area, a check should be made to be sure that the remaining net concrete area can sustain the initial prestressing compression without crushing. This is a temporary condition and a factor of safety of 1.5 on ultimate should be sufficient.

Connection details have received a great deal of study by the Prestressed Concrete Institute and their Connection Committee, who have issued reports on the subject from time to time. Additional information is found in the literature of the precast concrete building industry. Joints are further discussed in this book in Sections 4.3, Precasting; 4.6, Epoxies and Other Polymers in Prestressed Concrete Construction; and 4.7, Welding in Prestressed Construction.

Special care must be taken with the embedment and fixing of the connections of precast numbers in structures subject to earthquake. This applies to shear connections, joint details, and restrainer anchors. The precast concrete fabricator and erector must both adhere rigidly to the specified tolerances. Since the critical zones will typically be very congested, concreting will be difficult but the proper placement and consolidation of the concrete will be essential. Welds will have to be made with proper procedures so as to ensure there will be no defects that could give rise to brittle rupture under shock loads.

10.3 POSTTENSIONED BUILDINGS

Posttensioned concrete is extensively used in flat and waffle floor slabs. It not only provides strength but permits accurate control of camber and deflection so as to produce a "flat" floor throughout its service life. This latter is often achieved by the "load balancing" design method, in which the upward force of the tendons exactly counteracts the dead load plus the predicted actual average live load. Thus long-term creep leads to neither sag nor camber growth.

Slabs are cast-in-place, so the tendons are laid out in a generally orthogonal pattern, concentrated in bands on the columns lines. Although adequate bond between slabs and columns can theoretically be obtained by deviating the tendons around the columns, it is considered best practice to run a portion of the tendons through the columns. This is required in seismic design.

Shop drawings must be carefully detailed to eliminate conflict between tendons, reinforcing steel, ducts for telephone, electricity, and plumbing. It may be necessary to bundle the bars in the columns in order to provide sufficient space to run tendons through them.

The tendons are typically unbonded, greased-and-sheathed strands. A single 15-mm (0.6-inch) strand is often employed. Although earlier installations used paper wrapping, it was subject to damage and water infiltration. Polyethylene sheathed tendons are now employed and give excellent durability.

These tendons are draped up and down across the slab, in accordance with the design. They are usually supported on dobe blocks of cement mortar, although plastic chairs are also used. Support distance varies from 1.3 meters (4 feet) for single strand tendons to 3 meters (10 feet) for multi-strand tendons.

Being unbonded, the anchorages are of vital importance. They must be located in block-outs in the edge of the concrete slab, and anchorage zone reinforcement installed to prevent local bursting and spalling. Space must be provided for the tensioning operations. After stressing the anchorage must be encased in mortar or fine concrete to give fire and corrosion protection. Jacking is normally carried out with two single strand jacks, each working from one end on alternate strands, so that half the total are stressed from each end.

Defects in the slab concrete, such as honeycomb in the slab edge, may cause the anchorage to deform and pull into the slab. The anchorage must be removed and the slab edge repaired before retensioning.

Personnel should be kept away from directly behind the stressing equipment.

To affix the anchorage, it is necessary to strip the sheathing and clean off the grease with a solvent. These stripped ends should not be left open for any length of time, especially in a coastal environment. This is especially true of paper-sheathed tendons, where the paper may absorb salt from the atmosphere.

Bonded tendons are also employed for slabs. Light gauge steel ducts, of rectangular cross section, are laid out, supported on the reinforcement and "dobe" blocks, so as to have the design profile.

Typically three to four strands are entered in the duct after the concrete has been placed and hardened. They are then stressed in a group and injected with grout.

Shoring must stay in place until the stressing is complete, but should be able to deflect slightly as the prestress shortens the slab. If shores have been removed to allow stripping of the forms, they should be re-installed to prevent creep.

Post-tensioning is also used in foundation slabs. The unbonded technique, with greased-and-sheathed tendons, is typically employed for flat foundation slabs and grade beams of small-to-moderate size buildings.

Post-tensioning has been used for the thick mat foundation slabs used to support high-rise buildings. In this case, the required force is provided by large multi-strand tendons, (or multi-wire tendons) in semirigid steel ducts, similar to the post-tensioning systems used for bridges.

The prestressing serves to offset thermal stresses developed shortly after concreting, as well as to distribute the concentrated loads from columns, so as to more evenly distribute them to the soil. They are especially useful in preventing "edge curling."

The principal problem with this application to mat foundations is obtaining access to the ends for jacking and anchoring. Because of the usually exaggerated vertical profile, it is necessary to jack from both ends; thus all four sides of the mat need access. Jacking is usually performed in stages, as the dead loads are applied.

Further, the vertical profile requires careful venting of the ducts at the tops of rises, and the selection of a grout and procedure to minimize bleed.

Provision must be made for shortening of the mat under prestress. A sand layer between two polyethylene sheets has been employed for this purpose. An asphalt layer, over a concrete blinding layer ("mud mat") has also been used successfully.

Post-tensioned beams and girders have been widely employed in multi-story buildings, using the techniques commonly associated with bridge girders. Bonded tendons are generally employed, with steel sheaths and multi-strand tendons. In high-rise buildings, access to the girder ends may be difficult. Staging should be planned and designed so as to progress upward with the forms and scaffolds. Provision must be made to hoist the heavy jacks.

10.4 COMMERCIAL BUILDINGS

Prestressed concrete has enjoyed its greatest growth in this field, because it has enabled structural and architectural functions to be economically combined. Shopping centers, for example, need moderately long spans to permit flexibility and alteration of partitions. External walls must be attractive as well as functional, leading to the use of precast panels or double-tees, either as plain concrete or with exposed aggregate and, occasionally, color (Fig. 10.19).

The predominant precast structural system comprises columns, inverted tee girders, precast hollow-core floor slabs or double-tee floor slabs, and double- or single-tee roof slabs.

Cast-in-place posttensioned floor slabs have been widely used, with both conventional and lift-slab construction.

Fire resistance is a major factor in the selection of prestressed concrete, and performance to date in actual fires, has been good.

Proper connection details are essential to ensure provision for volume change such as shrinkage and creep. Failure to allow for this has produced minor distress conditions at the supports in a number of the earlier commercial buildings. A multibay layout tends to make volume changes additive.

Provisions must also be made to ensure waterproof joints in the roof despite thermal and other volume change movements. Leaks cannot be tolerated in shopping centers.

Figure 10.19. *Prestressed wall slab for science building incorporates both color and texture in its surface finish.*

In erecting precast prestressed roof members, it is frequently necessary to match the sections to ensure approximately equal camber, and then to tie adjoining members together at the third or quarter points by bolted or welded connections.

Prestressed concrete has enjoyed some of its widest popularity in parking garages, because of its long-span capabilities combined with fireproofing. Rectangular layouts have utilized the double-tee most extensively, supported on inverted tee girders. Circular layouts have utilized pie-shaped members; for example, a double-tee at the outer end tapering to a single-tee at the inner end.

One popular concept for parking garages utilizes pretensioned single tees which are seated on precast columns and then posttensioned after erection to provide a full-moment connection (Fig. 10.20). A cast-in-place floor slab then serves in composite action. It, in turn, may be post-tensioned transversely with bonded or unbonded tendons, thus tying the whole structure together.

It is increasingly common to utilize the roofs of garages for either parking or

Figure 10.20. Parking garage under erection in New York uses precast columns and prestressed single-tee floor slabs.

playgrounds (Fig.4.1). This frequently calls for the roof to be black asphalt. Heat absorption causes accentuated thermal movements of the long-span roof slabs, and this must be taken into account in the connection details. As noted earlier, camber will tend to grow over time and the girder to rotate and shorten.

Columns and girders for prestressed concrete buildings are sometimes precast, sometimes cast-in-place. The design for either method must provide for lateral support and integrity, and for volume-changes of the slabs.

Posttensioned cast-in-place construction is also being increasingly employed for garage construction, often using structural lightweight concrete. The advantage of long spans and thin slabs is maintained. Joint problems are eliminated, but shrinkage and plastic flow may be accentuated. Shrinkage-compensating cement has been used for a number of such structures. The relative economies of the two systems (precast or cast-in-place) will continue to vary in individual areas due to the interplay of many economic and competitive factors. What is clear is the ever-wider employment of prestressed concrete in one form or another.

10.5 HOUSING (APARTMENT BUILDINGS)

Precast prestressed concrete has been widely adopted in Europe for apartment housing construction, but has been slower to gain wide acceptance in the U.S., primarily because of the ready availability of low-cost alternatives. It is believed, however, that there will be a continual increase in this application because of a number of concurrent factors:

1. Greater emphasis on fire resistance.
2. Greater emphasis on sound (acoustic) insulation.
3. Increasing costs of alternative materials and methods of construction as opposed to the prefabrication and assembly of large elements.
4. The developing mass market for precast prestressed concrete, which will result in more efficient manufacture and comparatively lower costs.
5. Ability to integrate mechanical and electrical services.

At various times over the past 30 years, "Systems Building" projects have received wide attention and a great deal of development effort. The dominant trend of these studies has been to emphasize the need for and advantages of prefabrication, using standardized precast pretensioned segments.

The hollow-core floor slab is widely employed for housing construction (Fig. 10.21). Experience to date indicates that the voids are of only marginal benefit for utility runs. The voids, therefore, are primarily a means of reducing weight, and reducing concrete and steel quantities. Greater emphasis on acoustic insulation is causing an increased interest in some parts of Western Europe in solid slabs.

Use of a cast-in-place topping permits shear ties to resist lateral forces of wind and earthquake. This can act in composite action with the prestressed slab, and provide the needed thickness for heat transmission during fire, for acoustic insulation, and greater rigidity. However, the use of a cast-in-place topping is more

Figure 10.21. Hollow-core floor slabs used in low-rise apartment construction.

expensive than if full-depth slabs are employed. Development work continues on economical methods of jointing full-depth slabs to insure adequate shear transfer. Welded shear connectors often cause black stains on the concrete, and in actual earthquake, some have failed either in the weld (brittle fracture) or by the anchors pulling out of the concrete (bond failure). Therefore welded shear connectors require great care in detailing. Cast-in-place joints, with overlapping reinforcing bars, are structurally satisfactory if of sufficient width. Unfortunately, all of the transverse shrinkage is concentrated at the joints, and cracks will appear, unless special means are taken.

Lift-slab construction continues to be widely used for apartment housing construction. It depends on prestressing to keep the floor slabs level under normal (service) loading. Many lift-slabs in the United States are post-tensioned with unbonded tendons. In other countries, bonded tendons are generally used, placed in ducts, tensioned, and then grouted. Arguments continue as to the merits of the two systems: the low labor requirement and speed of the unbonded system versus the better behavior at ultimate (failure) condition of the bonded system.

Lift-slabs can also be produced by pretensioning on site, using the basement walls as abutments. This is the contribution of a system developed in England.

10.6 SCHOOL BUILDINGS

Prestressed concrete is selected very frequently for school buildings, because of its long clear spans, especially for auditoriums, gymnasiums, and cafeteria spaces. It is fireproof and has excellent acoustic properties.

In seismic areas, connection details and shear connectors must be given special attention to insure their ductile behavior under dynamic loading.

Long, slender roof spans must be designed with attention to sag under rain and snow loading, so that they will always drain and not allow ponding to occur. Because of the long spans, particular care must be taken to provide lateral staying or guying during erection.

10.7 OFFICE BUILDINGS

Prestressed concrete is increasingly employed because of its long clear spans, permitting maximum flexibility in partitioning, and its fire resistance. Properly designed, it is suitable for the higher floor loadings that computers and office machines impose.

At the present time prestressed concrete is most extensively employed in the form of floor slabs, such as hollow-core slabs, and wall panels, the latter of architectural concrete, often with exposed aggregate or special coloring, and including the window openings. Proper detailing is essential to ensure waterproof joints despite thermal changes (e.g., the sun's daily movement) and wind. Connection details must be sufficiently strong, including the embedment of anchors into the concrete, to ensure adequate securing in wind and earthquake.

Lateral shear connection between precast floor slabs is more important because of the heavier local loads associated with offices.

10.8 INDUSTRIAL BUILDINGS

The utilization of prestressed concrete for industrial buildings has been very extensive in Europe, especially in Eastern Europe and the former USSR. Precast pretensioned concrete elements are employed. Typical systems consist of double columns, prestressed roof trusses, wall slabs, roof slabs, and prestressed crane girders. Wall and roof slabs are designed to give thermal insulation as well as structural strength. Three types of slabs are used: sandwich panels, all-lightweight concrete panels, and composite panels. The composite panels consist of a thin slab of prestressed normal or structural lightweight aggregate concrete, combined with a thick slab of aerated (cellular) concrete.

It is believed that the prestressed structural lightweight aggregate panels will emerge as the predominant roof slab and wall panel for this framed system.

The roof trusses are generally prestressed in the lower chord only, mild steel being used for the web and upper chord. Deep web girders are also used, with large holes in the webs for weight reduction and passage of utilities.

Throughout Europe, most industrial construction has followed a similar framed approach, although thin shell roofs are more common than trusses. These thin shells usually have prestressed edge beams for the longer spans. North-light shell roofs are extensively used in The Netherlands, for example.

An interesting roof shell has been developed in Western Europe. This is a hyperbolic paraboloid, permitting use of straight pretensioning strands to produce a thin shell roof element for medium spans.

In the United States and Canada, the use of prestressed concrete for industrial buildings has lagged, for a variety of reasons. Tradition, the ready availability of steel, the trend for incorporating more services into the building structure (such as lighting and air conditioning), and the lack of a mature industrial building concept for prestressed concrete have been partially responsible. Earliest inroads have been made by roof slabs, where the long spans, fire resistance, and excellent appearance have led to their selection. Some long-span barrel vault shells, with prestressed edge beams, have been used. The single tee, however, has been selected more frequently for long spans, with the double tee for shorter spans, i.e., 20 m (70 feet) and below.

Recently, storage buildings for a variety of bulk products, such as potash, sulfur and coke, have been constructed of prestressed concrete. These have mostly been A-frame buildings. The most promising type, however, uses very long (150 feet or more) single or double tees to form the A-frame, thus supplying roof slab and beam in the same element (Fig. 10.22).

The emerging pattern in the United States and Canada, therefore, has been the use of large elements, such as single or double tees, of maximum width, to form both the structure and the cladding (side walls). Connections are generally welded. Joints are grouted, with plain or epoxy grout, or filled with a waterproof expansion material, such as Thiokol, to permit differential thermal movements.

For special types of industrial buildings requiring large volumes of air circulation, hollow-box precast prestressed girders have been used to combine the function of a fireproof, sound resistant duct with that of long-span structural support.

Figure 10.22. Prestressed tee units cover sulfur storage facility in Canada

Industrial buildings, because of their size and function, must be designed with consideration for volume changes due to thermal movement, shrinkage and creep, and differential settlement of supports. Shrinkage may be somewhat more complicated than usual if the inside is maintained at high humidity, as, for example, is done with coke storage. Creep will be of a relatively high magnitude due to the long spans and highly stressed members, and also due to the increasing use of structural lightweight aggregate concrete, with its somewhat greater creep factor.

Differential settlement of supports is a problem in bulk storage, where the high loadings cause consolidation and possibly lateral movement of the soil. Ties beneath the building floor may be more harmful than beneficial, since any settlement of the floor tends to pull the base of the walls inward.

Heavy industrial floors in Europe utilize post-tensioning to provide both bending strength and shear strength, this latter to resist concentrated loads.

10.9 EXHIBITION BUILDINGS AND STADIA

Many spectacular structures have been constructed by means of prestressing. Some of these have incorporated precast elements, principally as roof slab elements, supported and integrated with the posttensioned girders.

Long-span trusses have been constructed in Japan and Europe, frequently employing a combination of precast elements and cast-in-place construction, and utilizing high-strength concrete (Fig. 10.23). The posttensioning is usually run through the nodes, creating a multi-axial stress condition that has excellent rotational ductility under overload. Mild-steel binding reinforcement is of course needed, again on all three axes.

Cable-stayed roofs, as well as prestressed trusses, have been widely used to cover aircraft maintenance hangars.

A few major problems have been encountered in service, principally due to thermal movements and corrosion. The constructor of prestressed concrete structures of these types therefore must be meticulous in execution of joint details and in positive sealing of all tendons to prevent moisture infiltration.

The International Conference Center in Berlin, constructed in the early years of prestressed concrete, collapsed due to corrosion of tendons at a joint where the rotation under thermal strains produced cracks, allowing water to penetrate to the tendon. Corrosion may have been augmented by the cyclic bending stresses imparted on the tendon.

10.10 CIVIC AND RELIGIOUS STRUCTURES

These structures are often designed to convey an architectural theme as well as to provide utility of function. They must be permanent, durable, of clean and attractive finish. Prestressed concrete is ideal for this use; the concrete can be molded to almost any shape and the prestress varied to suit the varying structural requirements.

One of the foremost of such monumental structures, the Sydney Opera House,

Figure 10.23. *Long-span roof trusses of prestressed lightweight concrete in France.*

uses precast segments to form inclined arch ribs. These are then prestressed to create a stable structural system and a favorable state of stress. Indeed, without prestressing, this form could probably not have been achieved and maintained.

A recent trend has been the use of standard single- and double-tee elements in a variety of combinations to achieve striking architectural effects consistent with the dignity and spirit of the theme. The stems of the trees may be turned in or out and architectural treatments applied to one or both faces.

Some of the finest of such structures have integrated precast and cast-in-place concrete, with both pretensioning and post-tensioning, thus showing a maturity of engineering design and construction skill.

10.11 SYSTEMS BUILDING (TECHNICAL ASPECTS)

The general philosophy of systems building has been discussed earlier. This section is concerned only with the technical aspects of design, manufacture, and construction of components for incorporation in systems.

The concept of systems implies the integration of many functions. The integration of architectural and structural function is inherent in prestressed concrete. Insulation of various types may be applied by spray or gluing on of sheets. Holes may be formed and inserts placed for mechanical and electrical systems; in a truly integrated system, most of the runs will themselves be embedded in the concrete. Concrete girders may be hollow to serve as air ducts or vents. Window and door frames may be placed and the windows glazed and doors hung.

All of this will be successful only as each and every operation is integrated into the manufacturing process. A basic process must be established. Either the product moves to each operation station in turn, or the operational processes move in succession to the product. In either case, adequate time and space must be provided for the operation. Thorough consideration to material flow and work simplification must be given to each operation, even those such as mechanical and electrical, which have not historically been concerned with these aspects.

By such a program it then becomes economically feasible to employ a highly skilled artisan (plumber or electrician) at the final fitting station. This practice may lead toward a solution of the problem that has plagued many previous efforts at systems building and prefabrication: the resistance of labor unions and building code officials to factory-built components.

Inserts must be rigidly held in forms so as to prevent their displacement during placing and vibration of the concrete. They may require covers or other protection to guard them from damage during handling of the product.

Certain processes, such as sprayed-on insulation, etc., require enclosures to protect both the product from contamination and the adjoining workers from hazards. Prefabrication should lend itself to better safety practices in this regard.

Completed products must be protected by lagging or paper from damage and staining during storage, transit, and erection.

Code marking of all units is an essential ingredient of system prefabrication. Products must be tied in with a working drawing, a definite production sequence, storage space, shipping schedule, and erection sequence. Such a program has been computerized in at least one plant in the eastern United States, which utilizes computer-drawn strand patterns as well as printed numerical instructions, coded to both product and forms, as well as to storage, shipping, and erection.

Whether such sophisticated programming will be useful in all aspects of systems building or not is irrelevant. The basic principles of carefully organized production are applicable. They do make possible the acme of mass production theory: the multiple use of standardized elements, with alterations and modifications so incorporated into the production process that the full benefits of standardization are retained while varying the completed product to serve its required functions.

10.12 SHELLS

Prestressing has provided an excellent means of countering the tensile stresses in the edge beams of shells. It permits a degree of control and assurance that would otherwise be unobtainable. Edge beams, therefore, are frequently post-tensioned, with the sequence of stressing carefully dovetailed with the construction of the roof. The edge beams of domes are usually posttensioned with internal tendons covering a 90° to 135° arc between buttresses. On some notable domes, the edge beam has been stressed by wrapping wire under tension, in a manner similar to a tank. With other shapes, simple internal posttensioning tendons are employed. Where such tendons lap at corners, adequate mild steel must be provided. Precast edge beams have been used on occasion, especially when they were separate from the shell, and posttensioned (e.g., as tie-beams).

With cable-supported roofs, the compression ring may be constructed, then the cables strung and stressed to their specified length. Then forms may be hung and the roof concreted in a predetermined pattern. This procedure produces a varying state of final stress in each tendon. If precast slabs are hung from the tendons, then the tendons may be stressed in several stages as the deadweight increases, with final bonding with mortar in the joints between the slabs.

10.13 SUMMARY

Buildings represent perhaps the largest potential area for the application of prestressed concrete. The nature of buildings varies widely. As pointed out in the Indian Symposium on Human Habitation, "habitations" includes all the structures which house human services and which serve human needs. To be truly useful, they must serve the whole man. Prestressed concrete has a major role to play in buildings. It will realize its full potential only when the technical knowledge of the prestressing engineer is fully integrated with the skills of the architect to serve the "whole man."

SELECTED REFERENCES

1. ACI 318, "Building Code for Reinforced Concrete," American Concrete Institute, Detroit, Michigan, revised bi-annually.

2. PCI, *Design and Typical Details of Connections for Precast and Prestressed Concrete,* second edition, Prestressed Concrete Institute, Chicago, Illinois, 1988.

3. PCI, *Recommended Practice for Erection of Precast Concrete,* Prestressed Concrete Institute, Chicago, Illinois, 1985.

4. *FIP Guide to Good Practice,* "Quality Assurance of Hollow-Core Slab Floors," Institution of Civil Engineers, 1991, London.

5. FIP Recommendations, "Precast Prestressed Hollow-Core Floors," Structural Engineers Trading Organization Ltd., London, 1988.

6. FIP Recommendations, "Practical Design of Prestressed Concrete Structures," Structural Engineers Trading Organization Ltd., London, 1990.

7. CEB-FIP Model Code 1990 (Final Draft), Comite Euro-International du Beton, Lausanne, Switzerland.

CHAPTER 11

Prestressed Concrete Piling

11.1 INTRODUCTION

Prestressed concrete piles are widely used for marine structures and building foundations throughout the world. Actually, the growth in their use and the constant extension of the fields of application have been phenomenal. The original impetus was based on their durability in adverse environments, especially seawater, their strength as a beam for handling and as an unsupported column, their ability to resist tension, and their relatively low cost. Further advantages emerged quickly, especially their ability to be driven to deep penetrations and high capacities, using large piling hammers. Wider use gave rise to lower costs of manufacture, which made prestressed piles an economical competitor for foundation piles on land. Their ability to withstand hard driving, and their good soil-pile interaction behavior has led to rapidly increasing use in this application. More recently their use has been extended to sheet piles, fender piles, and soldier piles.

Prestressed piles have been utilized successfully in a wide variety of environmental conditions, from subarctic to tropical to desert.

Prestressed piles have been used as end-bearing piles on rock; as friction piles driven (or jetted) into sands, silts, and clays; and as piles set and grouted in predrilled holes in hard pan and rock.

Prestressed piles have been as small in cross section as 200×200 mm (8×8 inches) and as large as 3.5 meters (11.5 feet) in diameter. The longest piles made and driven in a single piece, up to the date of this writing, are 900 mm (36 inches) diameter cylinder piles, 80 meters (260 feet) in length, in Lake Maracaibo, Venezuela. Prestressed piles have also been installed in increments, spliced during driving, to achieve final lengths in excess of 60 meters (200 feet). Composite piles, with a top section of prestressed concrete and a lower section of steel, have been successfully installed to lengths of 60 meters (200 feet).

Prestressed piles offer these advantages:

1. Durability.
2. Ability to be transported, lifted (pitched), and driven without cracking.
3. High load carrying capacity.
4. Excellent combined load-moment capacity.
5. Ability to take uplift and tension.
6. Economy.
7. Ability to take hard driving and to penetrate hard strata and debris.
8. High column strength.
9. Ease of splicing and connection.

Prestressed concrete piling can be driven with an underwater hammer below the surface of the water, or with a follower below the surface of the ground or water, thus permitting piles to be installed prior to excavation and dewatering.

High-capacity prestressed concrete piles are particularly advantageous for deep foundations with heavy loads in weak soils. They have been successfully driven through riprap, debris, and old fills, and even through hard strata such as coral, and have penetrated into soft or partially decomposed rock.

Prestressed concrete has enabled the designer to utilize the pile in an economical and practicable way as an integral structural member of the overall structure.

11.2 DESIGN

The design of the prestressed concrete piling for direct bearing (short column) loads has gradually evolved from an arbitrary, highly empirical percentage of concrete cylinder compressive strength (e.g., allowable capacity = $P = Kf'_c A_c$) to a more rational evaluation of the actual capacity, including the effect of prestress at time of failure. To this a suitable factor of safety is applied. For example, a current code formula reads $P = (0.33f'_c - 0.27f_p)A_c$. In the above formula, P = allowable capacity, f'_c is the 28-day uniaxial concrete cylinder strength, A_c is the concrete cross section, and f_p is the design prestress value. The constants used reflect an allowance for safety factors to cover such matters as variation of actual strength from cylinder strength, accidental eccentricity, etc.

Most current codes, however, do not yet adequately consider the importance of the secondary reinforcement (spiral) in confining the core and in preventing longitudinal cracking. This matter will be addressed in detail in a later section.

It should be always recognized that the structural system includes both the pile and the soil, and that almost always a failure under actual load occurs in the soil, not in the pile. Structural failures in prestressed piles have been very limited and have occurred only when not properly reinforced and then only under combined bending, shear and axial load, as for example, in earthquakes. Where piles have a substantial unsupported length, their behavior as a column must be investigated, so as to prevent failure through buckling.

When prestressed piles must resist both moment and direct load, the interaction requires careful analysis. The presence of direct load, up to about 30% of the ultimate short-column compressive strength, causes an increase in moment-carrying capacity. As the direct load increases above about 30%, the moment-carrying capacity is reduced (Fig. 11.1).

At the pile head, where combined moment and direct load are often critical, the favorable effect of the transfer length (prestress varying from zero to full value over the transfer length) may be taken into account. Additional mild steel may be effectively utilized in zones of high combined moment and load, such as the pile head.

The construction engineer is usually responsible for insuring that the pile is manufactured, delivered, and installed in its design condition; that is, without cracking, damage, or permanent deformation from overstress. Therefore, he must investigate stresses during these phases (Fig. 11.2). These installation stresses usually far exceed the long-term service stresses.

For the handling of prestressed piles, the concrete strength should be based on the ultimate compressive strength of the concrete at the age in question. Since piles are usually lifted from the forms after 14 hours or so of steam curing and often driven at an age of only one week, it is apparent that early strengths are very important.

Impact must be considered and may impose additional stresses of 25 to 50% of the static stress. During handling and transporting, tension up to 50% of the modulus of rupture may be allowed to resist impact. It must be noted that at early ages, the modululus of rupture (tensile strength) of concrete is somewhat lower in relation to its compressive strength than at later ages.

Many pretensioned piles are manufactured in fixed forms with slightly tapered sides. At the time of lifting of these piles from the forms, a great many factors must be considered:

1. The pile may not have been able to fully shorten when the prestress was released due to friction on sides and soffit. Thus, as the initial bending is imposed, the pile may not have its full prestress.

Figure 11.1. *Pretensioned prestressed concrete piles. Pile interaction curves. The interaction curves shown hereon are presented as an aid to the Engineer when designing structures using the ultimate strength method. The curves are derived from a computer solution in accordance with ACI's Building Code. A rectangular stress block was assumed in the concrete with concrete strain limited to .003 inch/inch. Steel stresses are based on average values of stress strain curves for $\frac{1}{2}$ inch and $\frac{7}{16}$ inch diameter ASTM A416 grade 270 strands. Concrete strength is 6,000 psi in 28 days. A strength reduction factor ϕ of 0.7 has already been included in the diagrams for axial loads in the range from 100% to 10% of ultimate axial load. Below 10%, the ϕ factor increases linearly to a maximum of 0.9 for the case of pure flexure and combined tension and bending. For section properties, effective prestress, allowable working design loads, and other technical information for these piles see Fig. 11.3a. Note: 1 kip = 1000 lbs. = 1.356 kN. (Courtesy of J. H. Pomeroy & Co., Inc.)*

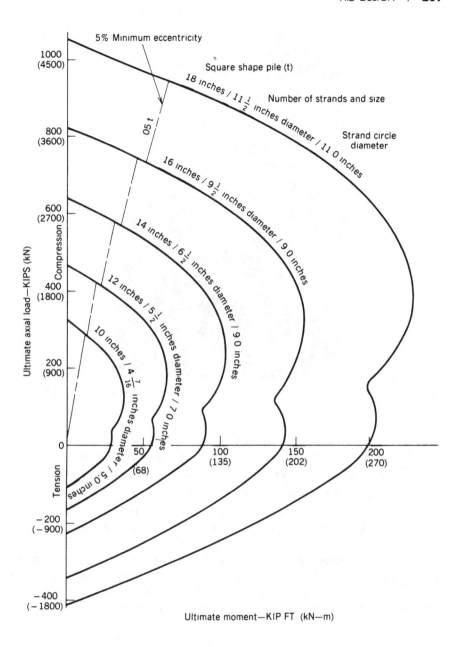

Ultimate moment—KIP FT (kN—m)

Figure 11.2. *Handling 34 meter (110-feet) long 250 mm (10-inch) pretensioned concrete piling.*

2. Lifting will have to overcome friction on the sides and suction.
3. Binding may occur at any irregularities, such as dents in the forms.
4. Dynamic loads from the crane or other lifting equipment must be allowed for.
5. The concrete is still young (green) and fully saturated, except at the top surface where tensile stresses may already exist due to drying and cooling.

To prevent overstressing during this condition, pick-up points may have to be spaced more closely, additional mild steel (or short pieces of unstressed strand) may be needed at the picking points, and it may be desirable to raise one end a few inches ahead of the other. Alternatively, the piles may be raised in the forms ("broken loose") by jacks.

Under seismic and wind loads a combination of axial load and moment may occur at the head of piles. This condition can best be provided for by means of auxiliary mild-steel bars embedded in the pile head, plus heavy spiral reinforcement in this zone. This matter of seismic design will be discussed further.

Prestressed piles have conventionally been designed with moderate values of prestress to meet the needs of handling and driving (Fig. 11.3 a,b). The increasing application of prestressed piles to resist combined direct load and moment requires that the designer be alert to the potential structural value and economy of using much higher values of prestress. An upper limit in the range of 30% f'_c is suggested.

11.2.1 Design for Resistance to Driving Stresses

The stress-wave theory explains many of the phenomena observed during the driving of prestressed piles. When a prestressed pile is struck by a hammer blow, the impact lasts from 0.004 to 0.08 second. The compression wave travels at the speed of sound in concrete, about 4000 m/s (13,000 feet/s) for a typical concrete used in prestressed work. (For structural lightweight aggregate concrete, this velocity is reduced 15 to 30%.) The wavelength is normally 15 to 45 meters (50 to 150 feet), depending on the duration of the impact. The maximum compressive stress in the pile head will typically range from 7 to 20 MPa (1000 to 2800 psi), but can reach over 30 MPa (4200 psi). (For structural lightweight aggregate concrete, the maximum compressive stress is reduced about 20%.) When this compressive stress wave reaches the pile tip, it reflects. If the tip is on hard bottom (e.g., rock), it will reflect as a compressive wave. If the tip is on soft bottom (e.g., mud), it will reflect as a tensile wave. (See Fig. 11.4.)

The toe stress may thus reach a compressive stress approaching twice the head stress in the extreme case of driving a short pile through water directly onto rock. Conversely, when driving the tip into soft material, a tensile stress up to 50% of the head compressive stress may be reflected back up the pile. Actual tensile stresses which have been measured by strain gauges can range up to 10 MPa (1400 psi) and even higher. Procedures recommended below should limit these to half the above value.

There are a great many variables (21 or more) involved in computing these stresses. Computer programs are available to handle these computations. The variables include weight and length of pile, modulus of elasticity, damping properties, weight and velocity of ram, stiffness of cushion, and soil resistance and resilience. Actual observations tend to confirm these mathematical solutions, although the values of maximum stress observed tend to be somewhat lower than computed, due to the dissipation of energy into the soil and into heat and sound and other inelastic behavior.

As discussed later under "driving," the values of maximum stresses can be effectively reduced by reducing the velocity of the ram during impact, and by increasing the amount of cushioning at the head of the pile.

Further, according to the stress-wave theory, with a short pile driven through water, or very soft soil, onto rock, a high compressive stress can reflect from the toe, travel up the pile to the head and, if the ram has left the head by this time, the wave will be reflected as a tensile wave. This is a very uncommon condition in actual practice, since soil embedment will absorb this energy in most cases. However, a few such cases have arisen, requiring modification in the driving procedure.

"Rebound" tensile stresses are resisted by the tensile strength of the concrete and

Standard Prestressed Concrete Foundation Piles

Pile Size Diameter (Inches)	Shape	Solid or Hollow Core (HC)	Concrete Area Ac (Inches²)	Weight per Foot (Lb)	Number of Strands (per pile)	Effective Prestress (to nearest 5 psi) psi
10	Square	Solid	98	105	4—$\frac{7}{16}$ inch	760*
12	Square	Solid	142	152	5—$\frac{1}{2}$ inch	830
14	Square	Solid	194	209	6—$\frac{1}{2}$ inch	730
15	Octagonal	Solid	186	196	6—$\frac{1}{2}$ inch	760
16	Square	Solid	254	273	9—$\frac{1}{2}$ inch	835
18	Octagonal	Solid	268	288	9—$\frac{1}{2}$ inch	790
18	Square	Solid	322	346	11—$\frac{1}{2}$ inch	805
20	Square	Solid	398	428	13—$\frac{1}{2}$ inch	770
20	Square	11 inches, H. C.	303	326	10—$\frac{1}{2}$ inch	775
24	Square	14 inches, H. C.	418	450	13—$\frac{1}{2}$ inch	730
36	Round	26 inches, H. C.	487	524	17—$\frac{1}{2}$ inch	820
48	Round	38 inches, H. C.	675	726	24—$\frac{1}{2}$ inch	835
54	Round	44 inches, H. C.	770	829	28—$\frac{1}{2}$ inch	855

(1) Nominal pile-size
(2) Holes for hollow-core piles are circular
(3) Reduction in area for chamfers on square piles has been taken into account.
(4) Tables are based on concrete of 155 lb/foot³ density
(5) Based on $\frac{1}{2}$ and $\frac{7}{16}$ inch diameter ASTM A416 Grade 270 strands with ultimate strengths of 41,300 and 31,000 lb, respectively. If different diameter strand is used, the number of strands per pile should be increased or decreased, in accordance with strand manufacturer's tables, to provide approximately the same minimum effective prestress shown in the table
(6) Effective prestress assumes a uniform distribution of strands resulting in a uniform prestress. Piles marked with * have effective prestress based on 60% of ultimate strand strength. All other piles have effective prestress based on initial prestress in strand of 70% of ultimate minus losses of 35,000 psi.

Figure 11.3.(a) Standard prestressed concrete piles.

Standard Prestressed Concrete Foundation Piles (Continued)

Moment of Inertia I (Inches⁴)	Section Modulus I/C (Inches³)	Radius r (Inches)	Perim- eter (Inches)	Allowable Moment		Allowable Loads	
				300 psi Tension (KIP Inches)	600 psi Tension (KIP Inches)	Based on f'_c 6000 psi (Tons)	Based on f'_c 7000 psi (Tons)
790	158	2.84	38	167	215	87	103
1,664	277	3.42	46	313	396	125	148
3,112	445	4.00	54	458	592	172	204
2,765	368	3.86	50	390	500	165	195
5,344	668	4.59	62	758	958	222	264
5,705	634	4.61	60	691	881	236	281
8,597	955	5.17	70	1,055	1,342	283	336
13,146	1,315	5.75	78	1,407	1,801	353	418
12,427	1,243	6.40	78	1,336	1,709	268	318
25,490	2,124	7.81	94	2,188	2,825	372	440
60,016	3,334	11.10	113	3,734	4,735	428	508
158,222	6,593	15.31	151	7,483	9,460	592	703
233,409	8,645	17.41	170	9,985	12,578	673	800

(7) Allowable bending moments listed are for a permissable concrete tensile stress of 300 psi with an effective prestress as given in the table, no external axial load, f'_c = 6000 psi and assuming a modulus of rupture of 600 psi. Allowable moments for earthquake or similar transient loads are based on a tension of 600 psi. Piles with both axial load and bending should be analyzed considering the effect of the sustained external load. When bending resistance is critical, the allowable moment may be increased by using more strands to raise the effective prestress to a maximum of 0.2 f'_c psi

(8) Allowable design loads are based on the accepted formula of N = Ac (0.33 f'_c − 0.27 f_{pe}), and are computed for f'_c = 6000 psi and 7000 psi.

(9) "Kip" = 1000 lbs. "f_{pe}" = prestress.

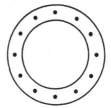

Figure 11.3.(b) Standard prestressed concrete piles.

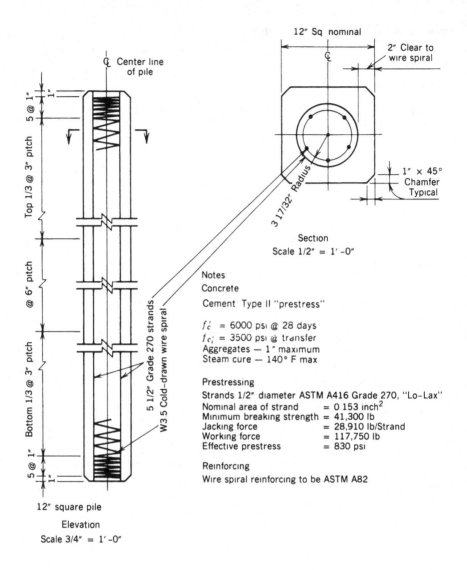

12" Sq nominal

2" Clear to wire spiral

Center line of pile

5 @ 1"

Top 1/3 @ 3" pitch

@ 6" pitch

Bottom 1/3 @ 3" pitch

5 @ 1"

5 1/2" Grade 270 strands

W3.5 Cold-drawn wire spiral

3 17/32" Radius

1" x 45° Chamfer Typical

Section
Scale 1/2" = 1'-0"

Notes:
Concrete
Cement Type II "prestress"

f'_c = 6000 psi @ 28 days
f_{ci} = 3500 psi @ transfer
Aggregates — 1" maximum
Steam cure — 140° F max

Prestressing
Strands 1/2" diameter ASTM A416 Grade 270, "Lo-Lax"
Nominal area of strand = 0.153 inch²
Minimum breaking strength = 41,300 lb
Jacking force = 28,910 lb/Strand
Working force = 117,750 lb
Effective prestress = 830 psi

Reinforcing
Wire spiral reinforcing to be ASTM A82

12" square pile

Elevation
Scale 3/4" = 1'-0"

Figure 11.3.(c) *Typical prestressed concrete foundation pile. (Courtesy of J. H. Pomeroy & Co., Inc.)*

the effective prestress, up to the point of cracking. After cracking, these stresses are resisted by the steel tendons, up to their yield point. When the stresses equal or exceed the yield point of the steel, the hammering ruptures the bond between steel and concrete and between paste and aggregate. The concrete in that zone will actually become hot. The tendon, with a wide stress range, soon fails in low cycle, high amplitude fatigue. Repeated driving at very high tip resistance (high compres-

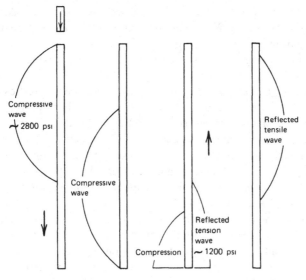

(a) Reflected tensile stress—wave   soft driving conditions at tip

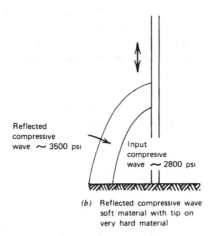

(b) Reflected compressive wave
soft material with tip on
very hard material

Figure 11.4. Dynamic stresses in driving prestressed concrete piles.

sion) may also lead to fatigue, with consequent disruption of the aggregate-paste and steel-paste bond, leading to eventual failure. These conditions are exacerbated when driving in water, due to the high pore pressures which build up in the pile and, after cracking, the hydraulic ram effects on the water which is sucked into the crack.

To prevent repeated cracking, the margin between yield strength in the tendons and the final prestress should theoretically require 1.0% steel tendon area. However, once the pile has cracked, damping reduces subsequent tension and in most cases in practice, 0.6 to 0.8% will prove satisfactory, especially if steps are taken to reduce the tensile rebound stresses. With piles less than about 25 meters in length, a steel area of 0.5% will usually prove adequate.

Means for minimizing the rebound tensile stresses during driving are presented in Section 11.8.

There is no general theory for design of the spiral binding for prestressed piles. In practice, steel spiral areas as low as 0.2% have been widely employed in moderate capacity land foundations without serious difficulty. (For this purpose steel area ratios are determined on a vertical section through the pile axis; Fig. 11.5).

However, longitudinal cracking has occurred with increasing frequency, especially with high capacity hollow-core and cylinder piles. The Concrete Society of Great Britain recommends a greater area of spiral for hollow-core than solid piles; values of 0.3 to 0.4% are recommended.

During driving, the bursting forces at head and tip are large. There are transverse tensile forces due to prestress and radial bursting forces due to the hammer blow. Under high amplitude cyclic loading, Poisson's ratio increases.

Specifications for the lateral binding of the end zones of girders generally set values for A_s/A_c of 0.4%, solely to resist the transverse tensile stresses in the end zone. A_s is the area of steel, A_c is the area of concrete. The additional bursting stresses due to driving are apparently of the same or greater magnitude. This indicates a total requirement of 0.8 to 1.0% spiral steel area in the top 24 inches. The Concrete Society of Great Britain recommends a minimum of 0.6% spiral over a length at the top and toe equal to three times the least dimension of the pile. This is based on observed (empirical) data.

The author has found in practice that the provision of 1% steel area in the top 12 inches of cylinder piles will effectively prevent splitting. Some engineers provide bands around the top for this same purpose. It is interesting to note that these bands

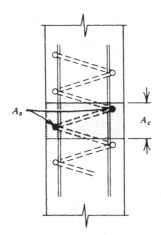

Figure 11.5. Vertical section through pile axis for computation of ratio of spiral steel area to concrete area. A_s = cross-sectional area of spiral steel. A_c = gross area of concrete on vertical section per pitch of spiral.

sometimes break during driving, indicating that the stresses are real. Hoops of mild steel are recommended in preference to bands, due to the favorable effect of the deformation. The hoops or bands can be placed very close to the outside wall of the pile, since cover for durability is usually provided by the pile cap.

Chamfering of the circumferential edges and top corners of the pile head will reduce bursting and spalling tendencies.

After installation, piles may be subjected to shearing forces and imposed curvatures due to lateral loads from soils and seismic forces. During driving, shear may occur due to restraint or obstructions. Therefore, some spiral should be provided throughout the length of the pile, not just at head and tip. Decreasing the spiral spacing is much more effective than increasing the wire size. Where the spacing becomes too close for concreting, the hoops may be bundled. High-yield steel has been reported as more effective than normal steel for spirals. Cold-drawn wire spiral is not only more effective but easier to instill. Epoxy coated spiral is not recommended since there is no adhesive bond and the cover is easily spalled. However, this could be overcome by using an epoxy to which sand has been added for adhesion.

In summary, therefore, the following amounts of spiral are recommended for prestressed piles:

Top 300 mm (12 inches)	1.0%
Next section [minimum 600 m (24 inches)]*	0.6%
Body of pile	0.3% (solid piles)
	0.4% (hollow-core piles)
600 mm (24 inches) above tip	0.6%
Tip 300 mm (12 inches)	1.0%

In zones of high seismicity, substantially more spiral is required. The International Conference of Building Officials, in the 1992 supplement to Building Standards, requires a minimum volumetric ratio of spiral reinforcement (ratio of spiral steel to included core of concrete) to extend over 120% of the length from first point of zero lateral deflection in the soil to the underside of the pile cap or grade beam. This ratio is set as $0.12 \times$ concrete compressive strength divided by yield strength of spiral.

11.2.2 Design of Raked (Batter) Piles

Where lateral loads must be resisted and the area is not subject to earthquakes or similar highly dynamic forces, raked (batter) piles are often employed. These piles must have the structural capacity to resist the design axial forces (tension or compression or both) and in addition the deadweight bending.

Since in typical cases, the majority of the pile length is below ground or water, the pile receives some vertical support from the soil or the buoyancy of the water.

---

*Since piles may not be driven to grade, this section should be lengthened as necessary to ensure that least 1.0 meter (40 inches) at the final top has this minimum spiral, after cut-off.

During installation, the pile will typically be in cantilever until it is firmly supported in the soil below. This produces high negative moment over the last support. To prevent excessive cracking, either additional prestress or supplemental mild-steel reinforcement may be required.

The pile head connection must be properly detailed to transfer the axial forces.

Because of deformations of the pile and adjoining pile, both moment and shear will be induced near the pile heads. Additional mild steel and confining spiral are required.

11.2.3 Design of Aseismic Piles

It is now generally recognized that the use of raked (batter) piles to resist lateral forces is not appropriate for areas of high seismicity (Fig. 11.6). The raked piles transfer the full seismic shear to the connection at the pile top, generally overloading the pile in moment and shear and/or punching through the pile cap or slab.

Figure 11.6. *Many batter piles failed at the head during Loma Prieta earthquake in California.*

Therefore, in most seismic regions, only vertical piles are used. The piles must have total longitudinal steel (both prestressed and passive) and heavy confinement in the form of spiral, to ensure that the pile has ductility and robustness. If it is also sufficiently connected into the structure above, moments and shears can be transferred and the frame will be able to resist the lateral forces and deformations in elastic and plastic response.

A typical ductile frame pile will therefore have, in addition to prestressing reinforcement, additional mild-steel longitudinal bars extending down $\frac{1}{4}$ to $\frac{1}{3}$ of the pile length to the calculated point of fixity in the soil, below the lowest stratum of soil of low shear strength. The pile will also have closely spaced spiral to confine the core over the length subjected to imposed curvature so that it can sustain high bending compression without disintegration from crushing. A steel pipe sleeve, with grout fill of the annulus, can also be used.

Recently, a high strength fiber epoxy encasement system has been developed and successfully tested at the University of California at San Diego. It is being applied as a means of retrofit of the columns of existing bridges in the state of California and appears fully applicable to new piles as well.

The fiber consists of glass and poly-aramid fiber. It may be applied under tension or, in less demanding cases, without tension. After wrapping, the fibers are encased in a high-elongation epoxy system.

The system enhances both flexural ductility and shear capacity.

11.2.4 Notes for Detailed Design

The following design notes relate to details for prestressed piling. Inattention to these details has led to difficulty in installation and subsequent behavior: like all mass-produced items, the basic details are of great importance.

11.2.4.1 PROVISIONS FOR LIFTING. Lifting holes through the pile are undesirable because they cause local stress concentrations. Piles should be lifted by slings around the pile, or by lifting eyes embedded in the pile. A loop of short pieces of strand, with each end embedded at least 60 cm (24 inches) into the concrete core, makes a good picking eye. Special picking inserts are also manufactured for this purpose.

If the picking eye or loop is permanently embedded in soil, no removal or corrosion protection is usually required, since there is insufficient oxygen available to cause other than local corrosion and since the loop runs perpendicular to the surface. If the picking eye is exposed, e.g., in the splash zone, then it should be recessed $\frac{1}{2}$ inch, then after use, cut off and coated with an epoxy patch. Epoxy putty should be used, as most epoxy mortar is permeable. Where lifting inserts are embedded, these are plugged to a depth equal to the cover over the spiral, since these have greater steel area.

11.2.4.2 PILE-TO-CAP CONNECTIONS. The following types have been successfully employed:

1. Mild-steel dowels are grouted into holes in the head. These holes may be preformed, with flexible metal duct, held in place during concreting by a mandrel. Removable tubes, such as hose or inflated rubber tubes, tend to leave too smooth a surface for bond.

Alternatively, the holes may be drilled in after the pile is driven, provided a small drill is used, so as not to break out, crack, or otherwise damage the pile. This however is difficult to accomplish and should preferably be limited to, say, four large bars, with long embedment, rather than attempting to drill six or eight such holes.

The bars should be embedded a sufficient distance (i.e., 30 or 40 diameters for deformed reinforcing steel) to develop full bond for pull-out under dynamic impulse. The bars may also be used to develop additional bending moment in the head of the pile.

Bars should be grouted in with cement grout of low W/C ratio. Dry pack is not nearly as satisfactory as grout. The grout should be designed for low shrinkage. Epoxy grout may be used, but is more expensive.

2. The prestressing tendons may be extended into the pile cap a distance of 500 to 750 mm (18 to 30 inches). They should be splayed. While such embedment will develop the full strength of the strand, there may be an excessive rotation of the pile-cap connection if the stresses allowed are greater than those used for mild steel of the same cross-sectional area. This is only of importance where the cap is small, as in high railroad trestles; for normal footings and caps, rotation is generally of little importance and may actually be beneficial under earthquake.

Electrical isolation of the cap or deck from the piles may be desirable in the marine environment, since the above-water portion, exposed to the air, may act as a very large cathode to cause corrosion at any anodic zone within the pile, e.g., at a spall or crack. This may be accomplished by coating the bars or extended strand of the connection with epoxy.

When piles must resist repeated tension, then strand embedment should be increased to 1.0 meters (40 inches) to provide safety against bond failure.

3. The pile may be extended up into the cap a distance of 500 to 750 mm (18 to 30 inches). The surface of the pile should be cleaned prior to concreting. It may also be roughened by bush-hammering or sandblasting. This provides excellent shear transfer; however it will not have much ductility under earthquake.

4. For hollow-core or cylinder piles, a cage of reinforcing steel or a structural steel beam may be concreted into the top portion of the core. To insure bond, the inside surface of the core should be cleaned and roughened, as by sandblasting, and epoxy bonding compound applied. The concrete in the plug should be designed for low shrinkage. Aluminum powder may be added to cause a slight expansion. The concrete plug must be designed for low heat of hydration, so as to minimize expansive force during hydration. The pile must have sufficient spiral through this area to develop the enclosed steel.

Precast plugs, containing the steel cage, are particularly useful as head plugs for large-diameter cylinder piles. The small annular space between the inside of the pile shell and the outside of the precast plug can be filled with grout. This essentially eliminates the problems of heat and shrinkage.

Note. When filling a hollow-core pile with concrete, be sure that there is sufficient hoop strength in the pile (spiral) to withstand the hydrostatic head of the fresh concrete. Rates of pour should be adjusted to prevent cracking or bursting the pile. Similarly, be sure there is enough spiral to withstand the expansion during heat of hydration at steel stresses below yield, so that any cracks will close.

5. Post-tensioning from the pile head to the cap is theoretically an excellent means of developing full moment, but has not been widely used except when tension under uplift must also be transmitted. If bars are used, these may be embedded and grouted into holes which are formed or drilled into the pile. Because the tendons to be stressed are very short, the anchoring system must be able to seat without loss, or else permit subsequent shimming to offset this seating loss.

Instead of multiple bars, a single large tendon may be embedded in the core of hollow-core piles and later post-tensioned through a sleeve to the cap.

The anchor within the pile will require sufficient reinforcement to transfer the prestress into the pile over a sufficient length to prevent cracking.

11.2.4.3 DISTRIBUTION OF TENDONS. The matter of distribution of the tendons within a pile has been given careful analysis in Japan and elsewhere. Current United States practice uses uniform distribution around a circular pattern, even for square piles. Swedish practice has been to group the tendons near the corner of the square. In the former USSR, short foundation pilings have been employed with tendons concentrated at the center. Elastically, and for concentric stresses, it seems to make little difference as long as the center of stress coincides with the center of gravity. After cracking in bending, however, the strand pattern does have an effect on the ultimate strength. Tests for the US Navy show that the highest ductility can be enhanced with a square pattern, properly confined. The Japanese tests have concluded that the ultimate bending moment is not affected appreciably as long as there are at least four tendons around the periphery.

A circular pattern, with circular spiral reinforcement, is used on round, hexagonal, and octagonal piles. With square piles, both circular and square patterns are used, with spiral shaped to suit. However, there appears to be no significant difference in behavior between the circular and the square pattern, and no evidence that the corners of square piles are more likely to break off when a circular pattern is used. Therefore, because of the savings, increased efficiency of circular spiral, and ease of manufacture, the circular pattern for square piles predominates in current practice.

11.2.4.4 PILE SHOES. Practice and opinion vary widely from country to country, based largely on early experience during the developmental testing of concrete piles. In the United States, pile shoes are generally not used at all, whereas in Scandinavia and Great Britain, pile shoes have generally been standard practice. The reason may lie in the geology of the regions. High local toe stresses occur mainly when driving through extremely soft materials (e.g., peat) onto hard rock (e.g., glaciated granite). Here a short steel dowel or plate may be useful in securing a toe hold in the rock, especially when the rock is sloping.

Extensive recent experience confirms that shoes are unnecessary when driving into sands, silts, clays, and even soft shales, etc. Pile shoes are of help when the piles must penetrate buried timber, coral (limestone) strata, and rock.

These shoes may take the form of plates, points, or stubs (Fig. 11.7). Plates should be sufficiently thick to withstand local deformation [e.g., 25 mm (1 inch) thick] and anchored into the pile tip by anchor rods of sufficient embedment to develop bond even under repeated loading (i.e., a minimum of 40 diameters). Reinforcing steel, if used for these anchors, should be mild steel, and preferably plug welded with full penetration welds to the plate. Intermediate or hardgrade bars often fail at the weld, although use of low-hydrogen electrodes can help. Fillet welds tend to fail under the repeated impact.

Fabricated or cast steel points may be used. These are useful in obtaining penetration, but tend to make the pile run off line; thus, pile alignment will not be as

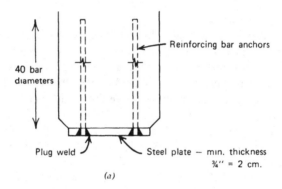

(a)

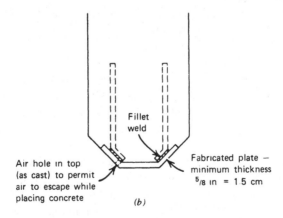

(b)

Figure 11.7. (a) Steel plate pile shoe; (b) "blunt" steel pile shoe.

accurate as with flat plates. A blunt point, therefore, is often chosen as a reasonable compromise.

Such points also serve to confine the edge of the concrete at its tip, preventing corner splitting or spalling. They need not be of as thick steel as a flat plate, since their shape serves to stiffen them; thus, 15 mm or thicker plate is usually satisfactory. In manufacture, some means must be used to prevent air from being trapped in the shoe while placing the concrete; an air vent is usually provided for this purpose.

Stubs consist of H-pile sections, solid dowels of diameters of 75 mm (3 inches) or more, or fabricated crosses. They may be welded to a plate, which is in turn anchored to the pile. More usually, they are anchored by embedment of the stub section into the pile, a distance of 1200 to 1500 mm (4 to 5 feet). In early applications, shear lugs were welded to the embedded portions; most recently, dependence has been placed on bond alone, with apparently satisfactory results. Weld heads can also be used. Again, means must be provided to permit placement and consolidation of the concrete around the stub and to permit escape of air; a hole through the web is usually provided.

Steel stubs must be of sufficient thickness, stiffness, and high yield to prevent their own distortion. (See Fig. 11.8.) One of the advantages of prestressed concrete piles in penetrating hard material is their rigidity and resistance to local deformation. Nothing is accomplished by installing a long, slender, steel stub if that stub bends, tears, or otherwise distorts. Cast steel shoes (tips) welded onto the steel H stub will minimize the tendency for distortion.

Stubs must usually be fitted into the strand pattern by forcing: even if this

Figure 11.8. Prestressed concrete bearing piles incorporate steel H-pile stubs to penetrate glacial till.

deflects the strands, it appears to have no ill effect as long as the stub is well confined.

Heavy spiral or stirrup reinforcement must be provided around the stubs to confine the concrete. The proper design of the tip assembly deserves careful consideration, so as to ensure proper force transfer in this critical section while permitting placement and consolidation of the concrete.

In Scandinavia, piles are often fitted with a steel dowel so that when they are driven onto sloping rock, the point will prevent sliding and pin the pile to the rock.

In Singapore, prestressed piles cast with a central tube have been anchored into rock by first driving them so as to seat on the rock, then drilling through the tube, inserting a prestressed tendon, grouting it throughout its lower (rock) portion, then stressing it, and finally grouting it into the hollow core. These piles have then been primarily designed as tension (uplift) piles.

The shape of the pile tip, like the shape of the shoe, may be varied to suit driving conditions. A square tip, with chamfered corners, is most common today in practice. In former times, points and wedges were formed. These tend to cause the pile to deflect during driving. High bending stresses are induced in the point, tending to break it off or to break the pile. However, rounded or heavily chamfered tips are useful when driving through old rock seawalls, etc. to prevent the edge or corner from catching on a rock.

Extended dowels may prove counterproductive when used with raked piles. In the Arabian Gulf, the point slid along the surface of a hard limestone stratum. A shorter, more compact, shoe proved more effective.

11.2.4.5 HOLLOW-CORE AND CYLINDER PILES. See Fig. 11.9. With hollow-core and cylinder piles, when driving through soft material, or under water, soil and water may be forced up the core, causing a hydrostatic head inside that may burst the pile. This is particularly serious if water completely fills the core so that the hammer blow causes a hydraulic ram effect.

In nonfreezing environments, vents may be provided to allow the water to escape. A series of such vents will be adequate for the hydrostatic head effect, but not for the case when the pile head is driven to or below the water surface. In this latter case, a driving head should be designed to provide full venting, that is, an area approximately equal to that of the core; otherwise, the hydraulic ram effect may destroy the pile.

Underwater vents through the walls are also useful in preventing internal hydrostatic pressures during sudden drops in the outside water level (as in a river or at extreme low tide). In freezing environments, the problem is very complex. The design engineer must give thorough consideration to all the potential effects of such vents: will they be properly located after the pile is driven, will they ever become partially blocked, e.g., with marine growth or frazil ice, etc. For these reasons, in coastal and harbor environments subject to freezing, it is preferable to use solid piles, not hollow core, wherever possible.

Hollow-core and cylinder piles used in marine structures should be filled solid in any zone exposed to impact from boats, barges, and ships. As with the case of head

Figure 11.9. Pretensioned cylinder pile for bridge across Napa River, California.

plugs, the use of precast plugs, set in place, and bonded by grout will generally be found preferable to casting concrete by tremie methods. The grouted precast plug eliminates problems of internal pressure and thermal expansion during concreting and curing.

11.2.4.6 COVER. Adequate cover must be provided over the tendons and spiral to insure durability. Current European practice calls for 40 mm (1½ inches) cover in fresh water and soil, and 50 mm (2 inches) in salt water. Current US practice is to use 40 to 50 mm (1½ to 2 inches) in fresh water and soil, and 50 to 75 mm (2 to 3 inches) in salt water exposures. Actually, there appears to be little reason for placing the strands very close to the edge, since the elastic bending resistance is unchanged and ultimate bending resistance is only slightly lowered; for example, with a 450-mm (18-inch) octagonal pile, increasing the cover from 60 to 75 mm decreased the ultimate moment by only 5%.

11.2.4.7 STRUCTURAL LIGHTWEIGHT CONCRETE. Structural lightweight aggregate concrete is sometimes employed for bearing piles, sheet piles, and fender piles. The weight for transporting and handling is reduced. The buoyant weight, under

Figure 11.10. *Pretensioned lightweight concrete pile for main line railroad bridge. Dunbarton Crossing, San Francisco Bay, California.*

water or soil, is reduced by almost 25% as compared with conventional concrete. This may give important benefits when large, long piles must achieve support in weak soils. By using lightweight concrete, substantially all of the frictional resistance of the soil is available for support of the superimposed structure (Fig. 11.10).

During driving, prestressed lightweight concrete piles behave approximately the same as conventional piles, the lighter mass tending to offset the lower modulus of elasticity and the greater damping.

There is an increased tendency for the heads of prestressed lightweight concrete piles to spall. This can be countered by providing increased spiral wrapping or bonding at the head.

11.2.4.8 VERY HIGH-STRENGTH CONCRETES. Such concretes, with strength in the range of 80 to 100 MPa (12,000 to 14,000 psi), have been used for piling to a limited extent. Their use permits a smaller concrete cross section, which reduces weight for transport and handling, and improves the driveability. In particular, it permits easier penetration of upper dense layers where such is required. Of course, the subsequent transfer of load from the pile to the soil must also be considered.

It is believed that such very high-strength piles may extend the use of prestressed concrete piles to soils where previously only steel piles were considered suitable.

11.2.4.9 CHANGES IN SECTION. Abrupt changes in section or properties (transformed section) should be avoided. When solid heads or tips are used with hollow-core piles, a gradual transition should be made, in the form of a cone (with rounded tip) or hemisphere. Additional mild steel, vertical and circumferential, may help to contain the stress concentrations across the transition.

Similarly, at splices or head zones where heavy longitudinal mild-steel bars are installed, the locations of the ends of the bars should be staggered longitudinally and extra spiral provided in this zone.

When an H-pile stub is embedded in the pile tip, a transition section should be designed, using heavy reinforcing bars welded to the stub, and extending them well back into the pile, e.g., by one meter or more.

11.3 MANUFACTURE

11.3.1 General

Prestressed concrete piles are manufactured by a wide variety of processes. The essential considerations of the manufacturing process are the following:

1. Uniform, dense, thoroughly consolidated concrete should be produced, of high strength.
2. Tendons should be in accurate location and be thoroughly surrounded by dense mortar.
3. No visible cracking, whether due to shrinkage or thermal contraction or restraint in the forms, etc. The manufacturing process should ensure that a minimum of surface tensile stresses are "locked into" the concrete.
4. The pile should not be cracked in bending during picking, handling, or transporting.
5. The surface of piles should be relatively dense and free from deep pockets or deep "bug holes" due to bleed. A typical allowance is 12 mm (0.5 inch) maximum depth; if exceeded it should be filled with epoxy putty.
6. The head should be truly normal to the axis and free from protuberances or ridges. A typical allowable tolerance for out-of-normal is 6 mm ($\frac{1}{4}$ inch).
7. The cross section should be sufficiently true to maintain the prescribed cover. A tolerance of ± 6 mm ($\frac{1}{4}$ inch) is usually considered acceptable.

11.3.2 Long-Line Pretensioning Method

This has been employed so far for piles ranging up to 80 m (260 feet) in length, and from 200 mm (8 inches) up to 1800 mm (72 inches) in diameter. Solid piles, hollow-core piles, and cylinder piles are extensively manufactured by this process. All shapes and cross sections can be made by this method (Figs. 11.11 and 11.12).

To the extent that the adopted cross section will permit the completed pile to be lifted out, forms are most commonly fixed forms with a slight taper. The precautions on lifting piles from the forms were discussed earlier under "Design," Section 11.2. In particular, the negative moment at each picking point is often critical and

Figure 11.11. Pretensioned concrete piling of square cross section.

Figure 11.12. Pretensioned concrete piles of octagonal cross section.

may require additional reinforcement and care at first breaking out. Removable forms are also used, and may be of the hinged, slide-back, or lift-off type. Removable forms are frequently used to form the upper half of octagonal or round piles (Fig. 11.13).

If the top sections are removed shortly after concreting, the top surface may then be finished to remove bleed holes and surface imperfections. Bug holes and bubble marks are primarily due to bleed and the escape of entrapped air, which has been trapped along the vertical and overhanging forms.

Horizontal sliding forms have been developed for the top surfaces of round and circular (cylindrical) piles: these are usually integrated with a concrete pouring machine which feeds the concrete into the forms. Such a sliding top form, when working properly, can eliminate the problems from bleed and entrapped air, and consequently produce a blemish-free surface.

For hollow-core piles, a long sliding mandrel is sometimes employed to form the core. This mandrel must have a device to keep the leading edge up: a notched wheel is sometimes employed which rides on a soffit, the notches being designed to ride between the spiral. Other systems guide the leading edge on the strand and then use removable lifting hooks to support the strand. The tail end support of the mandrel is supported in the freshly concreted pile behind. The mandrel is pulled along the line at a rate of about 300 mm (12 inches) per minute.

The mandrel may contain vibrators and heaters, so as to cause the concrete to take an initial set behind, without slumping or sloughing. A near frictionless surface for the mandrel is important: stainless steel has been used. The tendency for the

Figure 11.13. Removable top form for cylinder piles.

concrete to stick to the mandrel and than fall in should be minimized by wetting the mandrel or coating it with cement slurry just prior to entering the concrete. Usually the cement paste from the concrete then provides sufficient lubrication as the mandrel passes further down the line.

This process has worked well when equipment and workmanship were well done. However, it does require a high degree of coordination between concrete mix, temperature, rate of pour, and amount of vibration.

The most serious problem is sloughing of the top concrete under the strand and spiral. The strand and spiral hold the exterior up. This results in delamination which may not be readily apparent until later.

Proprietary extrusion machines have been designed to permit the "slip-forming" of both internal and external surfaces of hollow core piles at one pass.

An alternative method of forming the inside of a cylinder pile is to use mechanical, collapsing forms. These forms can be expanded to form the circle (Fig. 11.14). After the concrete has gained initial strength, for example, at the end of the curing cycle, the forms can be mechanically retracted, so that they can be pulled out endwise. The internal form must be held down, either by self weight or rolling supports, so as to prevent flotation as the concrete is vibrated. Typical tolerances in position of the core are 9 to 12 mm ($\frac{3}{8}$ to $\frac{1}{2}$ inch).

The mechanical system has proven to give a more reliable product although the labor costs are slightly higher.

With both systems, it is important that the forms be kept clean and placed in essentially perfect condition prior to the next pour.

Figure 11.14. *Pretensioned cylinder pile is manufactured with the use of a mechanically collapsing inner form and removable top forms.*

In the past, hollow cores have been formed by paper tubes and rubber tubes. Both require hold-downs at close spacing to prevent flotation during concreting and vibration: even then, there will be small upward deformations between hold-down points, so that the wall thickness will vary.

The typical production sequence for solid-core piles is:

1. Clean and oil forms.
2. Set out pre-cut bundles of spiral in the forms, one bundle to each pile location.
3. Pull strands down bed, through the coils. Take up slack, using a dynamometer to equalize length of each strand. Affix strand ends to stressing block, stress, and anchor. Alternatively, each strand may be tensioned individually.
4. Spread out coils to proper spacing, and tie them to strands.
5. Place segmental end gates at the pile ends and clamp them to strands.
6. Place concrete and vibrate.
7. Cover and keep damp for about three hours, then start steam-curing cycle.
8. At some point at or after the steam is turned off, uncover the piles, and release the prestress into the piles. Re-cover the piles and allow to cool.
9. Cut strands at pile ends.
10. Lift piles from forms to storage.
11. Employ supplemental water cure in storage (where specified).
12. Employ supplemental drying period in storage (where specified).

Supplemental water cure should be employed for piles which will be exposed to aggressive environment in service, especially for large, thin-walled, hollow-core, or cylinder piles, and whenever extremely drying atmospheric conditions will result in surface crazing.

The prestress should be released into the piles (Step 7) before exposure to the atmosphere; otherwise, the piles may develop horizontal shrinkage cracks as they cool. While most of these will probably close later when the prestress is released, the tensile strength of the concrete will be impaired, potentially aggravating the problem of tensile rebound cracking during driving.

A slow-cooling cycle is important for large, solid, or massive piles. If they are exposed to cold atmospheric conditions while the inside core is still hot, thermal stresses may cause cracking or crazing. Thermal and shrinkage strains are additive.

Steam covers should be sufficiently tight to provide protection from cold winds, so as to insure relatively uniform temperature conditions inside. If, because of wind, for example, there is a substantial difference in curing temperature inside, then when the prestress is released into the member, the strength, effective "age," and modulus of elasticity will vary, plastic flow will occur, and the resulting pile may have a permanent "sweep."

For piles which will be exposed to salt air or the saltwater splash zone, or to a freeze-thaw zone, if bleed holes ("bug holes") have occurred due to the trapping effect of vertical and overhanging forms, and if they are more than a nominal depth

Figure 11.15. *Finishing the top surface of cylinder pile.*

of 10 to 12 mm ($\frac{3}{8}$ to $\frac{1}{2}$ inches), then defects over that length of pile which will be exposed should be filled with epoxy putty. Because there appears to be no way to completely eliminate these air and water bleed holes from the upper surface of octagonal and round piles, the use of removable top forms followed by finish troweling appears to be the best solution (Fig. 11.15).

End-gate forms to separate piles in long-line pretensioning must be constructed so as to prevent the formation of transverse fins by mortar leakage. Any such fins will lock to the freshly cut concrete while the form is expanding under the rising steam temperature, and will cause transverse tensile cracks, which may later be aggravated under driving.

11.3.3 Segmental Construction

In this process, precast segments are fabricated, either by centrifugal spinning or in vertical forms. These methods of manufacture have evolved from precast concrete pipe technology. After the precast segments have hardened and cured, they are placed end-to-end, jointed, and posttensioned.

Spun segments offer the advantage of thorough consolidation of the concrete. The centrifugal force may be augmented by vibration or by an internal roller. Excess water is drained from the center. The outer surface is dense and smooth.

The spinning method requires that techniques be instituted to hold the duct formers in true position. They naturally tend to belly out from the centrifugal force. To overcome this, the duct formers are usually tensioned against the end forms and, after spinning, the duct formers are withdrawn, so that subsequent grouting of tendons will fill any voids.

Vertical casting of segments permits the economical forming and casting of any size pile, up to the largest diameters (Fig. 11.16). Shear keys may be easily formed in the top and bottom edges. Form vibration may be used effectively as a supplement to internal vibration.

Curing of these segments is very important, since the inside is hot and moist, while the exterior, if not protected, will be dry and rapidly cooling to atmospheric temperature.

After casting, stripping, and curing, the vertically cast segment must be tipped to horizontal. For very large diameter pile segments, this is usually done in a tilting frame, which allows rotation without imposing excessive crushing loads on the bottom edge. Another method is to set the segment in sand, then tip it while lifting. The sand serves to distribute the crushing load on the edge. A third method is to rotate it in air, using a double sling arrangement.

After segments are turned and accurately aligned on a horizontal bed, the joints must be made. If the ends as cast are very true, the joint may be made with epoxy glue. With one spinning process, true ends are achieved by using an absorbent form

Figure 11.16. Prestressed concrete cylinder piles are manufactured as vertically cast segments which are then joined longitudinally. Oosterschelde Bridge, The Netherlands.

liner (to draw off excess water and prevent bleeding). With vertical casting, the end forms are usually machined cast steel.

Wider joints may consist of cast-in-place concrete, with a width of 75 mm (3 inches) to 200 mm (8 inches). With joints thinner than 75 mm it is difficult to consolidate the concrete, with the result that stress concentrations and spalling may occur during driving. "Fine concrete," using a 10 mm ($\frac{3}{8}$ inch) coarse aggregate and high cement factor such as 450 kg/m^3 (750 lb/yd^3), is usually employed. The ends of the segments may be painted with an epoxy bonding compound prior to concrete placement in the joint. In any event, the surface film of mortar (laitance) should have been removed by light sandblasting or wire brushing. After pouring, steam curing may be beneficially employed to accelerate the gain in strength of the joint.

After jointing, tendons are inserted, stressed, and grouted (Fig. 11.17). In some systems, the end anchorage is removed after the grout has hardened, the tendon stress being transferred by bond alone. This is satisfactory, particularly with strands, up to a reasonable concentration of force per tendon. With higher forces, bond slip may be a problem, but more serious is the bursting force that may crack the thin wall outside the duct. Adequate spiral reinforcement, particularly at head, tip, and throughout the zone of possible cut-off, is essential. In the past, many of the large diameter cylinder piles [1.2 to 3.5 meters (4 to 11.5 feet) diameter] have had inadequate spiral reinforcement to meet all the circumferential tensile forces that develop. A minimum spiral percentage of 0.6 to 0.8% seems necessary to prevent vertical splitting in fabrication, installation, and service.

Sections of piles may be manufactured in standardized lengths, for example, 12

Figure 11.17. *Prestressed concrete cylinder piles, made of segments, are stressed to form 60 meter lengths. Oosterschelde Bridge, The Netherlands.*

meters (40 feet) and then spliced during driving. This method is extensively used in Japan, Sweden, and England. It is particularly well adapted to locations in which transport of long piles would be a serious problem, in which crowded conditions make the lifting (pitching) of long piles difficult, and in which the desired or available equipment for driving is limited in size and height.

Such standardized sections can be manufactured in a highly mechanized plant and stored in inventory. In driving, the lengths can be varied by simply adding or leaving off a section.

The sections themselves are manufactured by a variety of methods: centrifugal spinning with high-strength alloy steel bars posttensioned against the forms; centrifugal spinning with tendons inserted later and posttensioned; and horizontal casting with pretensioning.

11.3.4 Splices

Splice details must be strong enough to resist the repeated hammering during installation, as well as the design actions in service of tension, bending, and compression. They must be durable for the environment in which they will be embedded for service. Usually this will be underneath the soil or mud line. If the splice is below the soil and below the water table, there will normally be only surficial corrosion, since there will be a lack of oxygen. Exceptions are in old garbage dumps or where organic concentrations can cause high acidity: there special considerations are required, but even there, the lack of water movement will slow the initial corrosion process.

The Japanese have adopted a standard splice in which the tendons are anchored at each end of the pile segment. The segments are then joined by welding. An electrical connection is provided between segments when cathodic protection of the splice is planned.

Where splices are preplanned, it is very important that the fittings be correctly and accurately installed during manufacture, and that both mating ends be truly normal to the axis of the pile. One such method for insuring that the ends will properly match is to assemble the two steel splice plates, separated by two or three pile diameters, and held in a rigid frame that fits in the forms. Thus the end plates are held normal to the forms.

Holes for future dowel embedment may be best formed with corrugated steel ducting, into which a pipe mandrel is fitted so as to hold them rigid. The pipe mandrel, in turn, is supported by an external frame.

Although the corrugated duct may be pulled from the hole by unwinding, this seems unnecessary: it is best left as a hole former. Rubber tubes and paper tubes are not recommended for this purpose.

An alternative is to embed dowels, extending from the head. These dowels must be similarly temporarily fixed in a frame that fits into the forms.

In addition to welded splices, a number of mechanical connectors have been developed; these engage and lock mechanically, as by a screwed joint or wedge effect. These splices are, of course, very rapid in execution. The connectors generally are designed to transmit full compression and tension, but only a few are capable of transmitting moment as well. Improvements to these mechanical splices have been made in recent years, with the result that they are being ever more widely used. Long-term durability is generally considered not to be a problem if the final position of the slice is embedded in the soil.

Friction splices are also available, in which a sleeve is driven over a male casting, locking itself by wedging. These are generally fully effective in compression, but tension values are erratic, particularly under the first few hammer blows after a splice. Therefore, it may be necessary in soft driving to restrain the lower section so that the upper section may be driven onto the lower. This restraint has been provided by a timber clamp resting on the ground.

Experimental and developmental work has been directed toward a prestressed splice. To date, practical difficulties of tensioning have prevented widespread use. However, in the Netherlands, such splices have been developed for their very large piles of 3.5-meter diameter.

For splicing cylinder piles of diameters of 120 mm and greater, matching cast steel rings have been embedded in the two ends. They are anchored by multiple reinforcing bars welded to the ring, which distribute the driving forces back into the pile wall. The ring should be located in the center of gravity of the wall, which means its diameter will be slightly larger than the mean of the inside and outside diameters.

One ring has its edge pre-bevelled so that welding can start as soon as the sections are placed. In some versions a slightly curved back up plate is affixed to the lower ring: when the top is set, it automatically guides the top ring to its correct position.

Low hydrogen electrodes are used.

Epoxy-doweled splices have been very successful structurally and economically, but require a substantial time to set before resuming driving (Fig. 11.18). This may range from 30 minutes to 12 hours, depending on the setting characteristics of the epoxy and the temperature. Usually four dowels of deformed reinforcing bar are extended from the top section. When the top section is set, the dowels slide into corrugated metal tubes in the head of the bottom section. Bar lap (embedment) lengths of 30 to 40 diameters are recommended. The size of the four bars should be selected to give a steel area of 3% of the concrete area across the joint. Setting the four dowels in simultaneously is difficult. The operation can be facilitated and made safer if one dowel is longer than the rest, so it can be entered independently. Epoxy may be poured into the holes before inserting the dowels or injected afterward. Other proprietary compounds are also available. In Norway, unsaturated polyester resin has been used. The set is accelerated by internal electrical resistance heating. A small copper wire is wound around the dowel tubes. After the resin has been poured into the holes, the top section is set and an electric current from a generator set (about 24 V, 100 A) is used for five minutes to start the reaction. Once started, the reaction proceeds rapidly, even in cold weather, and sufficient strength is often generated to permit driving to restart within 15 minutes or so. Since copper in concrete is potentially a source of electrochemical corrosion of the prestressing

Figure 11.18. *Prestressed pile segments are spliced with epoxy-dowel splice for full-moment connection. Use of segments permits driving under high-voltage power lines.*

tendons, use of a steel wire conductor of proper size and electrical properties would appear to be a safer practice.

In the United States, steam jackets have similarly been clamped onto the pile at the splice and used to accelerate the reaction.

Because of the structural and durability benefits of the epoxy-dowel splice, developmental work is now underway along these lines:

1. Substitution of a single pipe stub in lieu of four dowels. If two nipples are run to the pile surface, steam can be circulated to accelerate strength.
2. Use of epoxy as an addition to one of the several mechanical splices now on the market: then commencing driving immediately, letting the epoxy set after completion of the installation.

It is important that with any splice, the sections be essentially center-bearing, and that the concrete immediately above and below the splice be contained with heavy spirals. If the splice is edge bearing, then the driving stresses will not be transmitted uniformly to the concrete, and a fatigue condition, under alternate compressive and tensile waves, may develop. With sufficiently rigid steel, as in the Japanese standard joint, this is apparently not critical; however, it is always desirable to keep the driving stresses as uniform and concentric as possible. Therefore a center bearing splice is preferred (Fig. 11.19b, d, and f).

11.3.5 Internal Jet Pipes

Internal jets, located on the pile axis, are increasingly used to aid penetration of piling through overlying dense soils, especially sands, where deeper penetration is sometimes required for geotechnical reasons. The advantage of the internal jet is that it discharges water at the tip, where it will be most effective in assisting initial penetration.

The entrance fitting (nipple and elbow) must be located below the head a sufficient distance that it won't be contacted by the driving head, even after the cushion block has crushed. In some cases, costs can be saved by locating the entrance well down the pile: this also reduces jet hose handling problems. These entrance fittings should be standard pipe sections. The nipple will permit subsequent removal and patching with epoxy putty.

As with all pipe bends, there is a radial thrust due to change of direction at the elbow. Therefore, this section of the pile requires adequate spiral confinement and should have an anchoring bar or bars.

The body of the jet pipe should be light gauge water pipe with screwed (watertight) fittings. To reduce costs, plastic pipes and conduit have been tried. While under ideal circumstances, these might work, especially if the jetting is carried out before the hammer is set on the pile, once driving is initiated, such nonwatertight piping is prone to fracture under driving stresses. For example, a tension rebound crack will reflect through the plastic. Then the jet water pressure can spread over the crack surface: under subsequent hammer blows, hydraulic ram forces will cause crack widening, paste erosion, and eventual failure.

The costs associated with even one or two such pile failures far exceeds any savings in materials.

The tip nozzle of the jet can be a swaged pipe fitting. While jet efficacy is not excessively sensitive to hydraulic phenomena, it is worthwhile to form the nozzle with some attention to entrance and exit losses.

In special cases, jet exit nozzles are constructed in the sides of the pile, at or above the tip, so as to give more lubrication against skin fraction.

Where the pile will be jetted down prior to replacing the hammer, then the entrance fitting may be merely a nipple at the pile head. It is important that this be recessed about 100 mm (4 inches) from the top. Otherwise, the hammer blow, directly transmitted to the fitting and jet pipe, may cause fracturing of the concrete.

Jet pipes in hollow-core piles present a serious complication. If for any reason the jet pipe breaks, e.g., under the impact of hammering, the resultant hydraulic pressure inside the hollow core will probably break the pile. Therefore, it's probably best to either change to a solid pile or use a free (extended) jet, inside or outside the pile.

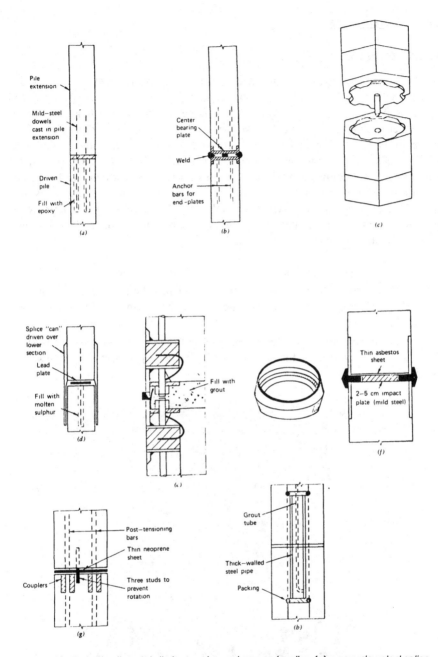

Figure 11.19. *Typical splice details for prestressed concrete piles:* (a) *epoxy-doweled splice—United States, Norway;* (b) *welded splice;* (c) *mechanical splice—Swedish patent;* (d) *steel splice sleeve or "can";* (e) *welded splice—Japanese patent;* (f) *"Brunsplice" joint—United States patent;* (g) *post-tensioned splice—Great Britain;* (h) *steel pipe splice for hollow piles—Norway*

11.4 INSTALLATION

11.4.1 General

Prestressed concrete piles are installed by a wide variety of techniques, including driving, jetting, drilling, vibration, and weighting. However, the great majority of prestressed concrete piles are installed by driving, either alone or in a combination with other techniques (Fig. 11.20). Well over 2,000,000 prestressed piles have been successfully installed by driving under the widest possible variations and combinations of subsurface conditions, ranging from penetration into silts, sands, clays, and soft rock, to founding on hard rock, and to penetration through overlying hard material into lower bearing strata. From this experience, some general rules have been developed to ensure successful installation. These rules have independently emerged in remarkably similar form in many countries, including the United States, Britain, Australia, Japan, Sweden, and Canada.

1. The pile and hammer should be held in accurate alignment.

2. The pile head should be square and free from protuberances or projecting wire or strand. Wedging or binding of the pile head in the driving head (helmet) should be avoided.

3. A hammer should be selected that will not require excessive (ram) velocity. Ram velocity should be reduced in soft soils, or when jetting in conjunction with driving.

Figure 11.20. *Driving prestressed concrete cylinder pile for reconstruction of railroad bridge.*

4. Adequate cushioning should be provided at pile head: 150 to 300 mm (6 to 12 inches) of softwood has consistently proven most effective in reducing internal stresses in the pile while facilitating maximum penetration of the pile. A new cushion should be provided for each pile. The Japanese have used corrugated paperboard packing; materials used elsewhere include a stack of crude rubber sheets, felt, sacking, asbestos fiber, coiled hemp robe, and plywood. Plywood is increasingly being used because it doesn't disintegrate: however, it has less than half the resilience of soft wood. Hence, its use is often inadequate, especially with piles over 20 meters in length. The cushion is intended to lengthen the duration of the impact and eliminate the sharp peak stresses, which can lead to rebound tensile cracking. Even soft wood, if dried out, as in arid regions, will become too hard: presoaking will restore its resilience. For further discussion, see Section 11.4.3.

5. Pile splice rings, or other fittings, should be protected with a ring cap.

6. Excessive restraint to pile, either in torsion or bending, should be avoided during driving. Once the pile is well embedded in the soil, its position and orientation cannot be corrected from a position at or near the head without causing distress. If sliding guides ("rabbits") are used to support the pile when it is in the leads, these should be removed as they approach the hammer, so as not to impose excessive bending moment.

7. For cutting off, preferably make a circumferential cut first with a diamond saw or small rotary drill. This will ensure a neat, unspalled head. Other means include clamping with a steel or wood band, then cutting with a small pneumatic hammer.

8. When predrilling or jetting, the tip must be well seated before full driving energy is applied.

9. The greatest driving efficiency is obtained by using a heavy ram, a softwood pile head cushion, and low-velocity impact.

10. With very long slender piles, guides or supports should be employed to prevent "whip," vibration, and buckling during driving.

11. With batter or raking piles, supports must be provided for the overhanging lengths—both those extending above and below the frame.

12. The lifting and handling points for handling and picking (pitching) must be properly marked and used. With long piles and multiple (two to six) pick-up points, the angle which the slings make with the pile affect the stress in the sling; pick-up points may therefore be different from those used for handling at the plant and in transport (Fig. 11.21).

13. In construction over water, pile heads are sometimes pulled into correct alignment. With long unsupported lengths, it is easy to overstress or even break the pile. Thus, specifications should limit the movement or force of pulling to ensure against overstress. Even the tidal current or waves may cause the head of a pile in deep water to move significantly. Specifications should be written in such a way as to be practicable of accomplishment and enforcement. The allowable amounts vary considerably, depending on the depth of water, soil conditions, etc. For example,

Figure 11.21. *Lifting long prestressed concrete pile for Escambia Railroad Bridge, Florida.*

with a standard 400-mm (16-inch) square prestressed pile at the outer edge of a wharf in 12 meters of water and a soft mud bottom, a force of 2.5 kN (500 lb) at the head will move the head about 75 mm (3 inches) and reduce the compressive prestress at a point 3 meters (10 feet) below the mud-line to about 1.5 MPa (200 psi). This may be acceptable in many cases.

In the same wharf, at the inboard edge of the wharf, in 2 meters (6 feet) of water and a rocky fill bottom, a force of 9 kN (1 ton) is necessary to move the pile 12 mm ($\frac{1}{2}$ inch). Thus, for such a pile, pulling the head into alignment is impracticable.

One practicable means that can be enforced is to specify that the tool used for pulling be set so it cannot exert a force greater than, say, 2.5 kN (550 lb).

14. After piles have been driven in rivers or harbors, they may require temporary lateral support against tidal current, wave forces, etc., and, in the case of batter piles, against deadweight deflection.

15. Diesel hammers are widely used to install prestressed piles. Because of generally light ram and high velocity of impact, particular care must be taken to provide a proper head cushion. Fortunately, the inherent performance of a diesel hammer prevents development of full ram velocity in soft driving, thus automatically reducing dynamic stresses. However, the first blow, the start of the driving, generally requires the ram to drop the full height. Thus the first blow may initiate cracking which later propagates.

16. Jetting is often employed to aid driving. This may be pre-jetting (pilot jetting) to break up hard layers or it may be jetting during driving. In the latter case it is hazardous to jet at or below the tip during driving, as this may create a cavity and produce a "free-end" condition, leading to excessive tensile stresses and cracking in the pile.

17. Predrilling, dry or wet, with water or with bentonite slurry, may be effectively used to aid penetration through hard and dense upper layers. This method is

becoming increasingly used due to the fact that soil engineers are requiring deeper penetrations for settlement control.

The pulling and removal of prestressed concrete piles is very likely to produce cracks because of inadvertent bending during pulling. It is very difficult to take a truly axial pull, even with a pair of slings which have equalized loops on both sides of the pile. Extensive jetting should be performed on all sides of the pile prior to pulling. The pull should be gradual and maintained, since soil tends to fail slowly in friction under a sustained pull.

Sometimes prestressed concrete piles are damaged during or after installation by accidental impact, etc. If the damage consists of spalled concrete, it may often be satisfactorily repaired by one of the following methods:

1. Underwater-setting cement, which does not contain calcium chloride, may be placed below the water surface by a diver. This is mixed in small batches, placed in a sealed can, and lowered to the diver, who then forces it into place.

2. Larger voids or spalls may be patched with tremie grout. A metal form is placed around the pile, and grout is poured into the void, with the pipe always immersed in the fresh concrete. The form should be overflowed to get rid of any laitance.

3. Patching with underwater-setting epoxy. If the void to be filled is larger than 50 mm in dimension, then the epoxy may be filled with graded sand on a 1:1 basis.

With all of these methods, the spalled area should be thoroughly cleaned of damaged and broken concrete, marine growth, silt, etc., and then, if possible, a key should be chipped by either a hand chisel or a very small air chisel. (Use of a large chisel may cause further cracking.) Patches should be protected from wave erosion, etc., until they have hardened.

Cracks may best be sealed with underwater epoxy, although in some cases underwater-setting mortar can be forced into the crack. With a deep structural crack, the epoxy injection method may be used. The crack is first sealed with epoxy, then a small tube is drilled into the crack and underwater epoxy pumped in until it exudes from the crack. Underwater epoxy is an epoxy containing a hydrophobic agent which dries the surface of the crack so the epoxy may bond.

The above repairs have been generally successful when carefully performed for the correction of structural damage. They are not necessarily applicable nor successful for repair of durability failures, i.e., disintegration due to corrosion or chemical change. This chemical and electrochemical degradation usually requires special, far more extensive techniques (see Chapter 25).

11.4.2 Examples

For the installation of very large and long piles, special construction techniques are often employed. For example, on the Oosterschelde Bridge in the Netherlands, the large cylinder piles were sunk through silts and sands by a combination of internal dredging and weighting. The derrick barge was equipped with an outrigger that was clamped onto the head of the pile (see Fig. 11.17): by means of rigging the barge was then lifted so as to impose several hundred tons of downward force onto the pile. The internal dredging was then performed by a cutter-head on a special articulated arm. After the proper tip elevation had been reached, a bottom plug of underwater concrete was placed. While initially the piles were unfilled, provision was made for later filling with sand, if needed, to dampen vibration. Freezing of the inside water was prevented by means of vent openings near the sand line, well below the ice zone.

On the San Diego–Coronado Bridge in California, 135-cm (54-inch) diameter cylinder piles were installed in three successive stages through clay and sand strata to the desired dense sand bearing stratum. In stage 1 they were pre-jetted with high-pressure, high-volume jets, then driven to within 2 meters (6 feet) of design tip. The jetting was controlled so that it did not affect the sands within this lower 2-meter zone. In stage 2, only side jetting to relieve friction was permitted, keeping the jets well above the tip, but the penetration was achieved by driving with a large hammer. This driving was continued until the pile reached a point about 0.6 meters (2 feet) above design tip. In stage 3 the hammer alone was employed to drive the pile to final bearing. The specifications required at least 200 blows in the last stage, to insure consolidation of the sand around the pile. After all piles in a pier footing had been driven, the hammer was again used for a specified number of blows (50) on each pile to insure that the overlying sands were re-vibrated and consolidated. Piles were then cleaned out, down to the hard plug of sand that had been driven up into the tip, and a tremie concrete plug was placed.

In driving prestressed piles on some of the Arkansas River projects, involving friction piles in sands, jetting plus driving was found to be most effective. It was found that driving should be continued for 20 to 50 blows after jetting ceased, even though no significant further penetration was achieved, in order to ensure re-consolidation of the sands. In these sands the jetting actually proved to be beneficial to the piles' bearing capacity because it washed out the fine silts, giving greater frictional and lateral support.

In driving 50-meter (160-foot) long piles 500 × 500 mm (20 × 20 inches) in cross section for a railroad trestle in Florida, a very dense stratum of sand was located at a depth of 20 to 25 meters (70–80 feet). To penetrate this, extensive external jetting was employed. Then the pile broke through into weaker silts below, eventually having to be seated in the 50-meter (160 foot) deep stratum. For the high design capacities specified, a high blow count was required for final seating, namely 240 blows for the last 300 mm (12 inches). In a number of cases the pile suddenly failed, apparently rupturing some distance above the tip. It is believed that initial cracking occurred when the pile broke through the upper stratum. Under prolonged driving on the cracked pile, the crack eventually widened and fatigue caused rupture of concrete and tendons.

Underwater, reinforced and prestressed concrete builds up pore pressure under cyclic loads, eventually causing cracking. As the cracks open and close under the hammer blows, the water is sucked in, then forced out. A hydraulic ram effect is generated, as well as erosion of the cement paste. The tendons now cycle beyond yield and soon fail in brittle fracture.

Recent research has indicated that this phenomenon where fatigue is accelerated underwater can be minimized by using very impermeable concrete, such as that obtained by the addition of microsilica to the concrete mix.

In particular, tests show that high-strength lightweight concrete with the addition of microsilica can withstand twice the number of high amplitude stress cycles as similar normal-weight concrete. This appears to be due to the secondary crystallization at the surface of the lightweight aggregates as a result of the microsilica.

11.4.3 Rebound Tensile Stresses under Driving

In the sections above, considerable discussion has been included regarding rebound tensile stresses. When these become excessive, they manifest themselves in a rather "mysterious" fashion, usually occurring in the soft driving, and especially in the first few blows when the pile breaks through the surface and runs. Cracks show up as puffs of cement dust, usually about one-third of the length of the pile from the top. If driving continues, other cracks will form about 600 mm (24 inches) apart. They usually go all the way around the circumference and are characterized by surface spalling that looks much worse than the actual crack. Repeated driving produces fatigue in the concrete next to the crack; the concrete gets hot, and the aggregate-paste bond is destroyed. Eventually the tendons fail in brittle fracture, due to high amplitude, low cyclic fatigue.

This type of failure is made harder to diagnose by the fact that the first crack usually occurs at a weak point, such as lifting eye, insert, form mark, or honeycomb. Such a local stress concentrator may determine the location of the first crack, but is usually not the cause of the crack.

The theory and preventive design for rebound tensile stresses were discussed earlier. However, should the rebound stresses prove to be excessive and cracking be noted, then one is faced with a situation in which the prestressed piles are already manufactured and a given piece of driving equipment is on the site. Several steps can then be taken to reduce rebound tensile stresses and prevent further damage:

1. Increase the thickness of head cushion and reduce the stiffness. The best material is unused, green softwood, made up of rough boards 25 to 50 mm (1 to 2 inches) thick to a total thickness of 300 to 400 mm (12 to 14 inches). These provide the maximum cushioning during the early, soft driving, and become much stiffer (twice as stiff or more) during driving. For this reason, they should not be reused, even if still in apparently good condition. Cushions which have dried out in storage should be pre-soaked. To hold the cushion together, pieces of plywood may be nailed on the top and bottom.

2. Reduce velocity of impact. If the driving energy must remain the same, use a heavier ram with less stroke. Usually, however, the existing hammer may be used by reducing the velocity of impact during the early, soft phases of driving. With a

differential-acting hammer, it is often only necessary to throttle down, so as to slow the hammer. With a single-acting hammer, a small reduction can be accomplished with throttle; significant reduction requires the use of an adjustable slide bar to reduce the stroke in the zone of soft driving. A diesel hammer may be changed to a lower setting. An hydraulic hammer can be adjusted to cushion the blow.

Note that for many diesel hammers, it is necessary to raise the ram all the way [typically 3 meters (9 feet)] for the first drop. This first drop may cause a crack if there is no resistance at the top.

3. Change jetting practice as necessary to prevent any tendency to wash out a hole or soft spot under the tip while the sides are still wedged in hard strata or gripped by side friction. Limit jetting during driving to the relief of side friction.

4. Some of the worst cases of rebound tensile fractures occur when driving through an overlying fill, into a very soft mud stratum. If problems persist despite steps 1 and 2, break up the overlying hard layer by predrilling or pre-jetting, so that the pile will push through the soft strata under is own weight plus the weight of the hammer.

5. With a diesel hammer, initiate the driving by raising and dropping the ram 4 feet or so, i.e., by using it as a drop hammer for three to four blows to seat the tip in firmer material. Then raise it to the minimum needed to start the diesel ignition.

6. Ensure that the pile head is not binding in the driving head, that it is free to turn. Any torsional and bending stresses are additive to the rebound tensile stresses and permit the start of tensile cracking on the faces of the pile.

11.5 SHEET PILES

Prestressed concrete sheet piles offer the advantage of durability, rigidity against local deformation, and excellent appearance (Fig. 11.22). They do not necessarily represent a savings in the first cost over steel piles, because sheet piles must take both positive and negative moments. To provide for this, prestressed concrete sheet piles must be essentially uniformly prestressed to a fairly high degree (Fig. 11.23). However, if a computation of design moments indicates greater negative than positive moments, then a degree of eccentricity of prestress may be used.

The amount of eccentricity that may be tolerated without affecting the ability to withstand driving stresses will, of course, vary widely, depending on the many variables in soil, pile, and driving equipment. However, eccentricities which give a variation of 1 to 2 MPa (150 to 300 psi) in the concrete have been successfully employed, as long as there is a compressive stress over the entire section.

Alternatively, mild-steel reinforcement can be added at the points of peak moment so that these zones will be "partially prestressed."

Since most prestressed concrete sheet piles are used as a seawall or bulkhead, an unique opportunity exists to post-tension the sheet pile after installation. For this type of structural system, the bending is positive in the portion exposed to air or water, negative at the top support or tie back, and reversing to both positive and

Figure 11.22. *Prestressed concrete sheet piles serve as a seawall.*

negative below ground, but to lower values of moment. Thus unbonded post-tensioning tendons, such as sheathed and greased strands, can be embedded on a profile designed to counteract the bending. After installation, the tendons can be stressed and anchored from the top. There is little danger of overstressing, since the deformations under prestress are increasingly resisted by the passive pressure of the soil. Thus highly efficient structural response can be obtained.

Special care is needed to provide corrosion protection to the top anchorage: for example, sloping the surface to prevent water from sitting on the top, and coating with epoxy or polyurethane to prevent chloride penetration. Alternatively, the top may be encased in reinforced concrete if the joint is thoroughly bonded or sealed against water infiltration.

Sheet piles are frequently installed by jetting. With jetting, it should be possible to use greater eccentricity of prestress. Another method of installation is vibration, with or without the aid of jetting. With vibration, both compressive and tensile stresses are traveling up and down the pile. We do not know the minimum prestress required to withstand the tensile stresses due to vibration but, presumably, they are similar in degree to those for driving, that is, 5 to 6 MPa (700 to 850 psi).

In the design of sheet piles some tension may be permitted to exist, provided the ultimate strength requirements are satisfied. Tension up to one-half the modulus of rupture is commonly permitted. A limit condition in design is cracking; obviously it is essential that the sheet piles not be cracked in the saltwater splash zone.

The details of lateral mild-steel reinforcement of sheet piles are very important. The mild steel serves the functions of confining the concrete in order to insure transverse bending strength, of providing shear strength, and of making the tongue

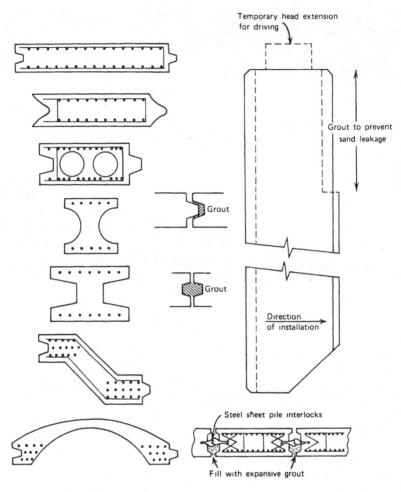

Figure 11.23. *Typical cross sections and details of prestressed concrete sheet pile.*

and the wings of the interlock grooves function as an integral part of the pile. These wings in particular tend to break off during driving, due to the wedging effect of the tongue of the adjoining pile, or due to the wedging effect of the soil. For this reason, it will generally be found best to drive the sheet piles with the tongue leading, so that there is no soil plug formed in the groove. When a double groove joint is used, a steel pipe is usually inserted in the leading edge of the groove as a temporary tongue, then later withdrawn. During such withdrawal, grout may be injected. Because of wedging stresses, it is therefore desirable to detail the mild steel with bars extending into the wings.

In driving prestressed sheet piles, it will often be found best to use auxiliary jetting, either several internal jets, or an external gang jet. To keep the piles vertical, a roller may be placed over the leading edge of the sheet pile being set, and pulled back against the preceding piles; this will tend to keep the pile up tight at the top. By sniping off the leading toe of each sheet pile, the toe also will be wedged back against the preceding sheet pile, keeping the toe tight at the bottom.

To drive each pile to the same grade, a special driving head may be needed, with a slot so it can pass by the head of the preceding pile. Alternatively, a short, narrow extension may be cast onto the head of each pile to serve as a driving block for the hammer. After driving, this extension may be cut off, and the tendons tied into a capping beam; or the extension may be left on and the slot between adjoining extensions used for tie-backs. As noted above, the head should be encased in concrete or otherwise protected to prevent corrosion.

Prestressed concrete sheet piles are utilized for waterfront bulkheads, breakwaters, cut-off walls, groins, wave-baffles, and for retaining walls.

In some recent building foundations, the prestressed concrete sheet piles were installed by a combination of predrilling and driving. After excavation, the joints were welded and filled with nonshrink grout, so that the sheet-pile wall served as the permanent foundation wall of the building.

When prestressed sheet piles are used for waterfront bulkheads and cut-off walls, the joints must be sealed. The most common method of sealing is to fill the joints with grout. In sand, a pipe may be advanced by jetting, which washes the sand out of the joint, then grout may be pumped through the pipe as it is withdrawn. In grouting through water, a polyethylene, burlap, or canvas bag may be pushed down the joint by means of a pipe, then the grout pumped into the bag as the pipe is withdrawn.

The grout used should be a rich, cohesive, workable mix. For most cases, Portland cement, sand, and water form the mix, preferably with an admixture to promote cohesiveness and reduce the W/C ratio.

A number of interlocks have been developed for prestressed sheet piles, to give both structural strength and sand-and-water tightness. The ordinary tongue-and-groove interlock transmits shear but not tension. Steel sheet piles can be cut in half and embedded in the prestressed sheet piles. These will, therefore, be as tight and have the same tensile strength as the steel sheet-pile interlock. A polyethylene interlock has been developed, which is embedded in the concrete and which acts as a sand and water-stop (Fig. 11.24).

Prestressed concrete sheet piles of different sections have been designed and tested in both the United States and Japan. The latter have developed a series of standards which include deep arch and Z-piles.

11.6 FENDER PILES

Prestressed piles have many advantages to offer as fender piles, dolphin piles, and other piles primarily resisting lateral forces. Prestressed fender piles are durable and economical. They can be designed to maximize deflection and energy-absorption by keeping the moment of inertia low in relation to strength (by choice of section and prestress levels).

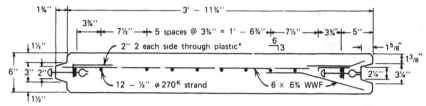

No 2 bar reinforcement for 6—in pile

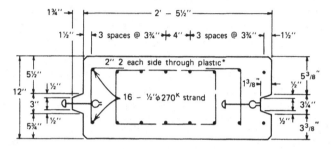

No. 2 bar reinforcement for 12—in pile

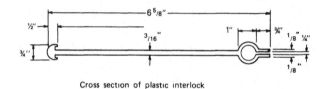

Cross section of plastic interlock

Figure 11.24. Plastic interlock for prestressed sheet piles.

One use of the fender piles is for wharves and piers, where ships must constantly berth without damage to ship or fender system. Prestressed fender piles have been installed on major cargo wharves in Kuwait, Singapore, Los Angeles, and San Francisco and have given very successful performance in service (Fig. 11.25). A serious fire at one installation destroyed the adjoining treated-timber fender but left the prestressed fender piles undamaged, except for slight surface spalling. Those fender piles installed in Kuwait in 1959 were still in service in 1989 (Fig. 11.26).

The US Navy has had an extensive research program on prestressed fender piles from which a series of standard sections have been developed. They are generally utilized with floating camels which protect the underwater attachments affixed to the ships' sides. Extensive tests by the Navy showed the need for confinement in order to give ductility.

Prestressed fender piles are usually designed for a maximum tension of up to one-half the modulus of rupture under service loading. It is important to prevent cracking that will reduce durability; on the other hand, it would be wrong to make

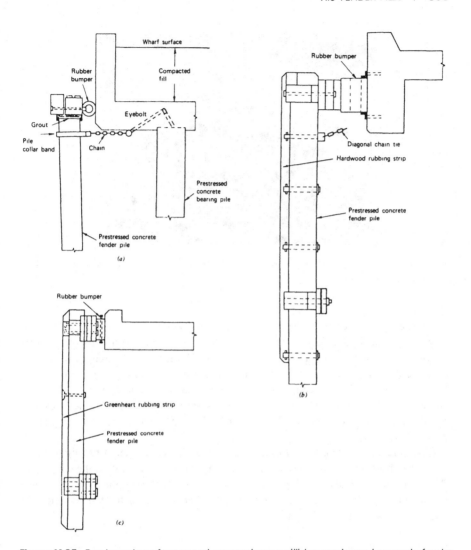

Figure 11.25. *Fender systems for general cargo wharves utilizing prestressed concrete fender piles: (a) Port of Los Angeles, California; (b) Port of Jurong, Singapore; (c) Port of Kuwait.*

the piles too stiff. The design should emphasize ultimate energy absorption. As regards durability, they can usually be economically replaced at intervals during the life of the structure itself, so they do not require as great a factor of safety as do foundation piles.

On most of the wharf fender piles which have been installed to date, it has been felt necessary to provide a timber rubbing strip so that the ship would not abrade and spall the concrete fender pile. Alternatively, timber camels or hanging timber pile butts have been used for this purpose. Since the energy absorption of the timber in this case is relatively small, its main purpose has been to prevent local spalling.

Figure 11.26. *Prestressed concrete fender piles at Port of Shuwaikh, Kuwait have given over 30 years of successful service.*

Another type of fender is the protective fender for bridge piers (Fig. 11.27). This fender is primarily designed to protect the pier: the prevention of damage to ships and the cost of repairs are a secondary concern. Here, ultimate strength plus durability are paramount. To increase the ultimate strength under excessive lateral load such as collision, unstressed strand or mild-steel bars should be added. Alternatively, the number of strands can be substantially increased, with the stress per strand reduced proportionally. Once again, a high percentage of confining spiral steel is required.

Fire resistance is of great importance in the protective fenders of bridges, especially steel bridges, where a fire in the fender might weaken and drop the steel span. This is a major advantage of prestressed concrete for this application. For these reasons prestressed concrete fender piles have been selected for the fenders of a number of major bridges in California.

11.7 ECONOMICAL EVALUATION OF PRESTRESSED CONCRETE PILES

The construction engineer is particularly concerned with the economical evaluation of alternative piling materials and systems because industry practice frequently offers or permits piling alternatives to the contractor or, otherwise, makes him responsible for their selection. The true economy of piles, as with other structural elements, is, of course, their ability to perform their assigned structural function, with adequate ultimate strength for overload conditions, for the period of the design life, and at minimum cost. Thus it is necessary to consider the structural behavior of the pile, the durability, and the costs of transporting, handling, driving, and connection, as well as the costs of manufacture.

Prestressed concrete piles have proven to be the most economic solution for a

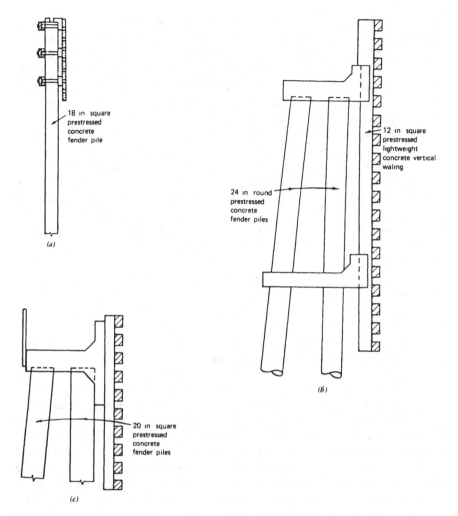

Figure 11.27. Bridge pier fenders utilizing prestressed concrete fender piles. (a) Eureka Slough Bridge; (b) Sacramento River Bridge; (c) Benecia-Martinez Bridge.

wide range of piling installations, especially when the inherent structural properties can be fully utilized, when durability or fire resistance are important, and when there is sufficient total volume in a geographical area to justify a proper manufacturing set-up and mobilization of proper driving equipment.

For marine structures the criteria usually include a life of 40 to 100 years, good column strength, high load-bearing capacity, strength in bending, ability to be handled and driven in long lengths, and low first cost. For all of these parameters, prestressed concrete piles excel and, thus, they have been widely adopted for marine installations throughout the world.

For foundations the usual criteria are low first cost, high capacity as a short column, and ability to be driven to the required penetration. In general, what is sought is the lowest cost per ton of carrying capacity. Since there are many types of piling available in the market for use as building foundations, a careful analysis should be made, comparing the load-carrying capacity of the pile, the soil capacity for the sizes and the penetration under consideration, the costs of furnishing, transporting, and driving, and the size, depth, and cost of footings. Time required for installation should also be evaluated.

These last two items, footings and time, give great weight to the choice of high-capacity piles. In concrete piles, high structure capacity can most economically be obtained by use of the highest-strength concrete available. In most localities, building codes establish a formula for computation of the maximum permissible bearing load, which is usually directly proportional to the concrete strength. However, high capacity is of no use if the pile cannot develop a corresponding capacity in the soil. Soil support is furnished either by side friction and adhesion, or by end bearing, or by a combination of both. Side friction–adhesion is primarily a function of the surface embedded in the bearing stratum. [In cohesionless soils of uniform density, such as beach sand, pile shape (e.g., taper) may also be a factor, as may surface roughness in some sands.] Prestressed piles are usually of constant cross-sectional area; therefore, they offer a large surface-friction area. Thus, for the great preponderance of friction piles, standard prestressed concrete piles of uniform cross section have the highest bearing capacity in the soil. In many soils, such constant cross section prestressed piles will develop the required bearing value at substantially less penetration than timber, steel, or tapered concrete piles.

For special cases in which friction may be very critical, for example, when it is desired to found the piles in a rather thin sand stratum between two clay layers, the friction support may be increased by special techniques. One of these consists of corrugations on the surface of the pile in the bearing zone. Special types of corrugations have been developed to maximize the frictional effect. Another technique consists of increasing the cross-sectional area of the pile through the bearing zone. This is much practiced in the Netherlands. This enlarged tip may be cast monolithically with the rest of the pile, or may be spliced to the column portion of the pile during manufacture. A third technique is to inject grout through an internal grout pipe, after installation. In this case, the pipe is usually also first used as an internal jet to aid in installation and to keep the pipe clear for grouting. This method has been successfully utilized to increase both ultimate bearing capacity and tension (uplift) resistance. The difficulty is in controlling the process, so that reliable values can be obtained. Experience in sands which are slightly permeable to grout shows an increase of 100% in capacity. Use of a cement grout with high penetrating ability is desirable; there are a number of admixtures made which reduce surface tension, and one process which uses colloidal mixing to aid flowability. At the present time this technique is primarily adapted to increasing the ultimate (overload) capacity in tension and bearing of piles which have already developed their design capacities by more conventional means.

End-bearing piles depend primarily on the size of the tip, for which constant-section prestressed piles are well suited, and on the ability to be seated into the material and mobilize the resistance of the soil. In soft rock, hard clay, etc., the soil capacity, as measured by deformation under load, is improved by the precompression from the pile being forced into it. Prestressed piles have demonstrated this ability of penetration and mobilization to a high degree, because the high-strength high-modulus concrete being held in a homogeneous (uncracked) condition by prestressing transmits energy efficiently to the soil.

The author has found it useful to prepare tables of cost-per-ton of load carried, evaluating piles of different types and capacities with prestressed piles of several different sizes and capacities. Such a table will quickly point up those factors which predominate. Above all, it will destroy the all-too-commonplace illusion that piles should be selected on the basis of the lowest first cost per foot.

Prestressed piles turn out to be very competitive in many such analyses. They must, of course, be compared with other forms of concrete piles. As compared with cast-in-place concrete piles, they offer high soil capacity, greater durability, better combined moment-load capacity, and usually, reduced footing size. As compared with conventionally reinforced (mild-steel) precast concrete piles, they offer major savings in column strength, durability, ability to be driven to deeper penetration, and savings in steel cost. Prestressed piles require about one-sixth as much steel weight as conventionally reinforced piles; since the unit cost of prestressing steel is double, the net cost is only $\frac{1}{6} \times 2 = \frac{1}{3}$. Due to this steel economy, and manufacturing economies made possible by mass production in modern plants (usually long-line pretensioning with multiple forms), the first cost of prestressed concrete piles is very favorable in comparison with other types of concrete piles.

For prestressed concrete sheet piles, the requirement for double-bending, that is, positive and negative moment-resisting capacity, offsets, to a large degree, the other economic advantages of prestressed concrete, as compared with steel. A recent study indicated that prestressed sheet piles of the Z or arch section are less costly than commercially available steel sheet piles of the same structural capacity, whereas the more conventional rectangular sheet piles are more costly. However, they are increasingly justified on the basis of resistance to corrosion and fatigue, i.e., lowest life-cycle cost.

Prestressed concrete fender piles are higher in first cost than treated timber piles, but give much longer service, especially when subject to abrasion, as from a floating camel. They may have a higher cost of installation. Their selection depends on an evaluation of their durability and performance. In evaluating durability, consideration should be given to the high cost of replacement of fender piles, timber or concrete, and the out-of-use cost of the wharf; on such a comparison, prestressed concrete will often be found very favorable.

11.8 PROBLEMS AND REMEDIES

11.8.1. General

Proper design, manufacturing techniques, and installation practice can be achieved only by a careful analysis of problems and failures. These represent an extremely small proportion (less than 1%) of the total number of prestressed piles which have been successfully installed. These known problems are presented as specific cases, with assigned cause, and corrective action taken.

11.8.2 Horizontal Cracking under Driving

This is the most common problem. Puffs or "dust" are seen about one-third of the length from the top, and horizontal cracks appear at $1\frac{1}{2}$-foot intervals, with considerable spalling at the surface. If driving continues, the concrete will grow sensibly hot just above the crack and will disintegrate in fatigue failure. Eventually, the tendons will break in brittle fracture.

Cause: *a.* Tensile rebound stresses from free-end condition at toe (soft driving).

b. Tensile rebound stresses from free-end condition at head when driving short piles through water or soft material to rock.

Cure: 1. Increase amount of soft-wood cushion.

2. Reduce height of stroke (fall) of ram, especially for first few blows.

3. Do not jet excessively below tip of pile during driving.

4. Increase weight of ram, while keeping the energy constant.

5. Predrill through overlying fill.

11.8.3 Inclined Cracking with Extensive Surface Spalling

Cause: *a.* Torsion plus rebound tensile stress.

Cure: Steps 1, 2, and 3 of Section 11.8.2, plus:

4. Make sure driving head cannot restrain pile from twisting.

5. Do not restrain pile in leads or template.

11.8.4 Apparent Breakage of Pile Tip Under Sustained Hard Driving

Cause: *a.* Fatigue under high amplitude–low cycle blows causes break-down of bond with aggregate and steel, especially when driving in water. Initial cracking may have occurred as described in 11.8.2.

Cure: 1. Reduce number of blows at very high resistance, where acceptable geotechnically.

2. Increase pile head cushion to reduce peak stresses.

3. Increase ram weight while keeping pile energy constant.

11.8.5 Vertical Splitting of Cylinder Piles or Hollow-Core Piles During Installation

Cause: *a.* Soft mud and water forced up inside to a level above outside levels, creating an internal hydrostatic head.

Cure: 1. Provide adequate vents in pile.

2. Increase spiral in piles which are not yet fabricated.

Cause: *b.* Hydraulic ram effect of hammer hitting water column.

Cure: 1. Do not drive pile head below water level or provide a special driving head with very large openings (greater than 60% of void area).

Cause: *c.* Shrinkage and thermal cracking in manufacture.

Cure: 1. Adequate water cure, especially for the first few days, starting immediately after conclusion of steam curing. Where water cure is not practicable, apply heavy coats of membrane curing compound at time of first exposure and again eight hours later. Insulate with blankets or tarpaulins.

2. Redesign mix to minimize shrinkage and heat of hydration.

3. Increase area of spiral.

Cause: *d.* Piles split while being filled with concrete, because of hydrostatic head of liquid concrete.

Cure: 1. Reduce rate of pour to ensure it will obtain initial set without exceeding tensile strength of pile concrete and also within acceptable strains of spiral, whichever may be lower.

2. Use precast concrete plug and grout annulus.

3. Increase area of spiral steel.

Cause: *e.* Thermal expansion of hydrating concrete fill due to hydration.

Cure: 1. Use grouted precast plug instead of fluid concrete.

2. Use low heat concrete mix.

3. Pre-cool concrete mix.

4. Increase spiral.

Cause: *f.* Vertical splitting at top during driving due to bursting stresses.

Cure: 1. Increase spiral at head or use a steel band at head.

2. Increase cushioning at head.

Cause: *g.* Vertical splitting at or above toe due to plug of soil wedging in void.

Cure: 1. Pre-jet or predrill to prevent plugs.

2. Jet or drill inside to break up plugs during driving.

3. Increase spiral at tip and/or band.

4. Use solid tip.

Cause: *h.* Jetting inside produces internal hydrostatic head in excess of external head.

Cure: 1. Provide vents in pile walls.

11.8.6 Vertical Cracking of Cylinder Piles after Installation

Cause: *a.* Freezing of water inside.

Cure: 1. Provide subsurface vent to allow circulation of water.

2. Place wood or "styrofoam" log inside.

3. Fill pile with frost-resistant material, e.g., coarse sand.

Cause: *b.* Piles cracked from logs, debris, etc., carried by river or boat impact.

Cure: 1. Fill with sand or concrete plug.

Cause: *c.* Thermal strains in service due to sudden drops in temperature.

Cure: 1. Provide jacket containing insulation around exposed pile surface.

2. Sheath exposed area by timber boards, banded on.

Cause: *d.* Delayed alkali-aggregate reaction.

Cure: 1. Install steel or concrete jacket and fill with grout, to confine original pile. Note: This cracking can extend underwater.

Cause: *e.* Corrosion of reinforcing steel.

Cure: 1. Cathodic protection. Underwater use sacrificial anodes.

2. Above water, use flame-sprayed zinc after first chipping off cover to expose reinforcing steel.

11.8.7 Excessive Spalling at Head Under Driving

Cause: *a.* Hammer impact on unconfined concrete.

Cure: 1. Chamfer head.

2. More cushioning.

3. More spiral or banding at head.

4. Make sure head is square and plane.

5. Make sure strand and other reinforcements do not project above the head of the pile.

6. Be sure driving head (helmet) does not fit too tightly.

7. Make sure driving head cannot become cocked and out of alignment with pile: it should ride in leads in axial alignment with hammer and pile.

11.8.8 Disintegration in Service of Corners of Prestressed Piles Below Water Line

Cause: *a.* Reactive or unsound aggregates.

Cure: 1. Jacket, with a steel or concrete sleeve and fill with grout.

Cause: *b.* Corrosion of spiral—see 11.8.6e.

11.8.9 Longitudinal Hairline Cracks Along Center of Faces of Solid Piles During Manufacture and Storage

Cause: *a.* Differential shrinkage of outside of pile plus differential cooling, especially in cold, dry, windy weather.

b. High radiant sun temperature during day, with rapid fall in temperature at night.

Cure: 1. Provide graduated cooling period during last phase of steam curing.

2. Cure in water immediately after removal from steam cure or insulate with blankets.

3. Additional spiral.

11.8.10 Transverse Cracks on Top Surface at Lifting Points

Cause: *a.* Negative moment in lifting exceeds concrete strength plus prestress.

Cure: 1. Provide mild steel at lifting points.

2. Use more lifting points.

3. Provide greater prestress.

11.8.11 Transverse Cracks Over Lower Support Points

Cause: *a.* With batter piles negative moment exceeds concrete tensile strength plus prestress, and is aggravated by rebound tensile stress during driving.

Cure: 1. Increase effective prestress.

2. Increase pile section modulus by increasing size. The pile section may be made rectangular to give greater section modulus.

3. Provide support for pile over a longer length (either above or below water or both), as by use of telescoping leads.

4. Make sure hammer is held at proper angle by slings or leads so that its weight does not induce additional bending on pile.

Cause: *b.* Excessive bending due to pile being held at three points: the driving head, in the soil, and at the support.

Cure: 1. Release gate and all rigid supports at bottom of leads as soon as pile has penetrated 5 to 10 meters into soil.

2. Follow pile head with leads and hammer so as to keep hammer aligned axially with pile. Note: Once the pile has penetrated 5 to 10 meters in the soil it is usually not possible to correct verticality, alignment, or position.

11.8.12 Breakage of Batter Piles After Release from Leads

Cause: *a.* Inadequate bending strength above point of support.

Cure: 1. Increase prestress.

2. Increase pile section.

3. Provide temporary support before releasing from leads.

11.8.13 Transverse Cracking During Manufacture, Apparent on Removal of Steam Hoods

Cause: *a.* Rapid cooling of concrete with forms restraining transfer of prestress.

Cure: 1. After transferring prestress, replace covers to retain heat until concrete has more tensile strength.

2. Make sure forms have no fins or other reentrant angles, for example, from cracks in forms at end gates. Provide soft rubber gasket where recesses or inserts are restraining movement.

11.8.14 Longitudinal Cracking at End of Curing Cycle, When Removing From Forms

Cause: *a.* Transverse thermal differentials: the outside is cooling faster than inside.

Cure: 1. Re-cover pile until it has cooled down further and gained more tensile strength.

2. Increased spiral.

11.8.15 Piles Drive Out of Alignment

Cause: *a.* Tip tends to wedge pile over.

Cure: 1. Change to square tip.

Cause: *b.* Excessive jetting on one side.

Cure: 1. Use internal jet.

2. Operate two free jet(s) on both sides equally.

Cause: *c.* Excessive driving where pile is essentially end bearing, causing pile to "walk." Upper sands may become liquefied.

Cure: 1. Reduce driving at high resistance, where geotechnically acceptable.

2. Provide steel stub to act as shear key in dense soil.

Cause: *d.* Uneven soil properties cause pile to deviate from alignment.

Cure: 1. Use square tip.

2. Pre-jet.

3. Predrill.

Cause: *e.* Driving piles through riprap and rock on or near surface causes pile to wander.

Cure: 1. Allow pile to seek its own path. Do not restrain. Otherwise piles will break. Note: "Spudding" before setting pile usually is unsuccessful as rocks fall into spudded hole as spud is withdrawn.

2. Provide for adjustment in pile cap and reinforcing to take care of out-of-tolerance results.

3. Drill, with casing or slurry through rock, then set pile.

REFERENCES

1. *FIP State of Art Report,* "Precast Concrete Piles," Structural Engineers Trading Organization, London, 1986.

2. Mehta, P. Kumar, *Concrete in the Marine Environment,* Elsevier Applied Science, London, 1991.

3. Marshall, A. L., *Marine Concrete,* Blackie, Glasgow; and Van Nostrand Reinhold, New York, 1990.

4. Naval Civil Engineering Laboratory, "Development of Prestressed Concrete Fender Piles—Preliminary Tests," NCEL 51-85-19, Naval Civil Engineering Laboratory, Port Hueneme, California, Sept. 1985.

5. Caltrans *Memo to Designers,* Interim MD 20-4, California Department of Transportation, Sacramento, California, 1992.

CHAPTER 12

Prestressed Concrete Bridges

12.1 INTRODUCTION

Many of the most notable successes of prestressed concrete have been in the construction of bridges. Its use offers advantages of low first cost, low maintenance, high durability, and attractive aesthetics.

Prestressed concrete has been successfully employed for bridges ranging from short spans to long spans, in all environmental conditions, from tropical to subarctic and desert to rain forest. It has been widely utilized in crowded city viaducts and for isolated structures in underdeveloped lands. Its concepts range from simple span slabs to cantilevered segmental construction to cable-stayed spans.

Prestressed concrete superstructures may be composed of precast or cast-in-place elements. When precast elements are used, the segments are joined with other elements for monolithic behavior by cast-in-place concrete or prestressing or both.

Prestressed concrete is also extensively utilized for pier shafts and columns and occasionally for abutments.

Prestressing is not limited to concrete alone: prestressing has been employed for long-span steel girder bridges and for composite steel-concrete superstructures.

Cable-stayed bridges involve many of the aspects and principles of prestressed concrete, such as high-strength steel tendons, and precast concrete segmental construction. However, they are not included in this book. The reader is referred to the excellent book in this same Wiley Series of Practical Construction Guides, by W. Podolny and J. P. Scalzi, *Construction of Cable-Stayed Bridges*.

12.2 BRIDGE SUBSTRUCTURES

12.2.1 General

Prestressed concrete has been extensively used for the substructure of bridges, both long and short span. Among the pier concepts which have made important use of prestressing are the following, listed together with the prestressed application most commonly employed.

1. Trestle-type bridges, with piles extending to just below deck level.
2. Piers constructed in cofferdams, with pile supports.
3. Major overwater spans, supported on large diameter cylinder piles.
4. Gravity-base caissons, seated on a prepared base.
5. Floating pontoons of post-tensioned concrete, both normal weight and light-weight.
6. Open caissons, sunk through overlying sediments to founding level, utilizing post-tensioned cutting edges and walls.

The above applications are partially described in the relevant chapters on "Piles" and "Posttensioning." However, certain additional aspects of special applicability to bridges will be addressed in this chapter.

12.2.2 Large Diameter Prestressed Concrete Cylinder Piles

These have been installed by combinations of jetting, dredging of the interior, weighting, vibration, and hammering. Bentonite injections may be used to reduce the skin friction in the upper zones. The dividing line between cylinder piles and caissons is one of degree only, as the two blend together under both structural and constructional considerations.

When high-capacity piles are employed for bridge piers, it becomes necessary to install them within a minimum tolerance to prevent eccentricity in the pier itself. Thus, there is a trend to the use of templates and guides in order to enable the accurate initial setting of the pile, or else drill the pile in.

Large diameter (3.5 meter) (11.5 feet) prestressed concrete piles 60 meters (200 feet) length were installed for the Oosterschelde Bridge in the Netherlands, through loose to dense sands (see Fig. 11.17). Techniques for the installation included internal excavation and weighting, the latter by an ingenious method of raising the crane barge by reaction against the pile. The pile was then plugged by grout-intruded aggregate concrete.

At Ju 'Aymah, in Saudi Arabia, a long bridge to an offshore terminal was constructed on cylinder piles up to 1.5 meters in diameter. These were installed by a combination of predrilling and driving with a heavy offshore hammer.

Similar piles have been installed by driving, accompanied by internal and external jetting.

For four of the five bridges of the King Fahd causeway, between Saudi Arabia and Bahrain, cylinder piles of 3.5-meter (11.5 feet) diameter were set in predrilled holes and grouted for fixity. The piles had a precast concrete plug that was grouted in place at the waterline to protect against boat or barge damage. Piles were extended up to support the superstructure (Fig. 12.1).

Splicing of prestressed concrete cylinder piles has been necessary when the lengths were excessive in relation to the water depth. One solution has been to embed matching steel rings in the top of the bottom section and bottom of the top section. These are made of thick steel plate, with the top one properly scarfed for welding. Care must be taken in installing the first section not to damage the ring: a protector is fitted over it and the first section is installed primarily by jetting.

Welding must be carried out slowly, with low heat, so as not to distort the ring nor spall the concrete.

The above splice system for cylinder piles is merely an enlarged version of the standard splices for hollow-core piles used in Japan.

The Dutch engineers have developed a prestressed splice, which gives full bending and axial capacities. If the two matching segments are match-cast during fabrication, the juncture can be made rapidly by the use of epoxy glue and short prestressing bars or couplers.

Jet pipes can be embedded in the wall of the cylinder piles, with discharge at the tip, if cutting action is desired. They can be arranged to discharge internally, if the purpose is to keep the inner core from plugging, or externally, to lubricate the pile wall and relieve the skin friction.

If desired, after the founding tip elevation is achieved, grout may then be injected through selected pipes to restore the skin friction and/or to consolidate material under the tip.

Figure 12.1. Prestressed concrete cylinder piles are extended up to support the superstructure of the bridges of the King Fahed Causeway, Saudi Arabia to Bahrain.

Splicing of jet pipes may be by screwed or drive fittings, but must be able to withstand the hydrostatic pressure without reliance on support from the concrete. Tapered sleeves, accurately machined, have been developed by the petroleum industry.

Splicing of pile sections when installed in deep sediments may be facilitated by first driving a casing, cleaning it out by jetting and air-lift, then setting the first section. The top section may then be set and the splice made. Then the whole pile may be driven and jetted to grade. Finally the casing is extracted.

12.3 PIER SHAFTS

The pier shafts for a number of bridges have been constructed of precast concrete segments, successively set on top of the foundation and the preceding segments. The joint has either been made by setting the subsequent section on a thin mortar bed, or, alternatively, by match-casting, with or without epoxy glue. The segments are then post-tensioned vertically (Fig. 12.2).

This latter requires use of specialized techniques.

In one system, short post-tensioning bars are set and the precast segment threaded over them by means of loosely hanging tubes. Then extension bars are coupled to the preceding ones and the process repeated. Care has to be taken that the screwing on of one extension doesn't tend to unscrew a lower coupler. The commercially available couplers are prone to this problem. Daubing the threads with epoxy before placing each coupler will prevent this.

A second system utilizes ground anchor hardware: after all segments are erected, a ground anchor with tendon is inserted, the anchor is grouted in place and, after hardening, the tendon is stressed.

Figure 12.2. Precast pier shaft segments are joined by vertical post-tensioning.

The third system embeds a U-tube in the foundation slab. The tube has a bell mouth on each upper end, so that when the tendons are later pushed or pulled down, they will enter the U-tube without catching.

As with all vertical ducts, care must be exercised to keep them covered, e.g., by plastic caps, at each stage, so as to prevent water entering (and subsequently freezing) and to ensure good grouting later. It also prevents debris from dropping in, an all too common real-life problem.

The special mix and procedures for grouting vertical ducts have been presented in Section 5.10.2.

12.4 CONSIDERATIONS AND CONSTRAINTS

Bridge superstructure spans, as long horizontal members designed to carry relatively heavy loads, are heavily prestressed as compared to most other structural elements. This prestress must be eccentric, to overcome the deadweight, and usually is designed to create a moderate upward camber, so that under design live load, the span will not have a sag. Hence shrinkage and creep are of great importance, not only because of the loss of prestress over time but because of the effects on the long-term camber.

Where the depth of the girder or slab is small, the prestress high, and the design live load infrequent, the camber may grow. In the more common case, where the span is relatively long, due to long-term losses of prestress, the camber diminishes.

For the longer bridge spans, shear becomes a major design consideration. It can be countered by inclined prestressing or, in very long spans, supplemental vertical prestress in the webs of the girders may be employed.

Because of the relatively high degree of prestress required for bridge spans, transverse tension in the end blocks becomes significant. These are the tensile stresses created between adjacent tendons or groups of tendons, tending to split the element. These are normally resisted by orthogonal grids of mild-steel reinforcing bars and/or both horizontal and vertical stirrups. These "between-tendon" stresses are in addition to the tensile stresses radiating out from the anchorage, which are usually resisted by spirals.

The result is typically a highly congested area of tendons, anchorages, spiral, bearing plates, two-dimensional stirrups, and face reinforcement (Fig. 4.6). It is difficult to detail these bars to prevent interference. It is difficult to fabricate the stirrups to the close tolerances required. It is even more difficult to place these multi-directional bars and to fit them in proper place while maintaining the prescribed cover. Finally, concrete placement and vibration is very difficult, so that in many places, external honeycomb and rock pockets occur, while in some cases, there is internal delamination or voids.

Several partial solutions have evolved. One is to ease the congestion by the use of short prestressing bars, so as to replace the many stirrups in one direction at least, by a few stressed bars. The other is to use mechanically headed bars instead of stirrups. Since these are not limited as to diameter for bending, a few larger bars may replace a large number of stirrups. Further, they can be accurately fabricated, so as to facilitate maintenance of tolerance.

Concrete mixes may be adjusted in the end block area so as to use smaller size coarse aggregate and to be more workable. For example, the coarse aggregate size may be reduced to 10 mm (⅜ inch), and the sand and cement contents increased slightly.

Internal vibration is essential, since external vibration is not powerful enough to reach through the maze of steel and thickness of concrete.

The practical problem with internal vibration is how to insert even a small diameter vibrator without encountering the steel.

Optimal locations may be marked on the forms, or short pipe mandrels preinserted. One good system for very deep girders is to preinsert a pipe mandrel, with a vibrator inside. As the concrete rises in the forms, the pipe mandrel is gradually withdrawn, with the vibrator following a short distance below.

Curved bridge girders are being increasingly used. Both pretensioning and post-tensioning systems of prestressing have been employed, the pretensioning using horizontally deflected strands. As opposed to vertically deflected tendons, whose upward force is resisted by gravity, the horizontal forces of the tendons must be resisted both locally (by stirrups) and globally, e.g., by the deck slab. Failure to recognize this and to provide for resistance to the radial forces has resulted in some serious failures during construction.

As indicated above, the upward force of deflected or eccentric tendons is usually resisted by gravity, with the corresponding downward force transferred to the piers. However, this is not always the case, especially during construction stages.

1. Precast concrete bridge girders during manufacture may be restrained from shortening by the forms or by the frictional component of their deadweight on the casting bed. Hence, the member as a whole is not yet pre-compressed and any large upward force may lead to bending cracks in the top.

2. Precast members during storage, transport, and erection may be supported at points removed from their final design supports.

3. On bridge construction by cantilever segmental or hammerhead systems, temporary overstress can result in excessive upward deflection of the extended arms, with consequent cracking.

In the three above examples, the results are usually limited to cracking. When properly calculated beforehand, using the prestress level (before long-term losses) and concrete strength at the age in question, the cracking may be acceptably limited by provision of additional mild-steel bars.

In the two following examples, however, the results can be catastrophic:

1. In one project, cast-in-place girders were cast on falsework supports which were designed to give uniform support. To provide for continuity of the completed bridge, with its multiple spans, the construction joints were located at the quarter

Figure 12.3. Precast girders being loaded onto rail cars. Note "hog-rodding" for support.

points, with the longer arm extending at the intermittent construction stage. As the post-tensioning was applied, the long arm cantilevered upward, raising off of the falsework except at the far end, which now had to carry about one-third of the total dead load of the span. This load, far in excess of the falsework capacity at that point, resulted in collapse, with loss of life.

2. When transporting or erecting precast prestressed concrete "I" or "T" girders, they may be tipped so that their web is no longer vertical. When the gravitational component of the dead load is reduced, the girder may "explode," as the concrete at the lower flange crushes. Thus during these stages, before the girders have been incorporated into the structure, they must be properly braced and supported (Fig. 12.3).

12.5 SUPERSTRUCTURES CONSTRUCTED OF PRECAST ELEMENTS

12.5.1 Precast Pretensioned Bridge Elements

These precast pretensioned girders have been standardized in the United States by the relevant user and industry associations such as the American Association of State Highway and Transportation Officials (AASHTO), the American Railway

Engineering Association (AREA), and the Precast Prestressed Concrete Institute (PCI). The standardized sections include I-girders, Tee-girders, bulb tees, double tees, channel sections, box girders, hollow-core slabs, and solid flat slabs. Other sections are also used where appropriate, e.g., M-sections, and for railway spans, through trough girders (Fig. 12.4a).

Precast pretensioned girders are generally employed for spans less than 35 meters (115 feet). When joined by a cast-in-place concrete deck, they act in composite action, as a highly efficient and economical bridge superstructure.

Newer sections, such as the bulb tee, have thinner webs, and employ higher strength concrete and shear reinforcement. As a result, they have extended the economical span range to 45 meters (150 feet) (Fig. 12.4b).

Where longer spans are required, this can still be accomplished with preten-

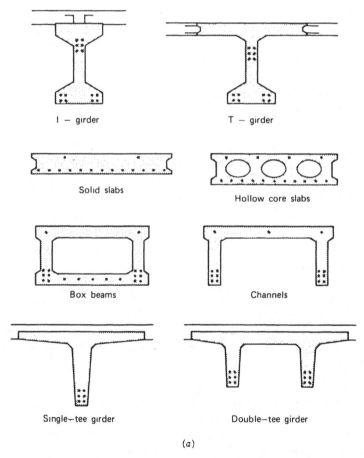

I – girder   T – girder

Solid slabs   Hollow core slabs

Box beams   Channels

Single–tee girder   Double–tee girder

(a)

Figure 12.4a. *Typical precast bridge girder sections for highway bridges (courtesy Prestressed Concrete Institute).*

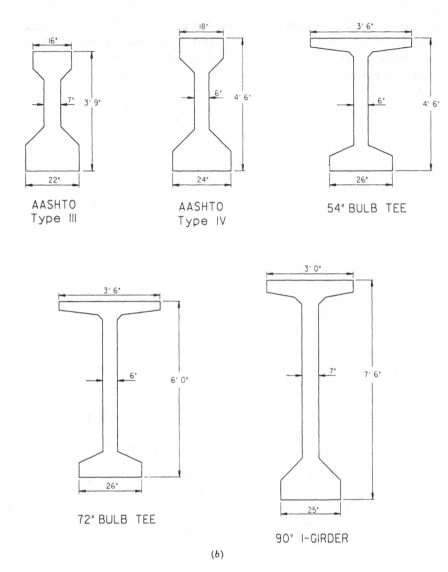

AASHTO
Type III

AASHTO
Type IV

54" BULB TEE

72" BULB TEE

90" I-GIRDER

(b)

Figure 12.4b. *Typical precast bridge girder sections for highway bridges. (Source: American Association of State Highway Officials, Washington D.C., and Prestressed Concrete Institute, Chicago, Illinois.)*

sioned members by use of the hammerhead-suspended span concept. Spans up to 60 meters (200 feet) have been attained by this combination.

Combinations of precast pretensioned segments of about 30 meters (100 feet) maximum length can be spliced by post-tensioning, for example at the inflection points, to extend the span range to over 50 meters (160 feet).

The manufacture of precast pretensioned girders follows the general procedure for manufacture as set forth in Chapter 6, with these important differences.

1. The density of prestressing tendons may be much greater, meaning that normal separation and clearances cannot be accommodated within the available section. In such cases, groups of two or even three tendons have been bundled together, separated at the ends only sufficiently to accommodate the grips for stressing.

2. Tendons are usually deflected at three or more locations along the girders (see Fig. 2.2). Typically some strands are run straight through while others are deflected in one or more groups.

3. As discussed earlier in this chapter, the tendon anchorages, together with the inserts for bearing plates and mild-steel bars and stirrups, create excessive congestion, making installation of the steel and the concreting exceptionally difficult.

It has been found especially valuable to first produce a full size mock-up of a typical end block before starting production casting. This mock-up may reveal spatial conflicts and may permit bundling or changes in the mild steel to facilitate placement and concreting. It is also excellent training for the workmen as they see the finished product from the mock-up, perhaps with defects such as honeycomb due to inadequate consolidation. The mock-up should be cut open to see if there are any laminations or excessive segregation.

4. With rare exceptions, precast pretensioned girders are designed for support at their ends only. Thus when picking (lifting), storing, transporting, or installing, the support points must be at or near the ends. Where it proves impracticable to achieve this, the support points may be moved inwards a short distance, provided that calculations show that cracking of the top will not occur as a result of the negative moment. Dynamic amplification must be added for picking (lifting) and transportation. It will often be found practicable to overcome small or moderate tension by the addition of mild-steel (passive) reinforcing bars at the top.

As noted earlier, tipping of I- and T-girders must be prevented by proper shoring and bracing.

12.5.2 Erecting Precast Girders

Precast girders may be erected by cranes operating from the land or cranes mounted on barges (derrick barges) (Fig. 12.5). They may also be erected by cranes or erection derricks mounted on the structure itself.

When using inclined slings, the temporary compression stresses in the girder must be considered, so as to insure against buckling. The lifting gear must be designed for the range of angles in which the slings will lead, both under static lift and when swinging in the air. Angles of force should be calculated for each position during the lift. Lifting loops must be suitable for all angles of lift, to prevent localized crushing or overstress.

Excessive swinging of the girder should be prevented and correct orientation

Figure 12.5. *Erecting prestressed concrete bridge girder.*

achieved by the use of tag lines. Tag lines can also be used to keep the girder from swinging into the boom.

Land cranes must have firm under-support, adequate for the concentrated temporary loads under their track, wheels, or outriggers. The position of the crane, angles of lift, and working radii must be plotted on working drawings and accurately laid out and enforced in the field.

When two cranes are used to erect a single member, each should have capacity to take at least 66% of the total load, and precautions should be taken to prevent undue swinging and side-pull on the booms, and to insure that the girder does not hit one of the booms during the successive steps of rotation of the booms.

Derricks or cranes mounted on the structure must be properly anchored down against uplift, and the temporary loadings imposed on the structure, including torsion, must be checked.

Care should be taken to prevent the boom from hitting part of the structure such as the cap beam, as this may cause the boom to buckle (Fig. 12.6).

Waterborne derrick barges should be checked for capacity during all stages of lift and placement, with due allowance for list due to load and wave action.

Also, the rotation of the revolving crane or derrick while the barge has a list puts added strain on the barge, the crane tub and roller path, and the swing engines. It

Figure 12.6. *Prestressed concrete girders being erected, Napa River Bridge, California.*

also produces torsion in the boom and added strain in the mooring lines. The list also increases the actual picking radius as the load drifts outward and, thus, may overload the crane. Therefore, before picking near-capacity loads in weight or reach, a thorough engineering check must be made for all phases of the pick.

Where reach or height make it impracticable to set a girder directly in position, it often may be possible to initially set it on the near edge of the cap beams, then slide it laterally on pads or rollers to its final position, then seat it by jacks.

This operation requires extreme care, not only to prevent sliding off the cap at one end but also to prevent tipping over of the girder. The latter may be prevented by a temporary bracing frame attached to the girder. Despite the obvious risk, this sideways sliding of girders has been successfully performed on a number of occasions.

As precast bridge girders are erected, they must be properly braced (Fig. 12.7).

12.5.2 Precast Posttensioned Bridge Superstructures

Some of the more spectacular recent bridge superstructures have been composed of precast members which have been posttensioned before or after erection.

One widely employed system embodies the sequential erection of precast segments by the method of progressive cantilevering (see Section 12.13). After each segment is erected and the joint has been made, it is post-tensioned back to the pier head.

In a second scheme, the precast segments may be erected in their final position on falsework or scaffolding. After adjusting for dead weight deflection to final

Figure 12.7. *Precast girders are temporarily braced prior to casting of concrete deck. Napa River Bridge, Vallejo, California.*

profile, joints are made and the span is post-tensioned, lifting the positive moment region off of the falsework. (See Section 12.9.)

In a variation of this, the precast segments may be assembled on a barge or other support, located immediately below the final position. They are jointed and post-tensioned. Then the entire span is lifted to position. The joints to the pier head are made and negative moment post-tensioning tendons run, locking the span in final position.

The King Fahed Causeway Bridges between Saudi Arabia and Bahrain utilized a system in which very large girders, representing all or a large portion of the entire span, are precast in a plant, then post-tensioned, transported, and erected by crane barges of several thousand tons capacity. (See Section 12.14.) Employed earlier in the USSR for simple spans of 100 meters length, the newer concept is based on the double cantilever scheme, which may be augmented by a suspended span between the ends.

This concept is now being used to erect 62 full width road girders and 62 double track railroad girders, each 110 meters long, on the Great Belt Western Bridge in Denmark and is being proposed for 250 meter long road girders on a bridge in eastern Canada.

12.6 CAST-IN-PLACE POSTTENSIONED CONCRETE SUPERSTRUCTURES

Cast-in-place posttensioned bridge construction methods are also extensively employed and are especially adapted to long-spans and complex and curved crossings typical of highway interchanges (Fig. 12.8). Cast-in-place methods have been extensively used with cantilever segmental construction, where successive segments are cast against the preceding member and prestressed to it before proceeding. Applying the current ability to control concrete quality and to progressively correct for minor deviations as the construction proceeds, long-span bridges have been brilliantly executed in this method.

The advantages of cast-in-place methods are:

- The ability to readily accommodate changes in alignment, section, and span length.
- The elimination of heavy transport and erection equipment and access for it.
- Ease of making adjustments and transitions.
- Construction joints that are generally more easily made.

Principal disadvantages are:

- Deformations as concrete weight is applied to supports.
- Thermal strains, shrinkage, and plastic deformation of young concrete.

Cast-in-place bridges are typically constructed on scaffolding (falsework). These may be box girders or tee girders. Box girders have especially favorable action for

Figure 12.8. Cast-in-place posttensioned box girder bridge under construction.

torsional resistance for bridges built on a curve. Attractive aesthetical results may be achieved by sloped webs (see Fig. 4.18).

12.7 JOINTING AND CONNECTIONS

Properly detailed and executed joints and connections are essential to the success of concrete bridge constructions.

Joints and connections must be so detailed as to meet the design requirements for transfer of bearing, shear, moment, etc., and, in addition, must be practicable of construction and inspection under the actual conditions at the site.

With stepped-end girders, as in cantilever-suspended span construction, high stresses develop due to the combination of direct bearing and anchorage stresses. Closely spaced stirrups are recommended and the tendon anchorage should be placed as low as possible.

Joints will generally be visible and thus will affect the appearance of the bridge. Well-designed and constructed joints may often be utilized and even emphasized, so as to enhance the appearance of the structure.

Consideration must be given to fatigue loading in all joint design, especially welded reinforcing steel splices and welded structural steel splices. Posttensioning tendon splices should have satisfactory behavior under cyclic (fatigue) loading as shown by manufacturer's tests and guarantees.

Joints in precast segments which are subject to a reversal of stress at any section under design loading should be detailed so as to provide adequate restraint against movement, or else the joint should be so detailed (with bearings, pads, or expansion joints) as to prevent hammering or other fatigue conditions.

Precast segments may frequently be combined with cast-in-place concrete to act in composite action as a monolithic structure. Provision must be made for horizontal shear transfer. The precast units may serve as forms for the cast-in-place concrete.

Consideration must be given to the various stages of loading to prevent over-stressing of the precast elements while the cast-in-place concrete is still plastic. The constructor must consider the effect of differential shrinkage and of different moduli of elasticity. See Section 4.4.

Transverse posttensioning may be beneficially employed in bridge decks to prevent sag, especially in cantilevered overhangs supporting a heavy curb (Fig. 12.9).

12.8 FLOATING-IN OF COMPLETE SPANS

Entire spans thereof are built or assembled on scaffolding on a barge, which is then towed to the site and moored in exact position. The span is then lowered onto the bearings.

Large single barges may be used, or multiple barges may be joined with trussing. The effect of hog and sag of the barge, due to waves, must be considered in its effect

Figure 12.9. *Both longitudinal and transverse prestressing being applied to long-span bridge deck.*

on the precast span. Wind forces on the barges and spans must be taken into consideration.

The deflection of the barge as the span is constructed on it needs to be countered by blocks and shims. Hence assembly of precast elements, with adjustment of profile before connecting, is preferable to casting the span directly on a floating barge. However, if the barge is first grounded on a prepared sand bed, then it will not deflect significantly as the span is cast.

Lowering may be carried out by utilizing the tides, flooding of compartments in the barges, or jacks. In flooding, the effect of the free surface on stability must be considered. This effect requires that the barges or pontoons be compartmentalized.

Stability must also be carefully calculated during transport on the barges because of the great weights involved and the height of center of gravity.

12.9 ERECTION OF PRECAST SEGMENTS ON FALSEWORK

A steel or aluminum truss may be placed in position and precast elements lifted one by one, for example, by crane, onto the falsework. When an entire span unit is erected, the precast units are jacked and shimmed to the exact profile, then joined and stressed. This method is especially adapted to the case of parallel girders, because after one girder is erected and stressed, the stressing automatically decentering the falsework, the falsework span may be moved sidewise for the next parallel girder.

Alternatively, the falsework truss span may be above the final girder location, the precast units being raised from barges into position by hoists, and held until jointed and stressed.

12.10 LAUNCHING GANTRY METHOD OF ERECTION

This method involves the use of a special erection or launching gantry, which may include means for moving itself forward as portions of the bridge are completed.

The launching gantry is typically a steel framed bridge of the cantilever-truss or cable-stayed truss type, which spans two bays. While a segment is being erected it is supported on a central pier, a forward pier, and a rear pier. After a span has been erected, the launching gantry slides itself forward, cantilevering over one pier, until it reaches the next one. It transfers lower legs onto this new pier, establishes a support by jacks, and the cycle recommences (Fig. 12.10).

Under this system, a precast segment is moved forward on rails or trucks, riding on the completed portion of the deck. The segment is then picked up by the launching gantry and carried forward. To enable it to pass through the supporting legs, it is usually rotated at right angles to its final position during movement, then turned back and set in its final position.

The individual precast segments are then jointed and stressed before the next segment is launched.

While the launching gantry has most often been employed to erect precast segments, it has also been extensively used with cast-in-place segments. The launching gantry typically supports the forms for two segments, one on each side of the central pier. Concrete is placed and, after gaining strength, the segments are stressed with post-tensioning tendons running from one new segment to the other, over the top of the central pier, to resist the negative moment (Fig. 12.11). When enough segments have been erected in the rear bay to meet the previous cantilever arm, the connection is made and positive moment continuity tendons are installed. Then the gantry can be launched forward for the next cycle.

When using a launching gantry to erect prestressed I-girders, provision must be made for lateral transfer of the girders after they have been moved into their span. Rollers, wheels on tracks, skidding with jacks, etc., are often employed to accomplish this lateral transfer. Positive stops must be provided to prevent the girder being moved beyond the end of the cap through accident.

Launching gantries are major steel bridges in themselves, subject to reversal of stress conditions as they are moved, and to impact as they handle the precast segments. Since the connections of the launching gantry are usually field bolted, it is important that provision be made for frequent inspection of all joints, and repair or strengthening of any members accidentally damaged. If high-strength bolts are used, a clear identification marking must be placed on them to prevent careless replacement by a conventional bolt.

Figure 12.10. *Launching gantry used for erection by means of cantilevered segmental construction. Bubaiyan Causeway, Kuwait. (Courtesy of FIP Notes and Buoygues.)*

Safe walkways and, where applicable, movable safety nets or platforms should be provided as an integral part of the launching gantry.

12.11 DIRECT LAUNCHING

This scheme has been employed to move precast girders lengthwise from a completed portion of the superstructure to their span location. A light steel or aluminum launching nose is overbalanced by a counterweight on the rear end of the girder.

Figure 12.11. *Launching gantry used to support cast-in-place prestressed cantilevered segmental construction of high bridge in Germany.*

Movement forward may be accomplished by jacking, rolling, tracked carriages, or cranes. The girder must be analyzed for temporary stress conditions as it is cantilevered forward and, if necessary, strengthened by external trussing or temporary post-tensioning. This method is particularly suitable for a single span in remote locations.

12.12 CANTILEVER—SUSPENDED SPAN CONCEPT

By this scheme, a prefabricated hammerhead (double-cantilever) section is installed on top of the pier and fixed to it, either permanently or temporarily. This can be done by floating in on two barges, which have been so strutted and lashed that they maintain their relative position during transport. The girder is supported on towers. After positioning over the pier, the barges are ballasted down.

Alternatively the hammerhead girder may be lifted at the center with one very high capacity crane barge and carried and set. On the Great Belt Western Bridge in Denmark, 124 hammerhead girders over 100 meters long, weighing almost 6000 tons each, are being set by a huge crane barge (Fig. 12.12).

Since the hammerhead girder is subject to maximum negative moment, its required final prestress force is very high. During this stage, before adjoining suspended spans are set, the tensile stresses in the bottom may exceed allowable limits. Stage stressing may be employed, or additional internal reinforcement provided in the bottom of the girders, or external structural steel beams bound to the segment.

Smaller hammerhead girders may be hoisted by one or two cranes lifting near the center.

Stability may be provided by stressing temporarily or permanently to the pier shaft or by an inclined leg support from the pier base or by falsework towers at one or both ends.

Figure 12.12. Giant Crane Barge lifts 6000-ton, 104-meter long, precast double-cantilever box girder for Western Bridge, Great Belt Project, Denmark.

Continuity of adjacent spans is attained by posttensioning across the mid-span joint.

Precast girders are particularly adaptable to suspended spans (Fig. 12.13). They may be lifted in with one or two cranes working from below or from the cantilevered ends of the superstructure, or they may be moved forward on a falsework truss or by launching gantry.

Suspended spans may be assembled from precast segments on barges, and then floated or lifted into place.

12.13 CANTILEVER SEGMENTAL CONSTRUCTION

This is a widely used useful method for construction of concrete bridges with precast or cast-in-place segments. As each segment is placed, it is jointed and stressed back to the completed portion of the superstructure. The sequence of erection is chosen to keep the partially completed superstructure balanced about a pier, in double-cantilever.

Since it is not feasible to lift or concrete the two segments exactly simultaneously, a step-by-step sequence is adopted, in which one segment is erected on one side, then one on the other.

This puts bending moments in the pier, for which it must be designed.

The unbalanced load is the load of the segment plus any construction equipment.

Figure 12.13. *Crane barge lifts 50-meter long suspended span for King Fahed Causeway Bridges between Saudi Arabia and Bahrain.*

This system may be employed with cast-in-place segments or precast segments.

With cast-in-place segments, each segment is directly cast against the preceding one (Fig. 12.14). The construction joint should be properly prepared by water jetting or other means.

Ducts have to be extended. These are spliced to the extensions of ducts from the preceding segments. Since this is a source of frequent difficulty, great care must be taken in splicing.

The ducts will be properly positioned at both ends but will tend to sag between. A mandrel of light steel pipe should be inserted in the extensions, until the concreting has been completed.

The forms for the cast-in-place extension should be set high at the forward end, to counter the expected deflection from the weight of the concrete.

Concrete placement should commence at the forward end and progress back to the joint with the previous segment (Fig. 12.15).

A schedule, often computerized, should be set up to monitor the deflection, lateral deviation, and twist, taking into account creep and modulus of the concrete at the various ages, loss of prestress, and temperature. Corrections can then be made progressively. Care must be taken to guard against over-correction, which would then result in an S-curve.

Figure 12.14. Cantilever segmental construction of cast-in-place long-span bridge girder. Columbia River Bridge I-205, Oregon.

Figure 12.15. Concreting the cast-in-place segment of a cantilevered long-span bridge.

Provision of extra unused ducts will permit installation of additional tendons should they prove necessary.

Ducts must be provided with vents at the high points and drains at the low points.

Precast segments are also employed in cantilever segmental construction (Fig. 12.16). These are match-cast segments, joined with epoxy glue. They enable very rapid completion of spans.

A segment is lifted (Fig. 12.16, 12.17) and pulled into fit against its mate as a check. The two are then separated and epoxy glue is applied to both faces. The segment is then pulled back into position and a temporary uniform prestress applied of about 0.3 MPa (50 psi) over the concrete cross section. Bars are usually the best system for applying this stress.

Epoxy is adversely affected by cold temperatures and water. Care must be taken in rainy weather that water doesn't drain on the deck to the joint while it is being made. "Hydrophobic" epoxies are available to enable application to damp, but not wet, surfaces.

Dowels can be used to help center a segment in exact alignment with the preceding one.

To seal each duct against cross-over to another at the joint during grouting, care must be taken to ensure that glue has been smeared all around each duct.

An even more positive system is to glue around each duct a 1- or 2-mm thick compressible seal or an O-ring, designed so they will allow tight contact of the concrete face.

Precast segments have one potential disadvantage as compared to cast-in-place segments and that is lack of continuity of conventional mild steel. This is especially a problem for box girders with extended flanges having no tendons in them. Dis-

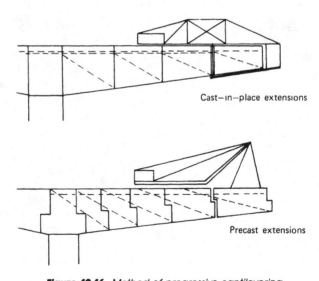

Cast–in–place extensions

Precast extensions

Figure 12.16. *Method of progressive cantilevering.*

Figure 12.17. *Precast cantilever segmental construction of approach spans. Columbia River Bridge, I-205 Oregon.*

placements and cracking may show up during construction or in service as a result of shear lag.

Provision of a duct in each flange, and insertion and grouting of a conventional bar can provide the needed continuity. Although this should be in the province of the designer, failure to provide it has caused excessive problems and costs for the contractor.

Temporary suspension of the cantilevered segments may be provided by external tendons, e.g., cables running on the deck or up to a temporary tower above the pier. Stability during erection of the cantilevered arms may be provided by temporary vertical stressing down to the pier, or by inclined legs or falsework towers.

12.14 SLIDING OF SEGMENTS

Precast segments may be slid forward to their position in the span, sliding on skids or rails or rollers over falsework trusses, or falsework girders. Similarly, they may be slid along temporary or permanent wire rope cables to their correct position in the span.

Special erection techniques, usually associated with building erection, are set forth in Chapter 10, Prestressed Concrete in Buildings. These include more detailed provisions for sliding and rolling in of girders, and provisions relating to the use of tower cranes, steel erection derricks, and helicopters.

12.15 INCREMENTAL LAUNCHING

In this scheme, a fabrication plant is set up at one or both ends of the bridge. A set of forms is mounted: these will remain in the same location and sequentially form segments of the bridge. Thus, the forms will usually be hydraulically or mechanically activated to ensure rapid stripping. A segment is formed, reinforcing steel and ducts placed and concreted. After stripping of the forms, tendons are inserted and stressed.

The segment is then jacked forward, sliding on Teflon which, in turn, is affixed to stainless steel plates on top of the piers.

A new segment is then constructed immediately adjoining the rear end of the first. Prestressing tendons are spliced, so the two segments are joined as one longer prestressed unit, and then jacked forward another length (Fig. 12.18).

The forward end of the bridge will be in cantilever. To aid in guiding it up over the next pier and in providing support, a light steel or aluminum truss is affixed to the forward end as a "nose."

The prestressed girder will thus be alternately in negative and positive moment. Thus, the prestress must be essentially uniform during launching. Once in final position, the prestress must be modified to conform to its design profile.

To handle these varying demands, three systems have been employed. In one, the final prestress is installed in internal ducts, and grouted. Offsetting prestress tendons are fixed externally to the concrete, inside or on top of the box, so as to result temporarily in uniform prestress. Once the girder is in final position, the temporary tendons may be cut.

The second system provides permanent uniform prestress in ducts at the time of manufacture. After the complete bridge has been launched and all spans are in their

Figure 12.18. *Jacks pull forward the girders successively in the incremental launching procedure of the entire bridge.*

final position, external tendons inside the box girder are stressed on an exaggerated profile, so that the combined result gives the design values. This is believed to give the most reliable and overall economical solution.

In the third system, external tendons, inside the box, are installed initially, located so as to give uniform prestress. Then when the girder is in final position, the tendons are jacked up and down and fixed on the final profile. This system, while economical of tendons, makes it difficult to ensure exact prestress values in each chord of the tendon.

Intermediate pier supports or towers may be used to give temporary support across longer spans, since the practical limit of span length for this type of construction is about 60 meters.

The coefficient of friction for such incrementally launched spans can be as low as 0.03, which means that the required jacking force is within practicable limits.

Bridges curved in plan can be constructed by this method as long as the curvature is constant. When launching from both ends of the bridge, different curvatures may be used, but this must be constant for each launching point. Similarly super-elevation must be constant.

When sliding on a curve over intermediate piers, especially when superelevation is involved, lateral guides may be necessary to keep the girder on course (Fig. 12.19).

After the launched bridge is in place, the bearing seats may be jacked up to remove the Teflon and the permanent bearings installed. Jacking points should have the required supplemental reinforcing to permit this application of force at a location offset from the permanent bearings.

Figure 12.19. *Teflon-coated shoes permit sliding while companion pieces guide lateral position.*

12.16 SCHEMES FOR BRIDGE CONSTRUCTION UTILIZING PRECAST SEGMENTS

Precast segments can be divided longitudinally, transversely, or horizontally. Transverse sections are generally, but not always, selected so that a fully stable cross section results. This may be the full transverse cross section of the bridge. Longitudinal sectioning results in girders, such as the solid or cored slab, channel, I, T, or box section (Figs. 12.20, 12.21), which may be joined in composite action by a cast-in-place composite deck. Alternatively, they may be jointed and stressed together transversely. Horizontal sectioning is generally used where precast flanges or deck sections are connected with cast-in-place webs, or conversely, where precast webs are joined to cast-in-place decks and bottom flanges.

Bridge concepts utilizing precast segments have usually been one of the following types:

1. Simple span.
2. Continuous girder.
3. Cantilever-suspended span.
4. Double cantilever.
5. Arch rib.
6. Flexible suspension.
7. External cantilever or rigid suspender type.

Figure 12.20. *Prestressed concrete T-girder will be joined by cast-in-place deck.*

Figure 12.21. *Pretensioned railway bridge girder for ballasted track.*

Trusses and framed construction are interesting solutions for moderate- and long-span bridges, being very light in weight. High-strength precast concrete web segments are joined with cast-in-place chords constructed by the cantilever segmental construction method. The precast segments are post-tensioned through the joints. Such bridges have been built successfully by Buoygues of France for viaducts in the Middle East and Europe. External tendons arranged to provide continuity are placed inside the truss (Fig. 12.22).

Precast deck sections may be assembled in such a way that the main longitudinal and transverse stressing is achieved by tendons located in the joints. The precast sections are first assembled on falsework or, in some cases, on cables. The tendons are then placed through the zone of the joint, the joint is concreted, and the entire assembly stressed.

12.17 ADDITIONAL REMARKS CONCERNING CAST-IN-PLACE PRESTRESSED CONCRETE BRIDGES

Cast-in-place prestressed concrete bridges are extensively utilized for medium- and long-span bridges, especially when the bridge alignment must incorporate horizontal curves. The cross sections most widely employed are T-girders, with the web and lower flange usually of the same width, box girders of rectangular cross section, and box girders of trapezoidal cross section. The latter is extremely efficient in its use of materials and lends itself to rather elegant and aesthetically pleasing solutions.

Most medium-span cast-in-place concrete bridges are constructed on falsework

Figure 12.22. Prestressed concrete trusses form graceful spans for the Sylhan Viaduct, France.

scaffolding (Fig. 12.23). Continuity is almost always employed for bridges greater than one span. Usually, such a bridge is made continuous for four spans or more. Because of maintenance problems and special costs associated with expansion joints, current practice tends to stretch out the length for continuity as far as possible, even around horizontal curves. In construction, therefore, a carefully planned sequence of tendon stressing must be followed, with the use of splices and overlapping tendons as detailed on the plans. Such a bridge lends itself to a sequential operation where, with careful planning, the prestressing crews may be utilized at a relatively constant level of work force over a substantial period of construction. The effect of expansion and contraction during construction must be considered for each stage (Fig. 12.24).

The problems in cast-in-place prestressed bridges are usually associated with shrinkage, creep, settlement of scaffolding, and differential moduli of elasticity. Shrinkage can be controlled by careful selection of the mix, a low W/C ratio (including use of a water-reducing admixture), and special attention to curing. Sometimes, to prevent transverse shrinkage cracks, a few tendons, perhaps 2 or 4, are initially stressed and grouted at an early age, such as 2 or 3 days. Then more may be stressed and grouted at 7 days, and full prestress imparted at 14 to 28 days. It is better to fully stress and grout a few tendons rather than partially stress all, due to the susceptibility of partially stressed tendons to corrosion.

Shrinkage can also cause horizontal cracks in the deep webs of girders. There is

Figure 12.23. Cast-in-place prestressed box girder bridge is supported on high falsework during construction.

always a tendency of the lower concrete to consolidate and draw away from the upper concrete, and cracks will usually form at a discontinuity boundary, such as a tendon duct. Low W/C ratio, thorough consolidation by vibration, and proper curing are the constructor's answers. Use of numerous small vertical reinforcing bars (stirrups) is the designer's answer. Very deep webs have been prestressed by vertical or inclined tendons.

Differential moduli of elasticity are usually due to variation in the W/C ratio and in the age at time of stressing. One solution used to overcome this and also to offset minor (elastic) deformations in the scaffolding is to provide 5 to 10% more tendon steel area than required and to then stress it to a point that raises the concrete spans to the required profile, using more or less prestress (within the limits provided in the design).

With long continuous spans and a complicated profile, friction losses during stressing may prove erratic. Use of a heavier steel duct will reduce the coefficients

Figure 12.24. *The Alsea River Bridge in the coast of Oregon utilizes a combination of segmental and cast-in-place construction, prestressed over five to six spans to provide continuity.*

of friction and wobble. Such tendons should be stressed from both ends. Additional water-soluble oil can be applied to the tendon as it is installed to minimize friction; it is then flushed out with the grouting.

The trend is to use ever more concentrated prestressing forces, thus putting greater loads on the anchorages. This results in a highly congested area for proper concreting and consolidation, and so in a number of cases, the anchorage block has been precast, then set in the forms and concreted into the cast-in-place concrete. Obviously, external tendons reduce this problem. A combination of 35% internal and 65% external appears optimal.

For medium- and long-span bridges, an overhead gantry may be used to support forms and each cast-in-place segment until it is cured and stressed.

With continuous cantilevering, all movements (elastic deflection, creep, shrinkage, and thermal) must be constantly computed for each step, then properly countered by varying the prestressed force within limits provided in the design. The profile must be constantly monitored, in order to prevent cumulative movements.

When tendon ducts come close together in the vertical plane, such as at points of maximum moment, attention must be given to the possibility of one duct squashing into the adjacent duct when the tendon is stressed. With round ducts, especially rigid ducts, the concrete which encases them is usually sufficient to hold them in shape, especially if the tendons occupy a large portion of the duct cross section. With very large ducts and a very sharp bend, the duct material should be made thicker, e.g., standard water pipe should be used at these locations (Fig. 12.25).

Figure 12.25. *Concentrations of ducts for tendons in negative moment areas over piers make proper concreting difficult.*

12.18 EXTERNAL TENDONS

External tendons are those which lie outside the main concrete members (webs and flanges), either inside a box girder or outside the primary girders (Fig. 12.26). They are typically deflected by saddles (deviators) placed at the points of high moments. External tendons also give the opportunity to strengthen bridges for higher loads and to replace tendons if they are damaged or deteriorated.

Being outside the concrete, they require independent corrosion protection. Typically this is provided by a polyethylene sleeve, through which the tendons are run and then grouted. Where the tendon passes a saddle, a curved steel casing is provided so as to ensure a proper length of contact and avoid "fretting" abrasion of the tendons. Minimum radius of the steel sleeve should be 3 meters (10 feet). The polyethylene is either sleeved completely through the curved sleeve or casing or is spliced to it. The steel sleeve should have slightly belled ends to prevent cutting of the polyethylene. The polyethylene must be hard enough to prevent the wires of the tendon from cutting into it while being stressed. The steel can be connected to the HDPE sleeves by welding, socketing with resin, or using thermal shrink sleeves. It is a very critical detail.

Saddles are members which transmit the vertical force to the webs of the girders. Since the tendons shorten during stressing and because of creep, etc., longitudinal forces may also be introduced which cause local bending of the saddles. Thus the design of saddles is three-dimensional. Adequate steel must be provided on all three

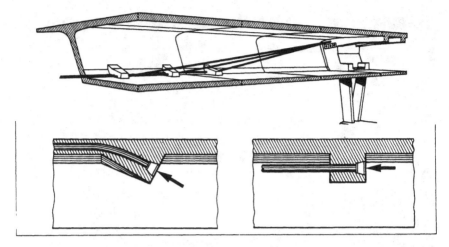

Figure 12.26. *External tendons are placed inside the box of a trapezoidal box girder. (From Post-Tensioning Manual, Courtesy of Post-Tensioning Institute)*

axes. Saddles or deviators should be spaced not further than 7 to 8 meters (23 to 25 feet) to reduce vibration.

The anchorages for external tendons are located on the end blocks. These again are very complex structural elements which also have to resist bursting during stressing, so the concrete must be thoroughly consolidated and compacted. Space is very constricted and adequate room must be provided for jacking. A 3-D layout of anchorages, jacking points, and deviators is recommended.

Because of shear lag and the need to provide proper crack control under overload conditions, a portion of the total prestress should be distributed to internal tendons, that is, tendons bonded to the concrete for the full length. A typical portion is 30 to 35%.

Other solutions include the provision of a few well-distributed ducts, through which unstressed tendons can be run and grouted, thus providing crack control as well as additional ultimate strength. This is especially desirable where precast segments are employed, as otherwise there will be no continuity of reinforcement across the joints of the segments.

One of the newest solutions for external tendons is to use individual greased-and-plastic-sheathed tendons, bundled into a group of 19 or so, and threaded through a larger polyethylene and/or steel duct. After stressing, they are grouted in the conventional manner. This reduces friction and gives double corrosion protection. Most importantly, it prevents "fretting" of tendons in the saddles or at deviators.

12.19 CONCRETING IN CONGESTED REGIONS

The concrete under the anchor blocks must be well-consolidated, as stresses are very high, yet it is at these locations where most honeycomb and rock pockets appear, due to the difficulty of placing and consolidating concrete through the congested reinforcement.

Therefore special means should be adopted. These may include:

1. Provisions for placing concrete through side windows.
2. Provision for external vibration.
3. Special arrangement for internal vibration. This is difficult because of the embedded reinforcing bars. Some of the means adopted are:
   a. Marking position of re-bars on the forms before placing concrete.
   b. Providing entry windows for internal vibrators.
   c. Setting pipe casings through the reinforcement before concreting. The poker vibrator is then inserted, the concrete placement begins, and the casing is gradually withdrawn.
4. Design of special concrete mixes for the end block regions, e.g., use of 10 mm = $\frac{3}{8}$ inch maximum coarse aggregate, and 50% sand, together with plasticizing admixtures.

12.20 CRITICAL AREAS FOR POSTTENSIONED BRIDGE CONSTRUCTION

12.20.1 Ducts (Sheaths)

For straight tendons, such as transverse tendons of deck slabs, ducts of high-density polyethylene or other plastic are often employed, as well as flexible steel ducting. The fiberglass (FRG) protects the tendons against chloride penetration as the decks are salted. Rectangular corrugated steel ducts are often used for the transverse tendons in order to reduce their thickness.

For longitudinal tendons, however, with curvature, only steel semirigid ducting should be employed. When the tendons are pulled through FRG or plastic pipe, they cut through the corrugations, resulting in leakage of grout. This is especially prevalent when the tendons are on a curved profile.

The steel ducting must have sufficient rigidity and wall thickness so as not to have excessive sag between supports ("wobble"). While in storage, it should be protected from corrosion. When flexible ducting is employed, a mandrel, such as thin walled conduit, should be inserted and left in place during concreting. This will prevent "wobble" and ensure against grout in leakage and duct blockage. The mandrel should be withdrawn shortly after initial set of the concrete.

Duct splices are an especial problem, particularly in segmental bridge construction where the splices of several ducts occur at the same location.

The ends of the duct should be cut by saw rather than by burning torch, and all burrs removed. Otherwise the end of the tendon will catch during insertion and as the pulling or pushing force is intensified, this catching will result in balling up the duct and a complete blockage.

Overlapping sleeves, sealed with waterproof tape, are currently widely used but have not proven 100% reliable. The result of deficient splices of ducts is cross-over when grouting the tendons.

One solution is to delay grouting until all tendons have been stressed. However, this exposes the tendons in an ungrouted condition for long periods of time, which is undesirable, especially in areas of coastal or industrial environment.

Heat-shrink tape appears to be much more reliable and is gradually replacing the standard tape.

Screwed couplings are excellent, but of course require thicker walled pipe.

Considering the costs of remedying duct blockage, it would appear that positive solutions, such as the screwed couplings or heat-shrink tape, are justified both from quality and economical viewpoints.

In cast-in-place segmental bridge construction, the rear end of the new duct length is supported by the extending piece of the previous duct. The forward end is supported by a stop bulkhead. In between, while there may be intermediate ties, the duct tends to sag. This results in a series of minor but significant "peaks" in the profile at each joint. Stressing of the tendons will not only encounter more friction but will exert a local downward force, tending to cause delamination of the concrete at the joints. Use of rigid mandrels through the ducts of the new segment will prevent this, and should be mandatory.

Similarly with match-cast precast segments, mandrels should be used to ensure not only proper location of ducts but also the continuation of their profile through both the old and newly cast segment.

Duct blockage can be caused by a number of problems. The catching of the tendon end on a rough end of a duct splice has already been mentioned. Another source is deep denting of the duct, for example, by supporting a thin duct on cross-bars of reinforcing. When the concrete is placed, the weight causes the bars to dent deeply into the duct.

With vertical construction joints, duct ends may be left protruding. These form a convenient "ladder" for workers, resulting in denting and distortion.

With horizontal construction joints and in slip-forming, the duct openings are exposed. These can be accidentally filled with aggregate or can become the receptacle for trash. On the Ninian Central Platform, two blockages were found to be Coca Cola bottles. Ducts should always be capped with plastic caps before delivery. This is good practice for all ducts, whether horizontal or vertical.

Duct blockage can also be caused by in-leakage of cement paste during concreting, e.g., where a duct splice sleeve has been dislodged. When a duct is grouted under pressure, any leakage at splices may communicate to an adjoining duct and block it. This is most prevalent at splices where two ducts are in close proximity.

Where tendons are coupled, an enlarged duct is required. This must be long enough and positioned correctly to ensure that the coupler can move without restriction as the tendon is stressed.

The transition between an enlarged duct and a standard duct should be tapered, so as to permit the grout to flow and completely fill the void.

Where ducts are concentrated, as for example, in the negative moment zones over the piers, special care must be taken in sizing, duct wall thickness, placement, and sequence of stressing and grouting, so as to prevent the tendons from "pulling through" from an upper duct to a lower one.

A concentration of ducts with little or no clear spacing between them can result in delamination. Normally the tensile capacity of the concrete is sufficient to transmit compressive and tensile stresses across the section, but when large ducts are placed side by side, they remove a major portion of the section. Delamination may result.

On the I-205 Columbia River Bridge, in Oregon, it was found necessary to install stirrups in the deck slab at relatively close spacing, so as to tie the upper and lower portions of the deck slab together.

Similarly, when many tendons are grouped, so that there is a high prestressing force on a section, the section's capacity should be checked for the net cross-sectional area, excluding the area occupied by the ducts.

12.20.2 Tendon Curvature and Anchorages

Where tendons are curved, they of course exert a local transverse force. Thus in horizontally curved web, when the tendons are stressed, they tend to burst out sideways. This actually occurred on a box girder bridge having a relatively short radius of curvature.

It also occurs when tendons are turned out of webs or slabs for intermediate anchorages. These anchors are typically located in bolsters (blisters) which protrude from the main member. The radius of curvature is usually quite sharp even though the length of the curved section is small. This results in very complex local stresses, where radial stresses combine with anchorage stresses, all trying to split the bolster from the member. Stirrups are used to tie the bolster to the member, but are often not correctly installed because of practical construction reasons.

These are very congested areas: longitudinal reinforcement in member and bolster, anchorage plates and trumpets, through all of which the stirrups must be placed and anchored. The critical stirrups are usually those nearest the anchorage, yet these are also the ones most difficult to force into position.

Mechanically headed bars may be found more practicable in many instances.

When longitudinal tendons anchor at intermediate locations along the girder, as for example, with continuity tendons, there is an abrupt change in stresses and strains. The stressing of the tendon tends to shorten the concrete section ahead, but also tends to pull away from the section behind. The result is tension behind the anchorage, i.e., in the zone away from that to which the tendon leads.

Thus in the zone ahead of the anchorage there are lateral tensile bursting stresses. Between adjacent tendons there are transverse splitting stresses. Behind the tendon, away from the stressed zone, are direct tensile stresses.

On the I-205 bridge, where ten high capacity anchors were anchored at one location, the anchorage zone ruptured, with a tensile failure behind the anchor, shear along the sides, and compression in front. The whole block suddenly moved 2 meters.

This phenomenon can be averted by anchoring such tendons in staggered positions. Behind each anchor, conventional reinforcing can be placed to transmit about

50% of that tendon's force. Obviously, these bars must be developed, i.e., bonded, in the compressed zone ahead of the anchor.

12.20.3 Construction Joints in Segmental Construction

These joints are the focus of many of the practical problems which have occurred. These problems can be overcome but require special efforts.

With precast segments, the problems are often inaccurate fit, nontight joints, and nonbonding of epoxy, where glued joints are used.

Inaccurate fit is intended to be overcome by match-casting each segment against its partner. It is of course important that no significant deformations or warping take place during curing. With steam curing, for example, the two segments, the old and the new, should be designed to minimize shrinkage: in particular, the W/C ratio should be kept low. Both the mold segment and the newly cast segment should be steam cured together, so as to have equal thermal response.

Alignment changes at joints are sometimes intentionally created during erection, to correct camber and lateral deviations. Usually stainless steel wire mesh is used for shims, correcting about 1 mm at each joint.

Poorly glued joints are often caused when the epoxy is improperly mixed or when the erection is done in the rain and water contaminates the epoxy. The epoxy should be hydrophobic, i.e., able to bond even when the surface is damp.

In any event, any openings or gaps must be sealed, as for example, by sealing with epoxy putty and then injection of epoxy.

Shear connections of segments are usually formed as keys. These keys are subject to high local stresses, especially when formed with polygonal joints. Multiple curved keys are best. In any event they should be reinforced near their extremity, to prevent cracks from developing. Also, if polygonal keys are used, a diagonal bar should be placed across each reentrant corner.

12.20.4 Construction on Falsework

Many short- and moderate-span concrete bridges are constructed on falsework, just as nonprestressed bridges are. There are, however, significant differences which relate to the stressing.

First, as pointed out by Prof. Fritz Leonhardt many years ago in his famous "Ten Commandments of Prestressing," there must be provision and ability for the girder to shorten when prestressed. The falsework must accommodate slippage or shortening without failing.

Ignoring this has led to collapse of the falsework, which, depending on the stages of construction, has, on occasion, also led to collapse of the girder.

Conversely, if the falsework is too rigid, there will be no prestress in the girder until the falsework is removed.

The second principle is that prestressing will raise a portion of the girder off the falsework. In New Zealand a fatal collapse of an extended girder occurred when the prestressing operation transferred its deadweight to the two ends, raising it off the falsework in the center. The result was to overload the falsework at the further extremity, leading to its collapse and falling of the girder.

12.20.5 Stages of Construction

Similarly, the erection of precast girders must consider all stages of construction, especially when the bridge is to be completed by cast-in-place jointing and composite construction. Half-depth cap girders may have to have additional reinforcing steel at their mid-depth. If half-depth girders are prestressed, they may have to be stressed in stages to prevent overstressing.

The important principle is to draw and check each stage of construction and erection. Most failures occur because of combining two or more steps in the programmed sequence.

An illustration of the latter was erection of a precast segmental bridge in Riyadh, Saudi Arabia. The principal problem was that during one stage of the complex erection procedure, there was temporarily a barely sufficient amount of prestressing tendons to carry the deadweight of the new segment.

When the temperature dropped at night, the combination of thermal strains and shear lag led to propagation of a crack completely across the member, with consequent collapse.

Other problems which have occurred during erection and construction have had to do with the sequence or erection of balancing cantilevers; the failure to recognize that the segments on both arms cannot be concreted simultaneously. Another frequent error is failure to include the weight of the construction equipment at the most extended location for that stage.

12.20.6 Control of Camber and Deflection

Precise calculations can be made, especially with the current availability of computer programs, for the camber and deflection control of prestressed girders. However, the validity of these calculations depends on the material properties which are input.

The long-term creep of concrete depends on numerous factors such as stress level and gradient, aggregate properties, W/C-ratio, strength, shrinkage, etc. Prediction of these values on the basis of short-time tests is always subject to miscalculation. In some cases, the laboratory values have been rendered inappropriate by changes in material sources or concreting practice in the field. For example, in one case, the creep and shrinkage of lightweight concrete were determined in the laboratory at 70% relative humidity while the site had a normal relative humidity of only 20%. This led to excessive loss of prestress and abnormal sag of the central span.

Usually such differences can be caught by careful field measurements during construction, but long-term variances will only show up at later dates.

Since usually the problem is one of sag due to excessive creep, it is advisable to include some spare ducts in which tendons can be inserted and stressed as required. These spare ducts will also prove useful in the event of duct blockage. An allowance of ducts to accommodate 5% additional tendon force is usual. A minimum of two ducts, one on each side, should be provided.

12.20.7 Closures of Cantilevered Girders

When prestressed bridge girders are cantilevered out from the piers, as they typically are with cantilever-segmental construction, or with the double cantilever concept, the long extended arms are subject to both long-term and short-term movements. A gap is therefore left between opposing arms and a waiting period of several months is left before closure, in order to allow the majority of creep and shrinkage to take place.

Prior to closure, these arms will move up and down with temperature changes and sideways with wind. When it comes time for closure, the two arms should be strutted and locked together, so as to prevent relative movement while the closure concrete is gaining strength.

Usually, heavy steel beams are secured over the deck and under the box, with rigid connections and diagonal braces.

After gaining strength, but before release of the struts and braces, the continuity (positive moment) tendons are stressed through the joint.

12.20.8 Expansion Joints

When expansion joints occur within a span, they usually will be located near the quarter point or, in any event, offset from midspan. This means that during construction, one arm must be progressed further than midspan.

When the expansion joint is only a short distance off center, it may be practicable to cantilever the additional distance by adding temporary prestress.

When the expansion joint is near the quarter point, it may be best to temporarily lock the joint segments together and erect the assembly as a typical monolithic segment. After final closure, the joint can be de-stressed and freed.

The temporary locking is usually made by installation of blocking and diagonal post-tensioning running downward across the joint.

12.20.9 Horizontally Curved Precast Concrete Girders

Longitudinal precast girders, reinforced by mild steel or by prestressed tendons, can be fabricated on a horizontal curve. Bracing must be employed to ensure torsional stability during erection and until permanently secured in the structure. Both box and T sections have been employed—the wider flanges facilitate this type of curvature.

With transverse segments, horizontal curvature may be achieved by varying the joint width progressively. For dry joints, the change in curvature can be accomplished in manufacture by adjusting the already cast segment to its new angle before pouring the adjoining segment.

Precast girders are relatively easy to erect when they are of moderate length and depth. When they become long and deep, stability can be critical, especially when on a curve or super elevation. Then they must be braced against overturning. Usually cross-bracing at the ends is sufficient, but for very long precast girders (over 30 meters), bracing may also be required at the ends and at mid-span.

This is a very difficult problem for the first girder, for which special frames and support may be required, especially with curved girders. In Germany, long curved precast girders were erected in pairs, braced together.

With cast-in-place box girders on a horizontal curve, the tendons will typically have both horizontal and vertical curvature. The tendons are typically located in the webs.

While it can theoretically be shown that, globally, the post-tensioning does not tend to cause the girder to "straighten out" nor change curvature, locally, within the web cross section, there is a radial force tending to pull the tendon out of the concrete. This must be resisted by closely spaced stirrups so as to mobilize the entire concrete cross section.

Tolerances on the placement of the tendons in the lateral position in the webs are very critical. While the proper performance is, of course, the province of the designer, the constructor may find it prudent to run an independent check to verify that sufficient web thickness and adequate stirrups have been provided.

12.21 INTEGRATION OF DESIGN AND CONSTRUCTION

The erection of long-span concrete bridges involves a full understanding of all factors involved and analysis of stress conditions at all stages from manufacture through transport, erection, concreting, and final design load conditions. Thus, regardless of how achieved, design and construction must be integrated. The designer must follow through all construction stages, and the constructor must be aware of the effect of each of his operations on the final results as they affect the design. Whether the designer and the constructor are in different organizations, as in conventional American practice, or in the same organization, as in European practice, there must be full collaboration if the best results are to be attained.

The most efficient use of precast elements can often be obtained by combining precast elements with cast-in-place concrete to work in composite action as a monolithic construction. By such an approach, the advantages of precasting can be employed in complex and variable-shaped structures. Such a combination approach requires consideration of the different moduli of elasticity, and the effect of differential shrinkage and creep, and analysis must be made of each stage and condition; thus, the designer is inherently involved and concerned with the constructor's methods.

The ingenuity of both design and construction engineers has been responsible for the development of outstanding concrete bridge concepts, in which prestressed concrete elements were utilized to their full advantage. This ingenuity has demonstrated itself in the matters of design, manufacture, transport, erection, and jointing.

Since erection concepts will often determine the size and configuration of the segments and since erection is primarily a responsibility of the constructor, he should be permitted as much freedom as practicable in selection of the size of segment within the general restrictions of the design. The importance of encouraging ingenuity on the part of the constructor should be given consideration during the preparation of plans, specifications, and criteria. This may be facilitated by the designer conferring with local fabricators and contractors during the design stage to discuss availability of erection equipment, special techniques, and cost comparisons. The thorough cooperation and coordination of both the designer and the constructor are essential if the maximum advantages are to be achieved.

REFERENCES

1. Podolny, W. and Muller, J., *Construction and Design of Prestressed Concrete Segmental Bridges*, Wiley-Interscience, New York, 1982.
2. *FIP Guides to Good Practice,* "Recommendations for Segmental Construction in Prestressed Concrete," Federation Internationale de la Precontrainte, 1978, London.
3. AFPC, "La Precontrainte Exterieure," *Annales de L'Institute Technique du Batiment et Des Travaux Publics,* Paris, 1991.
4. ACI Publication, "External Prestressing," American Concrete Institute, Detroit, Michigan, 1991.
5. Leonhardt, F., *Prestressed Concrete, Design and Construction,* 2nd edition, Wilhelm Ernst & Sohn, Berlin, 1964.
6. AASHTO, *Guide Specifications for Design and Construction of Segmental Bridges,* American Association of State Highway and Transportation Officials, Washington D.C., 1989.

CHAPTER 13

Prestressed Concrete Marine Structures

13.1 INTRODUCTION

Marine structures rank among the foremost applications of prestressed concrete. It was early recognized as the optimum material for harbor and coastal structures because it combined durability, strength, and economy. This chapter will discuss various aspects of this application. There is an obvious interrelationship with three other chapters, Prestressed Concrete Piling (Chapter 11), Prestressed Concrete Floating Structures (Chapter 15), and Bottom-Founded Concrete Sea Structures: Gravity-Base Offshore Platforms and Terminals (Chapter 14). Therefore, this chapter must be read in conjunction with the others for the full presentation of marine applications of prestressed concrete.

Prestressed concrete began to be generally applied to wharf and pier construction about 1955 and, within the short span of ten years, had gained acceptance for use in harbors throughout the world. It has been adopted as the principal material and technique for the harbor development of such diverse ports as Oakland, San Francisco, Los Angeles, Long Beach, Portland, Seattle, Galveston, Charleston, and Norfolk in the United States, and overseas, in Kuwait, Singapore, Indonesia, Malaya, South America, England, and Norway. The climatic environments have ranged in temperature from tropic to arctic, and the piles have been exposed to warm seawater, salt fog, and freeze-thaw seawater conditions.

Among the many types of marine structures for which prestressed concrete has been adopted are:

1. Container and general cargo wharves (Figs. 13.1, 13.2).
2. Petroleum terminals.
3. Overwater airport runways and taxiways.

Figure 13.1. Pretensioned precast haunched deck slabs span between pile cap girders to construct container terminal at Tacoma, Washington.

Figure 13.2. General cargo wharf at Jurong, Singapore, uses pretensioned half-depth deck slabs made monolithic by cast-in-place concrete deck.

4. Offshore platforms.
5. Trestle roadways and pipeways.
6. Bridge piers.
7. Navigation structures.
8. Dolphins.
9. Operational and protective fender systems.
10. Bulkheads and seawalls.
11. Coastal jetties.
12. Small boat harbors.
13. Groins.

Special features pertaining to these are discussed later in this chapter.

The complete structure for the preceding applications may consist of the following prestressed concrete elements:

1. Piles (see Chapter 11):
    a. Bearing piles.
    b. Batter (raker) piles (Fig. 13.3).
    c. Cylinder piles.
    d. Fender piles.

Figure 13.3. Fifty-meter (160 feet) long prestressed concrete batter piles support wharf in San Francisco, California.

    e. Sheet piles.

    f. Soldier piles.

  2. Wharves and quays:

    a. Half-depth deck slabs (for composite action).

    b. Pile cap girders.

    c. Marginal beams.

    d. Deck beams and girders.

    e. Firewalls.

  3. Bulkheads:

    a. Sheet piles.

    b. Curved slabs.

    c. Flat slabs.

    d. Ties.

    e. Struts.

  4. Miscellaneous:

    a. Craneway girders.

    b. Pipeway beams.

    c. Trestle approach spans.

    d. Catwalks.

  5. Substructures of bridges and offshore platforms:

    a. Piles.

    b. Piling, including large diameter cylinder piles.

    c. Skirt slabs.

    d. Gravity-base caissons.

    e. Pier shafts and columns.

  6. Intake and discharge pipelines and structures (see Chapter 14).

  7. Drydocks, floating docks, and dock gates (see Chapter 15).

  8. Offshore platforms (see Chapter 14).

  9. Concrete floating structures (see Chapter 15).

13.2 PREFABRICATION

Marine construction implies over-water construction, with all its inherent problems. Provision of temporary support is difficult; transport of and access for workers and materials is expensive and time-consuming. These challenges, frequently combined with saltwater splash below and hot drying (or freezing) conditions above, make it extremely difficult to obtain high-quality concrete possessing both strength and durability. Many marine structures are located in areas remote from the optimum location for construction. Local materials may be of marginal quality. Premium wages are exacted in some locations. Labor may be in short supply at other sites.

Figure 13.4. *The overwater extension of the La Guardia Airport in New York utilized precast pretensioned elements which were made monolithic by cast-in-place concrete and biaxial posttensioning.*

All of the above factors emphasize the value of prefabrication and precasting. Precasting is usually most efficient and effective when it employs prestressed reinforcement. Moreover, the connection of precast units so they will act monolithically can often be best done by posttensioning (Fig. 13.4).

Prestressing has another major advantage for marine structures, not recognized until recently, and that is the resistance to fatigue under cyclic loading from waves, and, in the case of driven piles, of cyclic stress reversals from piledriving.

13.3 DURABILITY

This feature has been a dominant reason for the selection of prestressed concrete for marine exposures. Chapter 4 treats the subject of durability extensively, so only special considerations relative to marine structures are examined here.

Prestressed concrete fully immersed in seawater is completely saturated. Oxygen contents are relatively low; therefore, most types of corrosion of reinforcement are minimized. On the other hand, chemical attack on the concrete is maximized. If the aggregates lack in soundness, their deterioration will be accelerated. Reactivity between the alkali in cement and reactive aggregates will be increased. The disassociation of NaCl leads not only to free ions of chlorine but also of sodium, a highly active alkali. With cements whose chemical constituents have been improperly selected, the cement constituents may be subject to replacement by softer compounds, leading to severe loss of strength. The incorporation of cement of extremely low C3A content may lead to early corrosion of the reinforcement. Therefore, for the underwater zone, particular care must be given to:

1. Soundness of the aggregates under sodium-sulfate tests.
2. Positive control of alkali-aggregate reactivity.
3. Use of a Portland cement, such as ASTM Type II having moderate C3A.
4. Low W/C ratio.
5. Inclusion of PFA or microsilica to maximize durability and impermeability.
6. Curing. Curing of prefabricated members will usually be by steam cure at atmospheric pressure, followed by a supplemental water cure and a drying out period. Properly carried out, this results in an optimal hydration of the concrete. For members cast-in-place, while water cure is probably the best, it is difficult to ensure thorough, complete, and continuous water cure, especially in windy and cold regions. Hence membrane cure is often the most practicable method of achieving satisfactory cure.

With dense concrete, internal curing will continue to take place after the termination of the water cure. Drying of the surface will improve the durability in seawater.

Near the sand line, concrete may be subjected to abrasion from moving sand and gravel. Abrasion can be partially countered by providing a dense concrete, of as high a quality as possible; that is, high cement factor, low W/C ratio, and inclusion of microsilica in the mix. Sharp corners should be avoided.

Aggregates should be selected for hardness and resistance to abrasive loss in the rattler test. It may be advisable to increase the cover in a zone of known abrasion.

Cavitation may occur in a surf zone. Cavitation is due to the collapse of bubbles of water vapor. It is very difficult to give complete protection against cavitation; a hard troweled finish will usually prevent serious cavitation loss. In extreme cases, or when repairs are required, an epoxy coating is recommended. Avoid sharp re-entrant discontinuities.

Attack of marine borers on concrete, such as mollusks and sea urchins, is usually of no consequence when a normally strong, dense concrete is employed. An exception is when limestone aggregates are used in certain extreme environments. Mollusks in the Arabian Gulf are able to penetrate a thin skin of cement paste and then bore into limestone aggregates, to a depth of at least 60 mm ($2\frac{1}{2}$ inches).

Above low water, in the so-called splash zone, lies the region of greatest exposure. In addition to the seawater itself, there is ample oxygen for corrosion of the reinforcement. Chlorides may be concentrated by evaporation in permeable concrete and lead to salt-cell electrolytic corrosion. This zone is also subjected to freezing and thawing, salt scaling, and in some cases, to ice abrasion.

Emphasis must be placed in this zone on:

- Low W/C ratio.
- High cement factor.

- Partial replacement of cement by PFA.
- Inclusion of 3 to 4% microsilica.
- Thorough consolidation of concrete.
- Good curing.
- Adequate cover over reinforcement.
- In freeze-thaw exposures, use of air-entrainment.

In the zone above the splash zone, often called the "atmospheric zone," (e.g., the decks of wharves) the concrete is still subjected to moist salt air (from spray or fog). The water evaporates from the surface, leaving the salts, which gradually become concentrated to an even higher percentage than the seawater itself. The most predominant form of attack is salt-cell electrolytic corrosion of the reinforcement. The concrete, therefore, must be designed for impermeability. Freeze-thaw attack must also be considered where applicable, especially on horizontal surfaces where ponding may intensify the deterioration.

Coatings and jackets may be applied in special cases, to seal the surface and to protect it. It should be emphasized, however, that prestressed concrete has been extensively and successfully employed in marine exposures in tropical, desert, temperate, and arctic climates with no coatings or jackets whatsoever.

The simplest and most common form of coating has been bitumastic paint, applied cold. As such, it serves primarily as a sealant. Hot asphalt coatings give a thicker and more continuous protection and have proven effective in completely sealing concrete from chloride penetration.

Epoxy coatings have been used in recent years, some applied to precast concrete while still at the manufacturing yard, others applied in place, using underwater-setting epoxies. Epoxy coats were applied to the cylinder piles of the King Fahd Causeway Bridges and have given seven years protection without allowing any chloride penetration. However, they do degrade with time under the ultraviolet rays and have to be renewed.

Water-based epoxy coatings have been very successful for over 13 years exposure on the shafts of offshore platforms in the Norwegian Sector of the North Sea, where they have resisted both waves and radiation.

Wood lagging has been applied for many years in the tidal range and splash zone in Western Sweden, to insulate it from freezing. The more recent trend, however, is to eliminate it and rely on the use of high-quality prestressed concrete with up to 8% air-entrainment.

Steel plates and wrought-iron jackets have also been applied in the past in areas of extreme ice abrasion. Where practicable, natural stone facing has been used similarly. The high cost of facing bridge piers with granite could be substantially reduced by preparing precast prestressed panels with embedded facing, then setting these panels up as side forms for the concrete, rather than using the stone-by-stone method now commonly employed. Considerable success has been had on dam facings by using precast panels of extremely high-quality dense concrete, again set up as forms and tied into the main body of the concrete pier. Some of these panels have been prestressed. This would seem to offer the added advantages of freedom from shrinkage, thermal, and handling cracks, the assurance of concrete continuously under compression, the ease of handling, and improvement of the tensile capacity of the entire structure.

Cylindrical jackets of precast concrete, sometimes prestressed, have been extensively used in the past, set around or over the previously constructed concrete. These have then been secured by grouting, either with tremie grout or by sealing, pumping, and grouting in the dry.

Unfortunately, conventional concrete jackets have sometimes proven less durable than the concrete they were supposed to protect. This is because jackets by nature are relatively thin. To place the concrete in such sections has required more workability, hence a higher W/C ratio, and more permeable concrete. Cover over the reinforcement may also be reduced.

The present trend, therefore, is to eliminate jackets, putting primary reliance in the quality of the basic concrete, and providing more cover in zones of high exposure.

13.4 REPAIRS

Where marine concrete structures have suffered corrosion, disintegration, or damage, considerable study must be given to the proper method of repair.

Localized damage, for example, from impact, is most easily repaired. If the damage consists of cracks, they may be injected with epoxy. The work is done by a diver, using a hydrophobic epoxy, that is, one which repels water and hence can bond and cure underwater. As with above-water crack injection, the crack is first sealed on the surface by epoxy. Holes are drilled on about 600-mm (24-inch) spacing. Starting with the lowest hole, epoxy is injected until it flows out the one above. Because of water dilution and contamination, it should be forced to flow out longer than is common with above-water cracks, until its color and consistency is the same as that being injected.

If the damage involves surface spalling, then the area should be thoroughly cleaned by water jet, assisted by wire brush. A thick coat of underwater-setting epoxy can then be troweled on. Piles supporting the bridge across the Columbia River at Astoria, Oregon, which had been damaged and scraped by a barge broken loose in a storm, were repaired by this method. The repair has reportedly performed well for over 25 years.

Prestressed concrete piles, especially hollow-core and cylinder piles, have suffered cracks during installation and service. The majority of these have been vertical cracks, as a result of circumferential stresses due to thermal strains or hydrostatic or hydrodynamic over-pressure on the inside, accompanied by an inadequate amount of spiral reinforcement.

Repairs have occasionally been carried out by epoxy injection, especially for solid piles, but that does not cure problems caused by thermal strains. For these, which are the majority of the cases, an FRG sleeve can be clamped around the pile and filled with epoxy containing tiny glass balls, so as to provide insulation. Such repairs have reportedly been very effective for cracked piles on the Atlantic coast of the United States and Canada.

For repairs which are not subject to major thermal changes, coatings of dense (solvent-free) epoxy or polyurethane can seal the surface against air penetration and thus restrict the rate of inflow of oxygen to the cathodic reinforcement. This will slow down the rate of corrosion but cannot stop it. The primary problem is that water vapor bubbles tend to form under the coating and burst out. Some proprietary coatings have the bond strength and elongation characteristics to prevent this.

Cathodic protection can be applied, either in the form of sacrificial anodes or impressed current.

Sacrificial anodes are best for the underwater portions, although the anodes must be installed or hung so as to "see" all surfaces, since the electrons travel in essentially straight lines. In sub-tropical areas, flame-sprayed zinc has performed well.

Impressed current cathodic protection is the only positive means of arresting corrosion above the water line. It requires the preparation of a smooth surface, by filling all the "bug" holes, bleed water holes, and water vapor blisters. Then the conductive layer, embodying a conductive grid, is placed. Finally a protective layer, usually of shotcrete, is installed. Obviously this is specialized work, time consuming and costly, but it can be fully effective in completely arresting the electrochemical corrosion. The problems with this method are both high costs and exposure of the protective layer to physical damage and thermal cracking.

Since prestressed concrete members usually contain conventional passive reinforcement near the external faces, corrosion often occurs earliest on this conventional reinforcing steel.

When this occurs in marine structures, repairs are best carried out by chipping out the spalled and surrounding concrete with a small chipping hammer, so as not to cause microcracks in the surrounding concrete. The concrete should be cut deeply enough to enable cleaning the corroded bars all around, back side as well as front. The steel should then be brushed and blasted to bright metal, then coated with a zinc-enriched epoxy. The author likes to clamp a thin bracelet of zinc at each end of the bar where it enters the adjoining concrete.

Then the surface of the hole is coated with bonding epoxy and the cavity is filled with fine concrete (10 mm or $\frac{3}{8}$ inch maximum size aggregate), having a low W/C ratio. Shrinkage gaps at the top are prevented by the use of a temporary "window box" for concreting.

For relatively small underwater repairs, an underwater setting cement may be used. For prestressed concrete this should not contain calcium chloride. There are several such products available on the market which have been used successfully in repairs. Underwater epoxies have been similarly employed.

Where concrete disintegration and/or extensive corrosion of reinforcement is occurring, the first step is to determine the cause and extent. In a few notorious cases (e.g., the original Hayward–San Mateo Bridge and an iron-ore loading pier in South America) the process had proceeded so far that permanent restoration was impracticable. In the first case, all of the concrete was permeable; it was in the splash zone, and salt cells had formed throughout the structure. Localized repairs consisted of chipping, replacement of reinforcing steel, encasement in shotcrete, and bitumastic coating. However, the anodes and cathodes just shifted and the electrolytic process continued.

In the case of the South American pier, the disintegration was below water and was occasioned by a combination of unsound aggregates, apparent alkali reactivity of the aggregates, and high alkali cement. Once again, attempts to apply coatings in localized areas were unable to stop the progressive disintegration within.

In an example apparently similar to the bridge, a wharf deck was found to be suffering salt-cell corrosion of the reinforcement. Extensive repairs were made by cutting out bad concrete, keying surfaces, replacing corroded steel, rebuilding the members by shotcrete, and coating the entire pier substructure with hot asphalt. Since the pier was covered, the concrete itself was basically dry, and the complete coating sealed off the electrolyte (water) from reentering. The repairs were completely successful. Another factor was the much better quality of the original concrete; thus, salt-cell formation had not penetrated so deeply.

For further discussion of repairs, see Chapter 25.

13.5 INCREASED SPAN LENGTHS

This has an important implication for marine structures, where live loadings are increasing all the time, and where economy depends so heavily on reducing the number of pile-to-cap-to-slab connections. Prestressing extends the efficient structural length of slabs and girders. This, in turn, permits the use of larger, more heavily loaded prestressed piles which, in turn, have greater bending strength, thus permitting the use of all vertical piles, which not only is less costly but ensures ductile performance in earthquakes. Post-tensioning has been used very effectively to provide for heavy live loads as well as to tie the whole structure together.

13.6 ECONOMY

Prestressed concrete has resulted in substantial economies in marine structures. These economies are due to two causes:

1. Greater structural efficiency (economies in design), as demonstrated by the following:
   a. More heavily loaded piles due to the ability to manufacture and install them to greater depths.
   b. Better column strength for deeper water.
   c. Greater lateral bending strength.
   d. Ease in handling and driving.
   e. Longer spans of deck girders, beams, and slabs.
   f. Better connections of precast elements for monolithic action.

g. Integration of structural function so entire structure acts to reduce lateral, vertical, and torsional loads.

2. Economies in production, arising from the following:

a. Prefabrication, especially precasting.

b. Plant manufacture—savings in mass production.

c. Standardization and modular design.

d. Economy of prestressing steel in relation to conventional mild reinforcing steel.

e. Ability to use higher strength concrete.

13.7 EFFECTIVE UTILIZATION

13.7.1 Wharves, Piers, and Pile-Supported Quays

13.7.1.1 GENERAL. A typical marine wharf is comprised of piles, both vertical and batter (raking), cap girders, deck slabs, and marginal beams. Certain wharves may have additional elements, such as pipeway beams, firewalls, catwalks, craneway girders, etc. Fender systems are employed to absorb the impact of berthing vessels, to minimize damage to ship and wharf, and to prevent vessels of low freeboard and debris from getting in under the wharf deck.

For efficient structural performance, the entire wharf should function as a unit to resist vertical and lateral loads.

Vertical loads arise from cargo, cranes, trucks, railroad cars, storage, snow, and people. Lateral loads generally arise from ship-mooring forces, waves and currents, earth pressures, including the lateral earth pressure due to surcharge behind the wharf, wind on moored vessels and buildings, and seismic forces on the wharf, the live load, and the earth slope. These lateral forces may cause torsion in the structure as a whole. A special loading condition may be produced by tsunamis and should be considered in those harbor locations where tsunami effects have occurred in the past.

A wharf deck, if properly tied together, can act as a huge girder in the horizontal plane, supported against buckling by the numerous piles. In considering the individual elements comprising a wharf structure, it must be emphasized that they usually are designed to perform as part of a complete structural system.

13.7.1.2 BEARING PILES. The special consideration pertaining to piles under wharves are that:

1. They usually involve long unsupported lengths, hence require consideration of column behavior.

2. They are exposed in the splash zone, where greatest attention must be given to durability.

3. Batter (or raker) piles must carry bending stresses as well as axial tension and compression.

4. Earthquakes or ship collisions may impose curvatures well beyond the elastic range.

In driving on a slope, it will usually be found best to set the pile a foot or so inboard of its final location, as it will tend to "walk" downhill during driving.

In driving through old riprap, etc., it has been shown best not to restrain the pile horizontally during driving. The pile should be held at the top only, while the tip tends to find its way through the obstructions, by displacing them laterally and downward. (Any attempt to provide a third point of restraint will cause high bending stresses and possibly break the pile.) This recommended procedure has been used satisfactorily in installing thousands of piles through old rock seawalls, and other unconsolidated rock fills, with a minimum amount of pile damage. However, this procedure will result in greater deviations in pile location, requiring that the pile cap be made wider, so as to permit a tolerance of the order of 300 mm (12 inches) or so.

Whenever piles are driven in a marine structure, some deviation in head location is inevitable. Common practice in the past has been to pull pile heads into true locations. This, unless properly guided by engineering analysis, has resulted in many broken and cracked piles. The cracks are usually below the mud line, at the point of fixity, thus unseen, but they have reduced the pile bending capacity significantly. On the other hand, the blanket prohibition against all pulling of heads into position may be unrealistic. The factors involved are the depth of water, soil properties, particularly near the surface, stiffness of the pile, and the pile's axial and confining reenforcement.

Enforcement of restrictions on pile head realignment is very difficult. The author has successfully used applied force as a reasonable solution which can be enforced. Pulling the pile heads is permitted only by use of a hand-operated jack, which is so set that the force applied cannot exceed the specified value, which has been tabulated for each range of water depth after consideration of all factors.

It is interesting to observe the flexibility of the typical piles in the outboard rows and hence deeper water, as they are displaced by the tidal current. In deep water, or when currents are swift or wave forces severe, it will be necessary to stay piles as soon as they are driven to prevent excessive deflection and breakage.

For marine structures, pile heads should always be notched, as with a saw cut, before cutting them off with pneumatic tools. This zone is under transverse tensile stress from the prestressing, so, unless care is taken, cracking and microcracking can extend below the point of cut-off. This, in turn, may lead to accelerated corrosion. For the same reason, it is desirable to embed the pile in the deck slab a few inches, to the extent that it is structurally feasible. Piles should never be cut by explosives or large jack hammers. Similarly, the drilling of holes near the head of piles in order to provide temporary support for the deck should not be allowed.

13.7.1.3 BATTER PILES (RAKERS). These are extensively utilized in wharf structures to resist lateral forces. During installation, these piles undergo reversals of stress, being cantilevered over a support during setting, then becoming a continuous beam during driving, and finally ending up as a simply supported beam in final position. These conditions must be analyzed. In deep water, with flat batters, it may be necessary to adopt one or more of the following steps during setting:

1. Set the pile on the bottom in a vertical position, then lean back to a support above water level.
2. Provide underwater support, such as below-water leads.
3. Hold in slings on specified batter, and set until tip rests on bottom and top rests on support.
4. Provide an above-water frame or leads to hold top of pile during installation.

Many batter piles, set properly, have been severely cracked or broken during driving by the weight of the hammer being placed on the overhanging pile head. The hammer must be constantly supported, either in inclined leads, or hanging in a bridle at the correct angle.

Batter piles in deep water, although properly installed, have been damaged or broken during the interval when they are released from the leads and support has not yet been transferred to the previously built structure. This is always a complex procedure that requires careful planning. Fortunately, the force required to support the pile head on release is generally very small, and can usually be provided by timber struts and/or wire lines to adjoining structures.

Batter piles are generally designed to function in both tension and compression with an adjoining vertical or opposing batter pile. The transfer of tension loads from a pile cap into a pile is usually accomplished by mild-steel bars, grouted into the pile in drilled holes to a depth sufficient to develop the full value of the bars in the fully prestressed portion of the pile, that is, below the transfer zone, which is usually 50 cm (20 inches) long. Thus, embedments of 1 meter (3 or 4 feet) or more are common. Drilling of such deep holes after installation is difficult and may damage the pile. Therefore, it is better practice either to form these holes in manufacture or to embed the mild-steel dowels during manufacture. In both cases there must be sufficient excess length to take care of driving tolerances.

Embedment of the extending prestressing tendons, (strands) from the pile into the cap, is adequate to develop strength. However, because of the small steel area, this may permit excessive rotation of the pile head. With wide wharves this may not be critical; in such a case the strand embedment is usually the least expensive and generally more satisfactory solution. Strand should be embedded a minimum of 600 mm (24 inches) into the pile cap.

Tension connections can also be made by use of a stressed tendon such as an alloy bar, anchored into the pile and stressed to the pile cap. This is a very effective structural means, but not extensively employed at present due to higher cost. It has many potential advantages, particularly with precast cap girders.

Typically prestressed concrete pile design has closely spaced spiral confinement extending only a short distance down from the pile head as manufactured. If the pile is then cut off, the spiral below that level is often unable to properly confine the concrete in the critical region of transfer. The proper design should therefore extend this closely spaced spiral so as to accommodate the potential cut-off and to confine the dowels and strand in this critical transfer zone. This is also the region where lateral displacement of the wharf under docking forces, and earth pressures, may lead to concentrated bending and shear.

Tension piles usually develop their tension in the soil by skin friction on the bottom portion of the pile. When driven into soft rock, such as shale or glacial till, steel stubs on the tips of the piles have effectively developed high pull-out values. Here also the embedded portion of the stub must be heavily confined by spiral. In Norway, steel dowels are similarly used.

Another scheme, particularly useful when driving onto hard rock, is to anchor the pile by prestressing. A pile with a hollow core or tube is driven to bearing and framed into the superstructure; in some cases, even the deck may be poured, leaving a hole formed over the pile. Then a drill, working off the deck or framing, drills through the pile and into the rock to form a socket. A prestressing tendon is then inserted, grouted in the rock, and stressed from the top, then grouted.

In recent years, as a result of the experience in earthquakes, the use of batter piles is being avoided in new construction because the resultant structure does not possess adequate ductility. Intersecting batter piles form a rigid point which initially provides the entire lateral support. The result is that the piles are often crushed or subjected to excessive imposed curvature. If the pile doesn't fail, it may cause the deck to fail in punching shear. It is today considered better practice to design the wharf structure as a ductile frame. Under such a design philosophy, the piles in the least water depth, i.e., those at the top of the slope, will resist the lateral forces by bending, transferring the load down into the soil. These master piles will generally have to be made larger and stronger, to resist the high forces. Where practicable, they are incorporated into the deck by a full moment connection. This can best be obtained by heavy reinforcing bars embedded into the upper portion of the pile and extended into the deck. Heavy confining spiral is essential.

13.7.1.4 CYLINDER PILES. Cylinder piles and other large-cross-section piles are being increasingly employed in wharf projects, with the lateral loads being taken in bending (Fig. 13.5). Usually the top of the pile is fixed to the pile cap girder by use of a heavy cage of mild-steel reinforcement. With a hollow-core or cylinder pile, this cage is set in the pile head and secured to it by a concrete plug. To insure that the plug will provide full shear transfer to the pile walls, several alternate techniques have been employed:

1. Roughening of the inside walls of the pile.
2. Provision of corrugations in the walls of the pile; these are formed in the top few feet of the pile during manufacture.
3. Painting the inside head with epoxy bonding compound prior to pouring the plug.

Figure 13.5. Prestressed cylinder piles for Port of Baton Rouge, Louisiana.

4. Use of a low-shrink concrete mix for the concrete of the plug.

5. Use of a precast concrete plug encasing the reinforcing cage; this plug is 20 to 50 mm (1 to 2 inches) smaller all around than the core. After placing, the bottom of the annulus is sealed with manila rope or paper pushed down, and the annulus is filled with grout. The precast plug may be a roughened or corrugated surface. This is the best system for large piles, as it minimizes thermal strains during hydration of the cement.

For cylinder piles 1 meter (36 inches) in diameter and larger, that is, having hollow-cores larger than 600 mm (24 inches) in diameter, the combination of techniques 1 and 5 is recommended wherever this can be incorporated into the structural scheme selected. For hollow-cores smaller than 60 mm (24 inches), the combination of techniques 1 and 4 is usually more practicable.

Where cylinder piles must take lateral loads in bending, special consideration must be given to shear and torsional resistance. Greatly increased shear and torsional resistance can be provided at minimum cost by increasing the number of spirals and, to a lesser extent, by increasing the diameter of spirals or hoops. In the typical cylinder pile with 125 mm (5 inch) thick walls, a maximum spiral spacing of 75 mm (3 inches) is recommended throughout the entire length that is subjected to bending and torsion.

Hollow-core and cylinder piles have unfortunately experienced a relatively large occurrence of vertical cracking. This is due to a combination of causes: shrinkage, water hammer when driving the head below water, wedging of soil in the tip, and filling the hollow core by fresh concrete, causing a bursting under the hydrostatic head. In other cases, jetting through the hollow core has caused excessive internal pressure. Water trapped in the core above the level of the outside water has not only exerted a hydrostatic head but has frozen.

However, by far the most common of the causes for splitting has been thermal stresses imposed by temperature differentials between the core and the outside. In northern environments, this has been caused when air temperature fell below freezing, often accompanied by dry winds, while the inside was moist and relatively warm. In the Middle East, the same phenomenon has been caused by the nighttime drop in temperature after an extremely hot day.

The preventive measures, which have proven successful, are to incorporate an adequate amount of spiral so that if the pile does crack from any cause, the spiral steel is still below yield. That is, the spiral across any vertical section should have a yield strength greater than the upper-bound tensile strength of the concrete.

13.7.1.5 PILE CAPS (PILE CAP GIRDERS). These are normally subjected to heavy shear forces as well as positive and negative bending. With comparatively short spacing between piles, therefore, prestressing is usually more or less uniform over the section (concentric). With long spans between piles, the center of prestressing force may follow the moment curve, with tendons raised over the piles and near the bottom of the cap between piles. Posttensioning after pile cap connections are made locks the cap to the piles, and forces the embedded portion of the pile head to act monolithically with the cap in resisting compression.

Because of the heavy bending and punching shears, pile caps usually have very heavy stirrups of mild reinforced steel. Mechanically headed bars may be effectively used to facilitate placement of steel and concrete.

Much modern design is aimed at providing continuity of the deck over the caps and at utilizing the upper portion of the cap as a part of the slab. This may readily be done with a monolithic cast-in-place deck. However, the provision of temporary supports for a heavy deck is costly, so an alternative, often practiced, is to construct the bottom half of the cap first, then set precast deck slabs on this cap, then pour the top half of the cap (Fig. 13.6).

Thus the advantage of precast deck slabs can be combined with the provision of continuity for the deck, and with common use of the concrete of the upper portion of the cap. The lower cap obviously requires minimum support during pouring, and the quantities of cast-in-place concrete are held to a minimum.

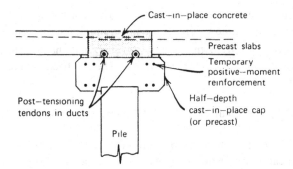

Half–caps, post–tensioned at mid–depth

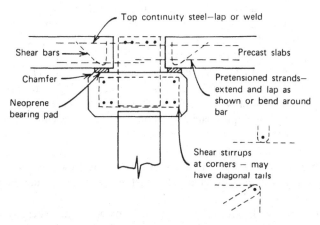

Figure 13.6. *Reinforcing and other details at supports of precast slabs or girders to resist shear and tensile stresses due to creep, shrinkage, and temperature, as well as bending moments.*

With this scheme, there is horizontal shear across the mid-depth of the cap. This means that the heavy stirrups must protrude from the bottom half-cap to tie the upper portion to it. The concrete surface should be as rough as possible, preferably with the cement paste sandblasted (or water-jetted off), so as to expose the aggregate. An epoxy bonding compound may also be used as an added insurance of bond. Shear keys may be stamped in the surface of the green concrete.

There is also temporary negative bending in the half-cap, so additional longitudinal reinforcing is required.

This bottom half-cap may be precast. The connection to the piles is then made in one of three ways:

1. Reinforcement protrudes below the cap and is inserted into the hollow core of the piles. A small hole permits pouring of the grout plug.

2. A large hole is left in the precast half-cap. The reinforcing cage is set through this hole into the hollow core of the pile and the plug is concreted.

3. A large hole is left in the precast half-cap and the pile head penetrates sufficiently far to develop the connection. The pile head is roughened so as to bond with the grout pour in the annulus.

When setting a precast half-cap over batter piles, an elliptical hole can be formed in the cap.

With precast half-caps, the bottom reinforcing must be in the side walls. Since the holes in the cap must permit tolerance for pile head locations, and the caps in turn must support deck slabs, this often requires widening of the cap beyond that required for structural reasons. While adequate side cover must be provided over the bottom reinforcement (prestressed or conventional), only bare minimum cover need be provided next to the hole, as this will be filled with concrete. In such a case, positive steps must be taken to prevent a shrinkage crack between the grout fill and half-cap. Widening of the cap is also an economical way to provide tolerance for setting of the precast deck slabs despite small inaccuracies in pile bent location.

When prestressed precast deck slabs rest on the cap, they tend to shorten due to shrinkage and creep and, thus, shear off the edges of the cap. The precast slabs, therefore, must be tied across the cap by means of extending strands or reinforcement. Stirrups should be provided to resist the shear in the caps (and the end of the slab). A strip of asphalt-impregnated fiberboard or neoprene will provide uniform bearing and permit rotation.

13.7.1.6 DECK SLABS. Precast prestressed concrete is at its best here: the economy of precasting and the structural efficiency of prestressing combine and augment each other in this application.

In most cases, simple precast prestressed slabs are used. These may later be made continuous by the use of mild steel over the pile cap. Lateral monolithic action may be provided by transverse post-tensioning or mild-steel reinforcement. Haunched slabs may be used to permit use of straight tendons and to enhance shear capacity. See Fig. 13.1.

Both continuity and transverse monolithic action are facilitated by the scheme of half-depth slabs. These are pretensioned slabs, usually having tie steel (stirrups) protruding for tying the two halves of the slab together. The surface is roughened by a jet of air and water while the concrete is still green, or by sandblasting. Mild-steel reinforcement is then placed for transverse load distribution and for negative moment over the cap; then the top concrete deck slab is poured in place.

This same continuity over the cap may be accomplished with full-depth deck slabs, by welding projecting reinforcement or by stepping or tapering back the top half of the slab for a few feet from the end. Transverse ties may be made by welding inserts in pockets in the joints between slabs, then grouting (Fig. 31.7).

With very wide but thin slabs, it is essential to provide a reasonably uniform seating by means of asphalt-impregnated fiberboard strips or neoprene.

A scheme which has been used in a few cases has been to set the precast slabs,

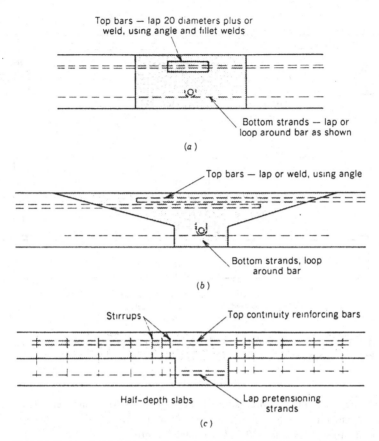

Figure 13.7. Methods of obtaining continuity over supports when employing precast pretensioned slabs.

then lay post-tensioning tendons in ducts in the joints between the slabs, raising these tendons in low saddles over the pile caps, concreting the joints, then stressing the entire deck longitudinally in sections of several hundred feet of length.

Another means of providing continuity was utilized on a large wharf in Kuwait. Full-depth slabs were desirable in this location because of the problems associated with the curing of large exposed surfaces in this environment. The slabs were made in double-span lengths, set alternately in checkerboard fashion so that every other slab was continuous over the support. Transverse post-tensioning tied the entire deck together (Fig. 13.8). This scheme was designed by T. Y. Lin.

These particular deck slabs were haunched, permitting the use of straight tendons, providing more shear next to the supports, and a greater effective depth for the prestressing to compensate for the fact that only half the tendons cross each cap.

Another scheme utilizing precast slabs for continuity is to provide diagonal post-tensioning in crisscross fashion. This is most readily achieved with half-depth slabs, with rigid ducts laid diagonally on top and stressed after concreting.

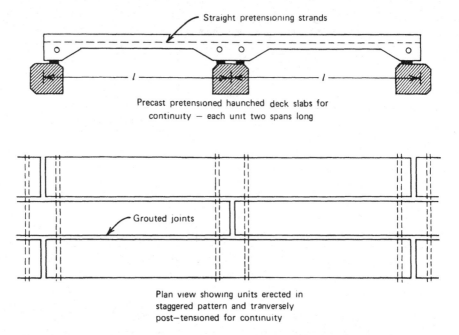

Precast pretensioned haunched deck slabs for
continuity — each unit two spans long

Plan view showing units erected in
staggered pattern and tranversely
post—tensioned for continuity

Figure 13.8. *Scheme for continuity using precast pretensioned deck slabs as employed at Kuwait Harbor Project.*

Where the wharf deck will be subjected to uplift due to tsunami effect, the slabs and their connections may require special design considerations. These are discussed more fully in Section 13.7.3, Coastal Structures.

13.7.1.7 COMBINED CAP-AND-SLAB DECKS. Much of the cost, time, and difficulty in precast systems for wharf decks is in the joints and connections. Prestressing is an extremely beneficial technique for making these joints structurally efficient. However, the cost and time remain. Therefore, considerable attention has been given to monolithic cap-slab deck sections, cast-in-place or precast in large combined sections.

With cast-in-place slabs, one such development is the use of a constant-depth thickness, that is, a continuous flat slab. This greatly simplifies the forming of the soffit. The cap is reinforced (or prestressed) in the line of the piles; it is, in effect, a wide, thin cap. The deck is reinforced (or prestressed) longitudinally right through the cap.

This same flat soffit can be maintained in ballast-deck (filled) wharf structures, even if structural requirements call for the cap to be a few inches thicker, by raising the top of the cap above the general level of the deck slab (Fig. 13.9).

For precast schemes one solution is to build the cap and a portion of the slab as one deck unit, allowing this deck unit to extend a few feet each way beyond the pipe bent into the adjacent bays. This permits the development of continuity. The struc-

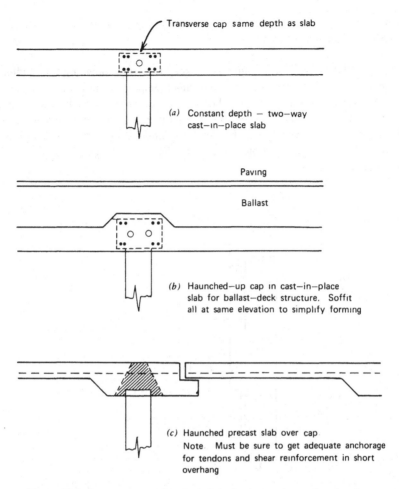

Figure 13.9. *Some schemes for deck slabs for concrete marine structures.*

ture can be made monolithic by splicing of the reinforcement and construction of a cast-in-place concrete joint.

An extension of this scheme is to widen the pile cap so that, in effect, it spans half-way to the next cap, thus becoming a double cantilever. Longitudinal post-tensioning and/or mild-steel overlap in the cast-in-place joint at the center of the span can be used to tie the structure together longitudinally. The soffit can be tapered in such a way as to improve the structural efficiency, and the main prestressing steel can be kept in the top of the slab, completely protected from the splash zone.

When this scheme is superimposed on a substructure of cylinder piles, with their wider spacing and the elimination of intersecting batter piles, then the precast element becomes a tapered wedge. A cast-in-place plug is used to fix the block or slab to the pile and all tolerance adjustment can be made in the cast-in-place joint at mid-span. Transverse and longitudinal post-tensioning can then be provided in the joints between the slabs.

13.7.1.8 FIREWALLS. Firewalls are utilized to keep fire from spreading longitudinally under long wharves and quays. They were much employed in the past with timber wharves where a fire could develop a horizontal draft chimney effect. Firewalls are still used in concrete wharves where oil or other petroleum products may catch fire on the water and the fire spread unchecked to adjoining ship berths, endangering the ships as well as any combustible portions of the concrete docks, such as timber fender systems.

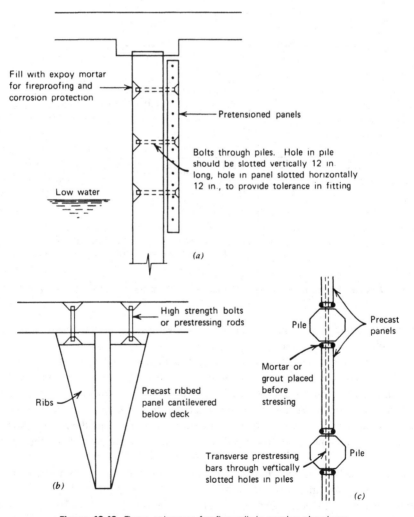

Figure 13.10. *Three schemes for firewalls in marine structures.*

Firewalls should extend below lowest low tide. Lightweight concrete is more effective in preventing heat transmission. Because of their location near and in the tidal zone, precasting is usually the most economical method of construction.

Firewalls tend to work back and forth in the waves; therefore they must be fixed properly against fatigue. In many wharves they are tied to the piles by bolting or with doweled and concreted joints. This puts an extra horizontal bending on the piles and may not be desirable structurally. An alternative method is to suspend the firewall as a cantilever from the deck, and to post-tension it vertically to the deck slab. Hunched ribs can be used to provide the necessary moment capacity and stiffness. (See Fig. 13.10.)

13.7.1.9 CRANEWAY GIRDERS. Many modern wharves must support large gantry cranes for container handling, ore loading, etc. Heavy, deep girders are required. Past design practice was to employ a large number of closely spaced piles under these girders, in order to minimize the shear and moment in the girder. With present wheel spacing and multiple wheels on gantry cranes, it may often be more structurally efficient today to use the same span as the remainder of the wharf, and make the craneway girder act monolithically with the rest of the deck slab, distributing the heavy wheel loads transversely to adjoining piles in the bents.

In some cases these girders have been cast in long lengths, heavily prestressed, and set across multiple caps. The prestressing force may be deflected up and down for positive and negative moment, or the soffit may be haunched (Fig. 13.11).

Particular attention must be paid by the designer to the reversal of moment that may occur with the crane wheels in an adjoining span, thus producing negative moment in the center of the span in question.

Similarly, the very heavy concentrated loads may produce axial shortening in the pile under the wheels, causing positive moment over the support. Mild steel should therefore be provided in the bottom of the girder over the support. Such long,

Figure 13.11. Haunched prestressed concrete girder for container crane rail.

prestressed girders are subject to shortening under creep. Particular attention should be paid to providing adequate stirrups in the caps under the girders and to chamfering the edge of the cap. A good detail is to provide an asphalt-impregnated fireboard or neoprene bearing strip at the edge of the cap to prevent spalling of the edge under concentrated bearing. At the joints where two girders join, there will be high forces trying to tear the cap apart. Longitudinal expansion should be permitted, as by a double bent, or else this jointing cap must be tied together with heavy mild steel.

When craneway girders are precast, they can be tied to the adjoining deck slab by transverse post-tensioning to distribute the load and to make the entire deck work as a whole.

13.7.1.10 PIPEWAY BEAMS. Many wharves carry extensive pipeways on trestle-type bents, connected and braced by pipeway beams. The forces on such beams include not only the usual loads, but also the longitudinal forces due to pipe expansion and contraction. These beams lend themselves to precasting because of the long spans usually employed; the problem has been how to connect them in a three-dimensional pattern with full moment connections.

Post-tensioning, usually with bars, has proven the most satisfactory method. Provision must be made to accommodate tolerances in pile location, either by moving the pile head to correct location (suitable for slender piles and deep water), or in the joint details. The jointing may be made over the pile, or a capping beam may be cantilevered a short distance out from the pile and the joint made there.

If the longitudinal pipeway beams are prestressed, consideration must be given to their shortening under creep and prestress. The usual practice is to take all longitudinal movement to an anchor bent, then provide expansion joints every 600 feet or so with a double bent, at which expansion loops in the piping can also be located (Fig. 13.12).

13.7.1.11 FENDER PILES AND FENDERS. Prestressed concrete is eminently suited for the fender piling of typical cargo wharves. Extensive installations have been in service for many years at Kuwait Harbor and the Jurong Wharves in Singapore (See Fig. 11.26). Smaller installations and test structures are in use in Los Angeles and San Francisco harbors. The US Navy has recently (1990) adopted prestressed concrete fender piles for many of its wharves, after an extensive research and testing program. These fender piles are discussed in more detail in Chapter 11, Prestressed Concrete Piling.

Although existing installations of prestressed concrete fender members have employed timber rubbing strips to protect the pile surface from abrasion, studies are being made on alternative ways. A heavy coat of epoxy or an anchored plate of polymer may prove to be a better solution than the bolted-on timber strip.

13.7.2 Bulkheads

13.7.2.1 GENERAL. Bulkhead wharves are much used in harbors and along inland waterways, serving to form a transition wall between the land working area on one

Figure 13.12. *The Liquefied Petroleum Gas (LPG) export terminal at Ju 'Aymah, Saudi Arabia, extends over 20 km into the Arabian Gulf. Both pipeway girders and a roadway deck are carried on prestressed concrete cylinder piles. (Courtesy of FIP Notes and Raymond International.)*

side and a navigable depth on the other. The explosive growth of container terminals worldwide has given great impetus to this type of construction. A great deal of engineering study has been directed toward analysis of the lateral earth loads, including those due to surcharge, and the failure circle which, in turn, determines the required penetration and the bending moment in the bulkhead. Having determined the loads and resistances, the structure and the slope beneath is then designed to be stable and to resist imposed lateral, longitudinal, and torsional, as well as vertical loads. The rigidity and ductility of the system will determine its behavior, especially under earthquake, and possible liquefaction of the soil.

The wall is exposed to impact and dynamic forces from the moorings and the environmental loads of the waves.

Prestressed concrete has been extensively employed for such bulkhead wharves because of its durability, strength in bending, and economy.

Ductile frames may be constructed using prestressed concrete piles with concentrated confinement in the form of spiral, with moment connection to the deck provided by additional mild steel reinforcement. These are very effective in resisting seismic action.

13.7.2.2 SHEET PILES. The design, manufacture, installation, and jointing of prestressed concrete sheet piles is discussed in Chapter 11, Prestressed Concrete Piling.

When used as a bulkhead wall, the point at which the tie acts determines the moment pattern in the sheet pile. Thus, the pile may require eccentric or deflected prestressing. Alternatively, mild steel may be placed in the relatively short region of high negative moment. Sheathed and greased tendons may be placed in the sheet pile, for stressing after installation.

To tie the sheet piles together along the wall, and to span between ties, some sort of wale (longitudinal beam) is required. If the tie is at the top, then a capping beam may be poured in place, either conventionally reinforced or, preferably, post-tensioned.

The post-tensioning will help to tie the wall together, so as to act as a continuous membrane, and also to keep the joints tight.

13.7.2.3 SOLDIER BEAMS AND SLABS. These soldier beams are usually of H cross section or else rectangular with suitable grooves. They may be steel or concrete. If prestressed concrete, consideration must be given to the points of positive and negative moment as discussed above for sheet piles, except that the magnitude of moment will be higher with the soldier beams (Fig. 13.13).

The connecting slabs serve as a wall to transmit loads laterally to the soldiers. These slabs may be prestressed planks or flat slabs, reinforced transversely, with vertical prestressing for durability and handling.

The slabs may rely on shell action to transmit the loads from the earth; cylindrical shells are sometimes used which have vertical prestressing for handling, durability, and edge restraint. They exert a side thrust on the soldier beams. This is resisted in the lower zones by the passive pressure of the earth, but at the top a longitudinal tie must be provided, such as a post-tensioned capping beam.

Figure 13.13. *Precast prestressed soldier beams are utilized to construct graving dock.*

13.7.2.4 TIES FOR ANCHORING BULKHEAD WALLS. Ties are used in tension to hold the wall back to the deadman anchor. These ties are typically exposed to corrosive action because of their usual location at about tide line in filled soil. They also must have bending strength to resist settlement of the fill, and heavy concentrated loads, etc. Steel tie rods, heavily coated and wrapped, have been much used in the past. In recent years, post-tensioning bars have been used. While very efficient, they have more elongation under load and less ductility. Both tie rod and bars require intermediate supports to prevent excessive sag. Highly stressed prestressed concrete tie beams appear to be optimal for many such installations.

Connections are always a problem with any material. With prestressed concrete, the connection at the wall may be by direct shear on a T-head, by doweling, or by post-tensioning to a capping beam.

Ties, being subject to both tension and bending, should be designed for no more than zero tension at the most highly stressed point. Additional ultimate capacity can be provided by embedded mild steel bars, which can also facilitate the end anchorage.

13.7.3 Coastal Structures

These are defined herein as structures built on beaches at the border between land and sea. Thus they penetrate the surf zone, passing from an onshore environment into an ocean environment.

Structures included are fishing and recreational piers, pipeline piers for intakes, discharges, and products; piers for the support of oil drilling and production, offshore loading terminals and facilities, groins and jetties for control of sand movement (erosion and deposition), and breakwaters. Completely submerged structures, such as intakes, outfalls, and pipelines, are discussed in Chapter 14.

The increasing emphasis on the coasts for recreational purposes, and the growing industrialization of the coastal areas have given a great impetus to the development of coastal structures. It is anticipated that there will be greatly expanded activity in the future. Great strides have recently been made in coastal engineering. Specific reference is made to the Conferences on Coastal Engineering and their published Proceedings.

Coastal structures must serve in a wide variety of environments, from tropical to arctic, and from gently sloping sand beaches to exposed rocky shores.

Environmental aspects to be considered include:

1. Waves, including both storm waves and the repeated action of all waves.
2. Sand movement, erosion, and deposition.
3. Current.
4. Seaweed (kelp) and marine growth (barnacles and mussels).
5. Mooring forces, especially the forces imparted by wind and waves on vessels moored to the structure.
6. Seismic action, with liquefaction of soils and slumping.

7. Tsunami, and storm surge, where applicable.
8. Ice build-up on structure.
9. Floating ice action against structure and moored vessels.
10. Freeze-thaw attack on saturated concrete.
11. Effect of saltwater on durability.
12. Effect of breaking surf, cavitation.
13. Wave slam and uplift.

Prestressed concrete is extremely well suited to the construction of coastal structures and has, therefore, been given preferential selection in many regions of the world. It must be emphasized, however, that the environmental aspects 1 through 13 above, present an extremely severe set of parameters, especially as many interact to intensify and augment each other. Judgment is often required to assess the probability of forces acting simultaneously. To design for the worst possible combination may be beyond economic feasibility in many instances. Similarly, an evaluation must be made of the consequences of failure, including contamination, shut-down of service, destruction of adjoining beach areas, and the effects on sand migration, currents, and wildlife.

Prestressed concrete used in coastal structures offers the advantage of durability, structural strength in bending, fatigue endurance under cyclic alternating loads such as those from breaking waves, and high column strength. In arctic installations, it offers excellent impact resistance at low temperatures, an especially important point under the repeated crushing of ice.

To resist storm waves, it is desirable to present a minimum area and a minimum volume to the waves. Use of highly stressed piles of circular cross section will reduce wave forces to the minimum.

Surf tends to produce cavitation on concrete surfaces. High density, well-compacted concrete, and provision of increased cover, will generally give satisfactory results.

Sand movement on the beach is extremely abrasive. Use of high-density concrete, in a circular or octagonal cross section (or a square pile with large chamfers), reduces the abrasive damage to a minimum. Lightweight concrete, made with high-quality expanded clay or shale aggregates, appears to give equal abrasion resistance to that of normal hard rock concrete.

Sand migration will also tend to give greater unsupported column lengths at times, followed by unequal build-up later. The high column strength and bending resistance of prestressed concrete can be beneficially employed to counter these changes in the sand line.

Seaweed and barnacles tend to increase the front presented to wave, surf, and current action. Hence drag forces are increased both by the increased size and roughness. In some areas, secretion of acids from barnacles is able to bore holes in high quality concrete made with limestone aggregates: use of igneous aggregates will prevent this.

The repeated action of waves, including those acting on a vessel moored to a structure, tends to produce fatigue-type stress reversals. Fortunately, prestressed concrete behaves excellently in resisting fatigue due to the low stress change in the steel tendons in relation to the change in loading.

However, stress reversal does sometimes have to be considered in sheet-pile groins, where the waves on the two sides of the groin are typically out of phase. This constant "working" has destroyed steel sheet-pile groins; prestressed concrete is, therefore, a more suitable material, but must be properly designed.

Fatigue is accelerated when the concrete is saturated and submerged. Thus under multiple cycles of repeated load the concrete may crack at about 50% of the stress levels associated with cracking under static loading. This in turn leads to pumping of water in and out of the cracks and eventual failure of the reinforcing steel or tendons. It has been found that high-strength concrete, highly impermeable, tends to resist these actions. Lightweight concrete, incorporating condensed silica fumes, appears to have enhanced fatigue endurance.

Prestressing, by eliminating or minimizing the number of excursions of the concrete into the tensile range, greatly extends the fatigue endurance.

Storm waves, as well as the rarer tsunamis, produce violent upward forces on the undersides of decks. They have torn apart many timber docks in the past. Some modern coastal structures are designed to lose their decks under such conditions; they are only slightly fastened down, with connections designed to fail, so that the substructure will remain undamaged.

A minimum of bracing and similar obstructions should be presented below the deck. This favors the use of prestressed concrete piles. Concrete deck slabs have inherent weight and can be designed to resist the upward force, provided they have sufficient mild-steel reinforcement to take the reversal in loading. Openings can be designed so as to minimize and relieve the wave and air entrapment in a particular area.

Flat undersurfaces lead to very high wave impacts. Even a small slope on the underside will dramatically reduce the impact force due to change in momentum of the water mass.

There has been a general tendency to neglect the consideration of tsunamis in design of marine structures, due to the unpredictability of occurrence and force. However, some harbors and coastal areas are notorious for focusing these waves and surges. A design for tsunamis can be based on a rational appraisal of probability, consequence, and practicability, including cost.

Ice build-up and collaring occur in areas of tidal change, and often include beach ice that is dislodged and wedged in the piling. The ice increases the lateral resistance to the current and adds weight. The downward weight on piles, particularly on batter piles, may be sufficient to break them. These ice masses also may slide down between batter and vertical piles and break them by wedging action. This destroyed a partially completed pier at Anchorage, Alaska. Thus the configuration of the structure must take ice build-up into consideration.

The force from moving ice is a major consideration for the designer and often exceeds all other lateral loads combined. There is a great difference in force, depending on whether the ice fails in tension or in compression. This, of course, is a function of the configuration of the structure and of the ice conditions. Prestressed concrete offers strength, rigidity against local buckling, and good fatigue endurance against continuous ice crushing. If adequately confined by stirrups, concrete slabs and shells have high ductility and robustness.

The installation of piles in sandy beach areas is frequently aided or accomplished by jetting. Internal jets have been found very effective in beach sands because they eject the water right at the pile tip. For most coastal structures, it will be prudent to use a standard steel jet pipe in the pile, with screwed or welded fittings; this prevents the possibility of cracking of the pile due to a leak in the jet pipe. Plastic jet pipes should not be employed. The compression-tension forces under the hammer blow crack the plastic, especially if the adjoining concrete cracks due to rebound tensile stresses. The water under pressure now has a large area on which to work. Repeated blows lead to hydraulic ram effects and fracture of the pile.

Care should be taken not to jet a large hole under the pile tip and then use an impact hammer to drive the pile, as this free-end condition will cause excessive tensile stresses and possible breakage.

In rocky beaches, prestressed piles are often concreted into drilled holes. After drilling, the hole may partially fill with sand. The use of an internal jet pipe may permit jetting in of the pile, and the pipe can then be used for grout injection.

Prestressed sheet piles are usually installed in a sandy beach; here the use of external "gang" jets, or several internal jets, are efficacious.

Many coastal structures in the surf zone are constructed "over the top," working progressively seaward out over the completed structure. Therefore the time required for completion of a span must be kept as short as possible. Precast elements may be temporarily connected by welding or bolting, and the construction equipment moved out over them. Later the permanent connections are made. Alternatively, prestressed connections may be employed, using short bars. Dry-fit joints between superstructure elements, with or without epoxy glue, are particularly expeditious. Provisions must be made, however, to accommodate tolerances in pile position and elevation (Fig. 13.14). Extremely clever schemes have been developed for this purpose, e.g., sleeve joints and jackets.

Use of extra-wide cap girders permits greater tolerances in pile location. Particular care must be taken not to overstress or break a pile by attempting to pull it into position.

Quick-setting grout and epoxy compounds help to speed completion of these connections.

Prestressed lightweight concrete may facilitate construction, permitting longer spans or allowing the use of smaller and lighter construction equipment.

Particular attention must be given to insuring the durability of joints and connections. These connections are frequently made under adverse conditions. The painting of all joints after concreting with epoxy is an excellent way of sealing them against moisture entry.

Furthermore, it is recommended that on the completion of a coastal structure, a detailed inspection be made of all connections to make sure of their completeness and tightness, for it is at these connections that most problems occur. Bolted joints

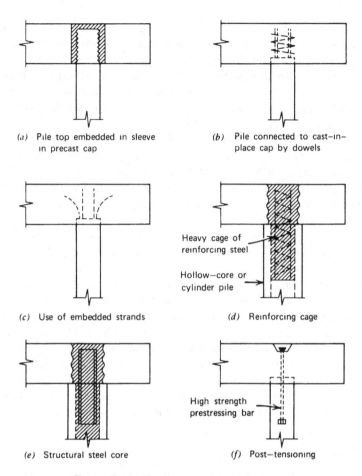

(a) Pile top embedded in sleeve in precast cap

(b) Pile connected to cast-in-place cap by dowels

(c) Use of embedded strands

(d) Reinforcing cage

Heavy cage of reinforcing steel

Hollow-core or cylinder pile

(e) Structural steel core

(f) Post-tensioning

High strength prestressing bar

Figure 13.14. *Connections of piles to pile caps*

can work loose under repeated wave loads and welds can fail in corrosion-accelerated fatigue.

Obviously, coastal structures are more subject than other structures to salt spray and the possibility of salt-cell electrolytic corrosion. Pile caps have been particularly vulnerable. Extreme care, therefore, should be taken to follow the procedures for maximum durability in the splash zones (Section 13.3, Durability; also Chapter 5, Durability).

13.7.4 Navigation Structures, Dolphins, and Bridge Fenders

Prestressed concrete has been extensively used for navigation structures because of strength and durability and because simplicity of structural concept permits prefabrication, simplified connections, and overall economy.

White cement and, on occasion, white quartz aggregate, have been used to give greater visibility. It would seem that panels could be made with exposed glass fragments or beads to improve the reflective power.

Such structures are frequently moored to by small boats, and may be hit by boats and barges. Thus the design must provide adequate ultimate strength to prevent the pile from snapping off. This can best be done by using additional unstressed strands or conventional reinforcement, and ample spiral.

Dolphins are subjected to horizontal loads only. The piles should be highly prestressed, and can use from 0.20 to 0.30 f'_c as effective prestress, while maintaining a balanced design. Such high prestress values require more spiral binding.

Dolphin piles should be selected, as fender piles are, to give maximum energy absorption. The cross section similarly needs maximum strength with minimum moment of inertia.

Prestressed concrete is increasingly used for the protective fenders of bridges (Fig. 11.27). This use differs in purpose from wharf fenders: for bridge fenders, the primary aim is to provide maximum protection to the pier. Prestressed concrete is fire-resistant, a very important fact to be considered when selecting the fender for a steel bridge. It is durable, resistant to abrasion and impact. As for all structures subjected to predominantly lateral loads, higher degrees of prestress, incorporation of additional mild steel, and heavy spiral are desirable. Hollow-core piles and cylinder piles may be filled to protect against local damage from debris and ice.

For bridge fenders, the criteria of minimizing damage to the vessel and ease of repair of the fender are secondary, since the cost of repair is usually paid by the ship. The primary purpose is protection of the bridge. Therefore, the top portions of the pile may be encased in heavy ring girders or horizontal slabs of concrete.

13.7.5 Overwater Airports

These may be constructed either as floating structures of prestressed concrete (Chapter 15, Prestressed Concrete Floating Structures), or as fixed structures supported on piles. The overwater extension of runways and taxiways at LaGuardia Airport, New York, is a notable example of the latter type (Fig. 13.15). The prestressed concrete deck structure consisted of precast cap girder segments, pretensioned half-depth deck slabs, and cast-in-place deck. The cap girders and deck slab were then post-tensioned to achieve monolithic behavior capable of supporting the extremely heavy design loading of jets crash landing, or alternatively, fully loaded at the moment of maximum thrust for takeoff.

The deck slabs were shallow, inverted double-tee sections, with holes through the webs, through which the post-tensioning tendons were later run. The cap girder segments were jointed over the pile heads, with a poured-in-place joint. Ducts crossing these joints were spliced and taped to prevent grout in-leakage. Stage stressing was employed; the first stage to support the deck slabs and the concrete topping, and the second stage to support the live load (Fig. 13.4). This meant that a considerable time interval occurred between placement of tendons and stressing and grouting. VPI powder was dusted on the tendons during their insertion and the ends

Figure 13.15. *Overwater Extension of La Guardia Airport, New York, was a pioneering triumph of the extensive use of prestressing.*

sealed. The precast cap segments were located just above the high-tide line, so prior to insertion of the tendons, they were flushed with fresh water. In the light of subsequent knowledge, this particular environment is so adverse for possible salt contamination and salt-cell formation that more positive steps would probably have been justified; however, no corrosion has been reported.

This combination of precasting, prestressing (both pretensioning and post-tensioning), and composite construction proved very satisfactory and showed its great value for decking large areas for extremely heavy loadings.

The interaction of piles and deck under heavy concentrated loads was of great importance in design, in minimizing individual pile loads, and in absorbing the impact energy under the design parameters.

Experience gained on this huge structure would indicate the following lessons for the future:

1. The precast concept is sound, economical, and rapid. It can be economically extended even to nontypical areas.

2. The combination of pretensioning for precast members with two-directional post-tensioning of the composite structure is a brilliant design technique and highly practicable.

3. The large tendons in the cap should be of a post-tensioning system that requires minimum size ducts; that is, the tendons should be capable of being inserted prior to placement of the end anchorage.

4. Special detailing and specifications should be directed to insure the utmost in durability.

5. Reinforcing steel detailing should aim at standardization and simplicity: on this project, the fabricating and tying of the mild steel was very labor-intensive.

6. Joints are the costliest item; therefore, the deck slabs should be as wide as possible in order to reduce the number of joints.

7. Adequate depth of concrete (cover) above the stems of double tees and reinforcing steel must be provided to prevent shrinkage cracks in the deck surface.

Other schemes of deck construction will undoubtedly be developed and used for similar overwater airports, such as the combination cap-slab units discussed in Section 13.7.1.7. This construction concept at La Guardia, however, represented a bold and successful step forward in construction concepts for overwater platforms.

13.7.6 Pile-Supported Offshore Platforms

Shallow water platforms have been constructed in Lake Maraciabo and along the shores of the Gulf of Mexico for the oil and gas industry, although the sulfur industry utilizes them also. As means are found further to exploit the resources of the continental shelves and adjacent coastal waters, platforms will undoubtedly be required for a wider variety of industries.

Most of these shallow water offshore platforms consist of piling, with jackets or other below-water framing, deck sections, and appurtenances. Prestressed concrete has been so far utilized extensively for piling in locations where design wave forces are low to moderate (e.g., Lake Maracaibo), and for deck sections of moderate size.

Pretensioned cylinder piles have been extensively utilized in Lake Maracaibo where durability was a major consideration, since steel piles corrode very rapidly. The strength of the prestressed piles permitted them to be picked and handled in lengths up to 60 meters (200 feet).

When prestressed concrete cylinder piles are used for offshore structures, they are of course subject to high bending under the action of storm waves.

Additional moment capacity can readily be provided at the deck by either an internal steel core or an external sleeve. At the mud line, however, increased bending resistance requires careful planning. Increased moment resistance can be provided for this zone by one of the following means:

1. Mild-steel bars added to the reinforcement of the cylinder pile wall, extending through and beyond the zone of maximum moment.

2. An inner steel core, lowered down the cylinder pile to proper elevation and grouted.

3. An external concrete or steel sleeve, jetted down around the pile and connected by grout.

4. Tapered pile sections. These have been extensively employed with large rectangular conventionally reinforced piles in Lake Maracaibo, but so far not with cylinder piles, mainly because the early cylinder piles were assembled from uniform cylindrical segments. Actually, the manufacture of cylinder piles by the pretensioning method lends itself to the use of tapered piles, giving maximum sections at point of maximum moment. Tapered piles could also readily be formed by the posttensioning together of precast segments, provided the segments are manufactured so as to fit the taper.

5. Installing a prestressed cylinder pile to moderate depth, then driving an inner steel pipe pile through and beyond it, overlapping the two in the zone of maximum moment, and connecting the two shells with grout.

Abrupt changes in pile stiffness must be given careful consideration in location and detail so as not to create a new weak point.

Prestressed concrete offers substantial advantages for the deck structures of these platforms where the equipment and operating loads are relatively fixed in location. The platform structure can be manufactured and assembled ashore, transported and set in segments, and connected by post-tensioning. This application would seem to lend itself to the use of dry joints with epoxy glue (Section 4.3). Connections can thus be effectively and expeditiously completed. Such deck segments can be highly refined in both design and construction; use of high-strength concrete in certain members combined with lightweight concrete in others is a promising approach. Deck girders may be framed or Virendeel trusses, deck slabs may be waffle-ribbed or hollow-core. Deep deck girders may be posttensioned longitudinally and, in addition, have their webs post-tensioned vertically with bars, to resist high shear.

To meet the restrictions of lifting capacities in construction yards and offshore, one system is to make up the deck structure in reasonable-sized segments of, say, 30 tons maximum weight. These can then be assembled and connected on a barge into units of approximately 500 to 1000 tons or more. Thus trusses may be formed of precast chord and web segments.

Final lifting and installation on site may utilize the capacity of one or more offshore derricks and crane barges.

Bracing frames of steel and of concrete have been used for providing support for the piles of offshore platforms. These are usually braced space frames, with sleeves for the piles. The connections are made by grout. The frames are typically used initially as a template through which the piles are driven.

The main problem with the use of concrete for space frames and bracing, other than weight for handling, is the difficulty in developing adequate connections for the bracing, whether truss or girder type. Posttensioning through the joints offers one solution which is particularly effective for truss web members. Another solution is to use steel stubs protruding from the ends of the truss members, so that the connections may be made by high strength bolting or by welding, then encasing the joints in concrete for protection.

13.7.7 Reconstruction and Strengthening of Steel Tubulars in Offshore Platforms

One major offshore construction firm (Brown & Root, Inc.) has developed a method for restoring the strength of old steel jackets, where corrosion and damage have weakened them. Working generally from inside the jacket's main legs, they install post-tensioning ducts in the bracing tubulars, then pump in a special high-strength grout-concrete. The main legs are reinforced as necessary at the anchorage ends, tendons are inserted, and the composite tubular member prestressed. Advantage is taken of the compressive strength of concrete confined in steel tubing. Connections are detailed so as to obtain the maximum reinforcement from multi-axial prestressing.

A variation of this scheme was used to upgrade the Ninian Central Platform's deck structure to carry heavier equipment.

This scheme for strengthening may indicate means of obtaining more efficient tubular members in new construction. With the present tubular design, the drag force from waves is primarily a function of the projected area. Relatively thin-walled steel tubes filled with concrete and posttensioned may offer a more economical technique in some cases for obtaining the maximum axial strength and most efficient connection details.

REFERENCES

1. Hasson, D. and Crowe, C., *Materials for Marine Systems and Structures* Academic Press, Inc. San Diego, California, 1988.
2. Mehta, P. Kumar, *Concrete in the Marine Environment* Elsevier Applied Science 1991, Essex, England.
3. Military Handbook *Piers and Wharves*, MIL-HDBK-1025/1, Pile Buck Inc, Jupiter, Florida, 1990. (Especially pages 205 et seq. "Prestressed Concrete Fender Piling.")
4. Bruun, Per, *Port Engineering*, 4th edition, Vol. I, Gulf Publishing Co. 1989.

CHAPTER 14

Bottom-Founded Concrete Sea Structures: Gravity-Base Offshore Platforms and Terminals

14.1 INTRODUCTION

The 1970's and 1980's saw the spectacular development of offshore concrete structures, standing in over 200 meters (650 feet) of water depth in the midst of one of the world's stormiest oceans, the North Sea. These gigantic structures, displacing up to 600,000 tons, are successfully withstanding the dynamic cyclic forces of waves approaching 30 meters (100 feet) in height; this is made possible by prestressing (Fig. 14.1).

Beneath the estuaries and rivers of coastal areas of Europe and Canada, underwater tunnels (tubes) of prestressed concrete carry multiple traffic lanes.

Prestressed concrete caissons form offshore terminals for gigantic ore carriers and cargo ships in Australia and France. A prestressed concrete breakwater wall protects an oil production platform in the North Sea (Fig. 14.2).

Off the coast of Newfoundland, a prestressed concrete caisson is now under construction; it is designed to resist the impact of icebergs from Greenland's glaciers.

Prestressed concrete caissons are being installed to serve as the piers of the Great Belt Bridge in Denmark. Sixty-two such piers are now under construction for the Western Bridge and a comparable number will be required for the Eastern Bridge.

Seafloor oil storage vessels are being designed. Conceptual studies for the crossing of the Strait of Gibraltar are based on prestressed concrete gravity-base caissons in water depths up to 460 meters (1500 feet).

In the Arctic Ocean, prestressed concrete caissons such as Tarsiut and the Glomar Beaufort Sea One (CIDS) have successfully withstood the Arctic environment with its multi-year ice floes containing embedded pressure ridges (Fig. 14.3).

Meanwhile, several thousand shallow depth concrete structures, some pre-

Figure 14.1. Offshore concrete platform, Statfjord "C", for Norwegian North Sea.

stressed and some conventionally reinforced, have been serving as compressor stations and pumping stations for gas production in the Gulf of Mexico and Nigeria.

Prestressing has had a major role in these developments, working in conjunction with conventional reinforcement to withstand the dynamic cyclic loadings of storm waves and continuous crushing of ice.

14.2 ROLE OF PRESTRESSING

While prestressing plays an essential role in these structures, they are nevertheless largely built with passive reinforcing steel in the conventional fashion. The concentrations of reinforcing steel are great, typically several times that required on comparable land structures. In particular, it has been found necessary to incorporate a large quantity of shear and confining reinforcement in the typically thick concrete walls.

Figure 14.2. Prestressed concrete breakwater caissons being positioned around the Ekofisk Offshore Oil Storage Caisson.

Prestressing has, however, been the catalyst which has made deep water structures practicable.

In addition to providing ultimate strength through its concentration of high-strength steel in critical zones, it provides control of cracking, which in turn assures fatigue endurance, durability, watertightness, and oil tightness.

Prestressing is utilized to ensure that the dynamic cycling of the great majority of waves lies wholly within the compressive mode, i.e., moderate compression to low compression but not into tension. This minimizes the development of cracks due to fatigue of the concrete and, ultimately, of rupture of the steel due to fatigue. The fatigue endurance provided by prestressing is especially valuable in marine structures since the submergence and the saline environment would otherwise render them vulnerable to the joint attacks of fatigue and corrosion.

Cylindrical and spherical configurations are much used in offshore and sub-aqueous structures due to the concrete's inherent ability to resist compression. However, these cylindrical configurations are ill-suited to the resistance of cyclic shear by conventional reinforcement, due to the fact that the orthogonal reinforcement is only partially effective in resisting diagonal cracks. Prestressing in one direction can change the inclined direction of the weak plane, the potential crack, to the orthogonal direction where the conventional steel may be more effective.

Specific areas and zones of offshore structures often see their most critical values during the construction stages. These occur during float-out of the base raft, during

Figure 14.3. *The CIDS platform successfully withstands the onslaught of the Arctic Ocean's sea ice with its prestressed lightweight concrete mid-section.*

reorientation of segments, and due to unequal loading from specific concrete placements. Prestressing can be used to minimize these forces and to counteract undesirable strains and deflections during construction.

Finally, by preventing and/or minimizing cracking during construction and in service, prestressing enhances the long-term durability of the structure.

Accidental loads may create tensile strains which would impair the continued serviceability of the structure. Prestressing is being used to counteract such forces from postulated accidents.

Prestressing may also be used to counteract ovalling from eccentric loads in service. For example, local impacts from icebergs and sea ice pressure ridges may create large tensions in the supporting structure, and prestressing may be beneficially employed to provide resistance and ensure serviceable behavior afterward.

Offshore structures typically have thick walls, 0.6 to 2.0 meters thick. Since high ultimate and early strengths are required, as well as durability, cement contents are high and substantial heat is generated during hydration and curing, with consequent thermal expansion. When subsequent cooling ensues, the contraction is often constrained by the configuration or by the friction on the supports, and thermal cracking ensues. Prestressing can be designed and its installation scheduled to resist such strains. At the same time, the influence of the changing weights of the concrete on the residual prestressing force must be taken into account.

14.3 CONSTRUCTION SCENARIO

Most gravity-base and bottom-founded concrete structures go through the following sequence.

1. Construction of the lower portion, i.e., the base raft, in a graving dock or basin. Prestressing of this portion of the structure (Fig. 14.4).
2. Launching and mooring of this lower portion in moderate water depth where it is protected from storms.
3. Completion of the structure afloat, including prestressing (Figs. 14.5, 14.6).
4. Installation of superstructure.
5. Tow and deployment to the site.
6. Installation on site.

Further details of this fabrication, deployment, and installation procedure may be found in B. C. Gerwick, *Construction of Offshore Structures,* Wiley-Interscience, 1986.

In initial stages of construction such as float-out, the base raft may be subjected to hogging moments which would cause large residual cracks. Prestressing is utilized to provide the resistance at this stage.

In some of the later stages of the construction, it may be necessary to pressurize

Figure 14.4. *The prestressed lightweight concrete base raft for the CIDS Arctic Ocean platform is heavily prestressed.*

Figure 14.5. Construction afloat of prestressed concrete offshore caisson "Condeep" in Norway.

the interior, thus creating tension in the outer shell. Prestressing can be used to resist these forces and to prevent cracking.

14.4 SPECIAL ASPECTS OF PRESTRESSING FOR OFFSHORE STRUCTURES

The installation, stressing, and grouting of the posttensioning tendons in offshore structures is usually more complex than in other typical structures, due to the varying cross sections, curvatures, directions of prestress, and conditions during installation (Fig 14.7). Many of the tendons have multiple curves, some in more than one plane. Anchorage zones are typically very congested due to the concentration of reinforcing steel and intersecting and overlapping tendons.

Field operations are frequently carried out afloat and portions of the ducts may be near or below the external waterline. It's important that salt water be kept clear of the ducts and, should any enter due to splash, the ducts must be washed out with fresh water.

Most of the concrete offshore structures constructed to date have been built in northern regions such as Norway and Scotland. During winter, water in ducts may freeze and the consequent expansion may cause multiple cracking of the surrounding concrete.

Even when the ducts have been kept clear of water and dry, the concrete itself

Figure 14.6. Gulfak's C platform, now serving in the North Sea in 220 meters (730 feet) of water depth, will soon be surpassed by Troll, under construction for 305 meters (1000 feet) water depth.

may become chilled below the freezing point. Grouting may be carried out when the air temperature is above freezing, but the cold mass of concrete may extract so much heat as to freeze the grout. Blowing warm air through the ducts for a period prior to insertion of tendons, stressing, and grouting may be necessary during periods of very cold weather.

Posttensioning tendons for offshore structures may have significant vertical runs (Fig. 14.8). Those for shafts and walls may be entirely vertical. As such, complete filling by grout becomes quite difficult due to the bleed and the wicking action of the strands. This phenomenon may display itself in the formation of a void at the top of vertical ducts.

Solutions to this problem have been by two means. One is to use a thixotropic admixture that causes the grout to gel as soon as pumping action ceases. The second is to arrange an empty hole in the anchorage plate to which a standpipe is attached.

Figure 14.7. *Vertical post-tensioning of peripheral ice wall for CIDS Arctic Ocean platform is combined with heavy circumferential prestressing.*

At the completion of grout injection, the grout is forced up into the standpipe. Any shrinkage will occur within the standpipe, rather than at the top of the duct.

Despite these dual efforts, field investigation still frequently reveals small voids at the top of very long vertical ducts, such as those in shafts and towers. Refilling from the top is needed.

Recent improvements in thixotropic admixtures should minimize or eliminate this problem.

Many problems have been encountered in maintaining the grout-tight integrity of ducts while keeping the inside clear so that tendons can be inserted and stressed with minimal friction and interference.

Splices are a weak point in the system. To prevent the tendons from catching as they are pushed or pulled past a splice, the ends of ducts should be cut clean with a saw and any burrs removed. The overlapping sleeve should fit snugly.

Horizontal ducts should be supported by saddles on the reinforcing bars so as not

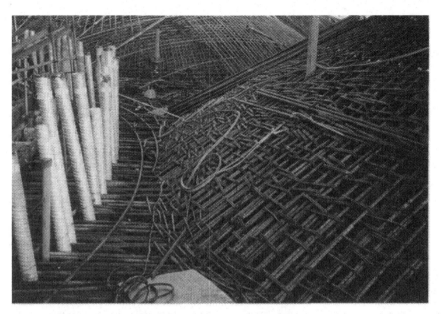

Figure 14.8. *The shafts of the typical offshore platform have heavy vertical post-tensioning while the dome roofs of the base caisson are densely reinforced.*

to be dented as the concrete is placed and vibrated. No person should be allowed to climb or stand on them for the same reason. This can, of course, also cause unwanted sag or wobble of the ducts during concreting.

Where sharp bends must be made in ducts, the ducting should be changed to thick wall standard pipe. In cases of very sharp bends, the pipe should be pre-flattened around the curve so as to give space for the individual strands to flare out and not bind on each other.

Where flexible metal ducts are used because of the sharp curvature, a mandrel should be inserted to hold them in true location during concreting (Fig. 14.9).

In insertion of tendons, both pulling and pushing will minimize friction as compared to either alone. The tendons can actually be worked back and forth a little before anchoring, so as to equalize the stresses in them.

Because of their length, the long periods when they must remain open, and the complex other operations that go on during construction, ducts in offshore structures have experienced more than usual blockage. Unfortunately, many cases have been due to the entry of foreign materials: short sections of reinforcing bar, and pieces of coarse aggregate. Thus duct segments should always be delivered to the site with red plastic covers over both ends.

The other common source of blockage is in leakage of cement paste or grout at splices. Daubing the ends with epoxy prior to slipping the sleeve over them will help to prevent pulling apart. Watertight or heat-shrink tape will prevent leakage in most cases, especially if the sleeve fits snugly.

Figure 14.9. The complex curvatures of the Frigg Manifold Platform require special provisions for the post-tensioning to minimize frictional losses. (Courtesy of DORIS Engineering.)

Attention to such details is extremely important, because the cost and delay of field corrections for a blocked duct, even assuming access is still possible, are enormous.

To provide a contingency solution in the event of blockage of a critical duct in spite of all precautions, extra ducts, e.g. 5%, may be provided.

Cracking problems often occur near anchorages, due to the complex three-dimensional strains introduced by stressing.

Where the tendon is turned out of the wall into a bolster for anchorage, radial forces tend to tear the bolster from its parent wall. Particular care must be taken to position the stirrups or ties so that they prevent the start of a crack, i.e., just under the anchorage plate and at the start of curvature. Yet these zones are also extremely congested by the anchorage plate, spiral encasement of the trumpet, and other reinforcement (Fig. 14.10).

Mechanically headed bars (T-bars) may be substituted for closed stirrups to minimize congestion and ensure placement in such critical locations, while reducing congestion and the physical difficulty of affixing bent stirrups and their tails.

These anchorage zones also develop tension behind them, that is, in the concrete structure away from the lead of the stressed tendon. Additional mild steel bars covering this zone can distribute the tensile strains and prevent cracks.

Similarly, transverse strains develop, especially where there are multiple parallel tendons, tending to allow shear along the outsides or cracks between. Here additional stirrups or T-headed bars will be useful.

Figure 14.10. *Posttensioning tendons will impart circumferential prestress around the edge of the roof dome. Where these ducts turn out from the dome, they will be tied to it with multiple stirrups*

The concrete under anchorage heads is highly stressed in bearing. Therefore it must be well-consolidated and free from honeycomb. Yet it is precisely here where access for concrete placement and particularly for vibration is often very restricted, to the point of impracticability. External vibration will help but cannot consolidate deeply.

Placement of one or two pipe sections (tubes) among the reinforcement and ducts will allow a pencil vibrator to be inserted. The tubes are withdrawn gradually, allowing the vibrator to work below their tip.

Grouting of long vertical runs, such as shaft tendons, perhaps 100 meters (330 feet) or more in height, requires high pressures. In the event of any leakage or breaks in the ducts, the bursting effect may cause major cracks. This is blamed for the widespread laminar cracking on the Ekofisk Barrier Wall, although there appear to have been other causes as well, such as thermal strains and transverse strains due to the high circumferential forces.

Therefore for these high and critical ducts, the use of relatively thick steel pipe (eg 2 mm = 0.08 inches) and screwed or belled sleeve couplings is recommended. Multiple stirrups will resist laminar cracking.

Tendons for most building and bridge projects are typically handled and installed under dry and protected conditions, whereas tendons and their appurtenances for offshore structures are usually installed near or over salt water. More than usual care is required in order to prevent saltwater contamination, e.g., from spray, during delivery to the site and installation. After installing and stressing of the tendons, but before grouting and prior to concreting the anchorage pocket, such zones should be protected from spray by plastic covers.

This is especially critical where the anchors are located externally, only a meter or so above the water lines. There small bolts can be incorporated in the forms for the anchorage pocket and later used to hold plastic covers. Obviously, if the anchorages can be relocated inside, this problem can be avoided.

14.5 CONVENTIONAL REINFORCING STEEL: SPECIAL CONSIDERATIONS FOR OFFSHORE STRUCTURES

As noted earlier, the reinforcing steel densities in offshore structures are frequently several times those in more conventional structures (Fig. 14.11). This is due to a number of reasons, including:

1. The concrete sections are made as thin as possible in order to reduce weight during the floating stages of construction and transport. Sections are usually reinforced in both tension and compression.
2. The need to control crack widths to extremely small values in order to prevent leakage of water and oil, corrosions, and fatigue.
3. The need to resist bending and buckling under the very high hydrostatic

Figure 14.11. *Conventional reinforcing competes for space with post-tensioning ducts in the base raft of a Condeep platform*

pressures encountered during construction stages, especially that of deck or superstructure installation.

4. The need to provide robustness, that is, maintenance of high resistance over large strain deformations, due to the impact of accidental loads such as barge or boat collision, and due to impact of sea ice and/or icebergs. This requires three-dimensional confinement of the concrete core as well as primary and shear steel.

5. The many loading combinations, typically omni-directional, which the structure may be called upon to resist during construction, installation, and service.

For these reasons, densities of reinforcement have run from 275 to 350 Kg/m³ (465 to 600 lbs/yd³) and in specific zones have reached twice that amount or even more. This is physically accomplished by the bundling of bars in groups of two, three, or four, and the close spacing of adjacent bundles.

This obviously introduces problems in concreting, requiring a very workable mix of small coarse aggregate size and intense vibration.

The problem is intensified at locations where the bars of bundles must be spliced. These of course are staggered, so that only one bar is spliced at a location. Since splices will typically have to function in compression as well as tension, the laps should be tightly bound with tie wire at both ends.

Use of larger bars and mechanical connectors should be considered as a means of reducing congestion.

The primary reinforcing bars are often curved and are then installed in a wall of single or double curvature. There's an obvious conflict between tolerances in fabrication of the bar, concrete cover over the bar, and termination of the bar at splices or at ends. Thus accurate fabrication is essential. Where practicable, the bars should be furnished slightly longer than required by the design to ensure adequate lap length at splices and development length in a compressive zone. Consider mechanical splices.

Care should be taken that correctly positioned bars are not subsequently displaced by workers walking or climbing on them, especially during concreting.

Offshore structures typically require a great many stirrups to provide transverse shear connections, to prevent pull-out of curved tendons and primary reinforcing bars under tension, and to provide confinement for impact and local overload. Thus the number of stirrups or equivalent transverse steel is far greater than in more conventional structures (Fig. 14.12).

Closed-loop stirrups are very difficult to install, especially in the congested zones of bundled, closely spaced primary steel. It is usually impracticable to turn the "tails" into the core compressive zone. Alternate arrangements, such as Z-shaped stirrups and U-stirrups, the latter overlapping in the concrete core, have been employed. However, such stirrups must have tight bends in order to encase the primary bars and not invade the specified cover, and this limits their diameter. Thus it has often been necessary to bundle stirrups, adding to the congestion.

The mechanically headed bar (T-headed bar) appears to be a viable answer to this

Figure 14.12. Bundled vertical and circumferential reinforcement is confined transversely with multiple stirrups while post-tensioning ducts are extended up to the shaft.

problem, especially when it is locked behind the primary bars. Then the bar diameter is not limited, so one T-headed bar may replace 6 or 8 stirrups. This helps to minimize congestion for concreting and significantly reduces the labor time and cost of installation.

Many hundreds of thousands of these bars have been installed on offshore platforms in the North Sea.

Because of the large size of concrete offshore platforms in deep water, with consequent high hydrostatic heads, and the design demands to resist such loadings as ice impact, the concrete sections are typically thick, e.g., 500 to 2000 mm (20 to 80 inches).

Thus thermal strains will develop as a result of concrete hydration. Later, in service, thermal strains may develop from hot stored oil. Long-term settlement and secondary deformations may also produce internal strains. Thus even for sections where the analysis shows compression under all loading conditions, a minimum amount of reinforcement should be provided on all three axes. This is especially important at reentrant angles. The bars should be sized to prevent crack propagation regardless of how the tensile strain originates.

The trend today is to use even higher yield strength reinforcement, even up to 550 to 600 N/mm², (80,000 to 87,000 psi). This of course reduces congestion and installation labor, but makes it even more important that locations, splices, and confinement be accurately located. It also means that the steel must be protected in storage and handled with more care to prevent damage. Crack widths will be more critical, so secondary reinforcement in the form of mesh may be more frequently employed.

The steel will of course have to have the requisite properties of elongation at failure and ultimate strength. In locations subject to potential damage, it should be weldable.

14.6 SPECIAL REQUIREMENTS OF CONCRETING

The demands on the concrete for offshore structures are frequently very exacting and require an optimization of material selection, mix design, including admixtures, and placement techniques (Fig. 14.13).

These demands include:

- High compressive strength (50 to 75 N/mm² (7000 to 11000 psi) cylinder strength).
- High tensile strength (5 to 6 N/mm²) to resist cracking and to develop shear capacity.
- Low heat of hydration to prevent development of excessive thermal strains.

Figure 14.13. Concreting of the roof domes of offshore platforms requires careful balancing of conflicting criteria of placeability, strength, and durability.

- Low permeability and diffusivity to provide durability.
- Good bond characteristics, to develop the reinforcing steel and inhibit fatigue.
- To these may on occasion be added light unit weight, to minimize draft and to aid floating stability.
- Arctic and subarctic structures may add abrasion resistance and freeze-thaw resistance.
- Fire resistance and thermal conductivity may be important in certain installations.
- Resistance to dilute acids may be required in those cases where anaerobic sulfide-reducing bacteria produce H_2S which, in the presence of seawater and air, may produce dilute H_2SO_4. Oil storage tanks and outfall sewers are subjected to this type of attack.
- Because of the chemical ions present in seawater—sulfates, alkalis, and chlorides—the materials and mix must have high alkali aggregate resistance and the aggregates must be sound in the presence of sulfates. The concrete must resist salt scaling.
- The concrete mix must be capable of being placed around, under, and between the congestion of reinforcing and prestressing steel and must have minimal entrapped air and bleed.
- Placement techniques must ensure thorough consolidation, a dense and uniform surface, with an absence of rock pockets and honeycomb. "Bug" and "worm" holes and surface cusps must be minimal and not penetrate more than 5 to 10 mm ($\frac{1}{4}$ to $\frac{1}{2}$ inch).

Despite these severe and often apparently conflicting demands, suitable concrete mixes and placement techniques have been developed and maintained with excellent consistency on the large offshore structures of the North Sea (Fig 14.4) as well as specific structures elsewhere, such as the CIDS platform in the Alaskan Beaufort Sea.

Among the solutions which have been found to achieve this are:

Cement: A relatively rich mix of a Portland cement having moderate C3A content or a blast-furnace slag cement with 70 to 85% BFS.

Coarse Aggregates: To be small, e.g., 10 to 20 mm ($\frac{1}{2}$ to $\frac{3}{4}$ inches) maximum, so as to promote workability and tensile strength.

Natural Sands: To be a relatively high proportion of total aggregate, e.g., 45 to 50%.

Fly Ash: To replace 15 to 20% of the cement.

Microsilica (Condensed Silica Fume): 3 to 6% of the cement.

HR WR Admixtures: Added at the appropriate stage.

Low W/CM Ratios: For example, 0.37%.

Minimum Silt Contents in the Aggregate: Rewashing and re-screening may be required.

Figure 14.14. *Ekofisk Oil Storage Caisson, shown here under construction, has been in service more than 20 years, with no evidence of any corrosion or loss of concrete durability.*

Workability: As measured by slump, of 150 to 250 mm (6 to 10 inches).

Air Entrainment: Where freeze-thaw attack is a consideration, air content should be 6 to 8% and the air void or pore spacing factor should be less than 0.23 mm. Sequence of adding the air entraining admixture and the high-range water-reducing admixture may have to be staggered, e.g., 50% of each at a time.

Thorough Mixing: Mixing times may have to be extended.

Pre-Cooling of the Mix: In hot climates or tropical waters this may be achieved by such means as evaporative cooling of aggregates, batching of crushed ice for the water, and injection of liquid nitrogen.

Design of Forms and Care in Joints: To prevent reentrant angles or fins that will restrain contraction under cooling.

Use of Insulated Forms and Blankets: To minimize through-thickness thermal strains. It is especially necessary to avoid exposure of uninsulated steel forms and fresh concrete to chilling winds.

Good Internal Vibration: To consolidate the concrete and prevent delamination and internal honeycomb, etc. External vibration may be used to supplement but not replace internal vibration.

Curing: Water curing is usually impracticable because of the size and the environment in which the structures are typically built. Membrane curing compounds are more practicable and positive. On thick members and in hot climates, at least two coats, four to eight hours apart, may be required.

Where the structure is built in the floating mode, so that recently cast concrete is submerged in salt water within a few days, consideration should be given to spraying on an epoxy membrane coating to prevent chloride intrusion, since the permeability of concrete is many times higher at these early ages. This would seem especially advisable where there will be air on the inside face of the wall, e.g., in an utility shaft, since over the years, chloride ions will diffuse to the inner face and concentrate there. With the ready availability of oxygen, eventual corrosion of the reinforcing steel is probable.

The excellent performance of concretes generally conforming to the principles of the above guidelines or their variants is attested by virtual corrosion- and problem-free performance of concrete offshore structures in the North Sea approaching 20 years (Fig. 14.14) and in the arctic and subtropics for over 10 years as of the writing of this book. It is also attested by the early nonprestressed forerunners of today's offshore structures, such as the lighthouses of the Baltic, Canada, Ireland, and the United Kingdom, the Mulberry caissons of World War II, and the Southampton lighthouse of World War I vintage.

At the same time caution must be continually exercised, especially in new regions, since alkali-aggregate reactivity, disruptive sulfate expansion, and freeze-thaw attack may occur many years after construction. Long-term satisfactory performance in land structures may not necessarily prove the same for offshore structures where water saturation is inherent.

14.7 GRAVITY-BASE CONCRETE STRUCTURES OTHER THAN OFFSHORE OIL PLATFORMS AND TERMINALS

The subaqueous tunnels ("immersed tunnels" or "tubes") of Western Europe are mostly constructed as huge rectangular boxes, prestressed on all three axes. The prestressing enables the structure to resist the external hydrostatic and backfill pressures and ensures watertightness (Fig. 14.15).

Since the loads on the roof of such a tunnel are not fully applied until the segment is sunk to grade, temporary external counter-stressing may be provided to prevent excessive cracking.

These tunnel segments, typically 100 meters (330 feet) in length, have very massive walls to provide deadweight and to resist design loads from water and backfill and accidental loads such as a dropped anchor or sinking ship. At the same time, they are much more lightly reinforced than the typical offshore platform. Therefore increased attention has to be given to preventing excessive thermal strains and consequent cracking.

In Japan, 12 prestressed concrete tunnel segments, each 130 meters (425 feet) long, 40 meters (130 feet) wide and 10 meters (33 feet) deep, were cast in a construction basin for the southern extension to the Tokyo Harbor Expressway. The

Figure 14.15. *Precast concrete underwater vehicular tunnel segments are posttensioned in stages to withstand hydrostatic pressures of submergence.*

sections are enclosed on the sides and bottom with steel plates and the tops are covered by a rubber membrane.

Caissons for bridge piers, such as those of the Great Belt Bridge (Fig. 14.16), do not experience the intense storms of the open sea, but must have a much longer service life, typically 75 to 125 years. Hence durability becomes a dominant criterion.

Recent experience has demonstrated the phenomenon that slip forms slightly alter the internal structure of the concrete, producing numerous microscopic laminar cracks parallel to and near the surface. While this does not appear to have had any adverse consequences as far as permeability is concerned, it would seem best to require temporarily fixed panel forms for concrete in the splash zone. These are required on the Great Belt Bridge.

Consideration should be given to epoxy coating the surface of the concrete in the splash zone in all regions except those exposed to freeze-thaw attack. The experience on the Saudi Arabia–Bahrain Bridge piers (King Fahd Causeway) in the Middle East environment has shown the efficacy of epoxy coating in preventing chloride penetration of the concrete. See Figs. 12.1 and 12.2.

Epoxy coated reinforcement has not been felt to be necessary or desirable for the offshore petroleum-related platforms but may be considered for bridge piers because of the long design life. Opinions vary on this matter because of the concern over adhesive bond and cracking of the cover. This is a special problem for the hoop steel in columns, where the sharp bends are vulnerable both to cracking of the epoxy coat

Figure 14.16. *The caissons for the bridge piers of the Great Belt Crossing in Denmark have post-tensioned base rafts to withstand the combination of global and local bending.*

and to spalling of the concrete cover. If epoxy coated bars are used on the bridge piers, it would appear advisable to increase the concrete cover to 75 mm (3 inches).

Similar concerns need to address the fatigue characteristics of epoxy coated large bars typical of bridge piers in submerged concrete. There is an indication that cyclic loading may adversely affect the bond.

The Oosterschelde Storm Surge Barrier, across the Eastern Scheldt Estuary in the Netherlands, is the culmination of the Delta Plan to protect the Netherlands from flooding from the North Sea. It is one of the great achievements of the 20th century (Figs. 14.17, 14.18).

It comprises 66 large piers which, in turn, support the movable gates. These piers are designed for a lifetime of 250 years, during which they must resist the North Sea storm waves and the corrosive sea environment.

BFS-cement was used for the concrete. The pier structures were cast in a temporary construction basin. Multi-axial posttensioning was used to supplement the passive reinforcing steel and provide a crack-free structure.

After completion in the basin, the piers were partially floated, partially lifted for transport to the site.

14.8 GRAVITY-BASE OFFSHORE TERMINALS

The emergence of very large oil tankers (VLCC), iron ore carriers, and coal colliers, with their deep drafts and large displacements, has led to the construction of a number of deep water terminals in the form of gravity-base concrete caissons.

Figure 14.17. Prestressed concrete piers for Oosterschelde Storm Surge Barrier are constructed in basin.

Figure 14.18. After the basin has been flooded, the piers of the Oosterschelde Barrier are partly lifted, partly floated, to site.

Figure 14.19. Precast concrete panels were assembled in a construction basin, then jointed with cast-in-place concrete and post-tensioned Hay Point Coal Terminal, Queensland

Figure 14.20. Towing the completed concrete caisson, with ship loading facilities mounted, to the offshore site.

Several of these have incorporated prestressing technology and a few have extended this to include precast elements which have been joined by prestressing.

The most notable of these is the Hay Point Coal Shipping Terminal in Queensland, Australia. Three large berthing structures support the ship-loading facilities. They, plus seven mooring dolphins, were constructed as gravity-base caissons, built in a construction basin, then completed inshore and towed to the site, where they were seated on a prepared foundation (Figs. 14.19, 14.20).

14.9 REPAIRS AND ALTERATIONS

The very nature of bottom-founded fixed structures, whether serving as operating oil platforms, surrounded by boats and barges and visited at frequent intervals by captive tankers, or serving as bridge piers in a crowded waterway, means that they inevitably will be damaged by minor collisons. In addition, operating platforms may be called upon to service adjoining fields, necessitating additional risers and penetrations into shafts.

Thus repairs and alterations will have to be made. As far as the concrete is concerned, it is a relatively straightforward matter of securing an external cofferdam, cutting out the concrete, replacing or augmenting the reinforcing steel, with welded splices and/or drilled-in and epoxy grouted dowels. The concrete is then placed, using a "window-box" in the inside forms to ensure complete filling and to allow bleed water to percolate upward. After concreting, the window box protrusion is carefully chipped off and ground smooth.

To ensure complete filling and watertightness at the top, a tube or tubes may be inserted there before concreting. Later, epoxy may be injected.

However, if a prestressing tendon runs through the area to be cut out and replaced, and if this tendon has been or must be severed, then it should be done with a burning torch using a yellow, low-heat flame so as to gradually anneal the cold-drawn wires and allow them to elongate without shock. After the initial cut, they may be cut wherever required within the cut-out zone.

Restoration of prestress is usually not required, there usually being enough excess in the original installation.

However, where this is not so, there are several alternatives, each representing a higher degree of restoration, as required.

The new prestress tendons, being short and requiring positive anchorage, are best comprised of bars.

In scenario number one, holes may be drilled in the concrete walls of the cut-out, bars inserted, and anchored by epoxy injection. Prestress may then be applied by torquing a male-female nut in the middle of the opening, aided by scissor jacks. Because of the short length, the degree of prestress will be very indeterminate.

Longer bars may be inserted in diagonally drilled holes in thick walls, running across the cut-out section. The angle should be selected so as to provide an adequate development length. Care should be taken to ensure that the stressing does not in turn create cracked or overstressed areas at each anchorage.

The last scenario applies where restoration of full prestress is essential. In this case, one or more external tendons are placed and anchored on new bolsters which have been doweled into the walls well above and below the cut-out zone. These bolsters may be post-tensioned to the walls by short bars in drilled holes through the wall thickness. To prevent cracking behind the bolsters, the anchorages should be staggered.

Finally, after all this has been done, including prestressing, the whole assemblage should be coated with coal-tar epoxy to protect against corrosion.

REFERENCES

1. API-RP2A, "Recommended Practice for Planning, Designing and Constructing Fixed Offshore Platforms," American Petroleum Institute, Washington, D.C. 1990
2. ACI 357 R-84, "Guide for the Design and Construction of Fixed Concrete Offshore Structures," American Concrete Institute, Detroit, Michigan. 1984
3. CSA Preliminary Standard, CSA-S-474-M1989 with commentary, "Concrete Structures for Offshore and Frontier Areas," Canadian Standards Association, Toronto, Canada, 1989.
4. Gerwick, B. C., Tr., *Construction of Offshore Structures*, John Wiley & Sons, New York, 1986.
5. FIP Recommendations, "Design and Construction of Concrete Sea Structures," Structural Engineers Trading Organization, London, 1985.
6. ACI 357-IR-85 "State of the Art Report on Offshore Concrete Structure for the Arctic" American Concrete Institute, Detroit, Michigan 1985

CHAPTER 15

Prestressed Concrete Floating Structures

15.1 INTRODUCTION

Prestressed concrete is especially well-suited to the construction of floating structures. It possesses the essential qualities of strength, watertightness, adaptability to shell and double-curvature configurations, durability, vibration damping, fatigue resistance, redundancy, repairability, and economy. Its apparently detrimental weight-to-strength ratio turns out, in many actual cases, not to be a serious drawback; in some cases it may even be an asset.

One of the principal advantages of prestressed concrete as compared to conventionally reinforced concrete is the high resistance to fatigue. Since cyclic loading is much more damaging to saturated and submerged concrete, prestressing is a requisite for successful utilization of concrete in floating structures.

Prestressed concrete has a favorable mode of energy absorption, even under accidental conditions, such as grounding, collision, or explosion. It develops local cracks and spalling, but is not subject to crack propagation, ripping, and tearing, as is typical of steel.

Maintenance of concrete in seawater is inherently minimal if the structure has been properly designed and constructed. Weeping and condensation have been virtually absent in concrete hulls in service. The high impact resistance and fatigue strength of prestressed concrete at low temperatures make it the preferred material for service in the arctic and subartic.

Most floating and submerged structures must resist bending loads which tend to cause buckling of the skin. Concrete, in the thickness necessary to provide the requisite strength, has rigidity to resist local buckling and excessive distortion. Shell action can be utilized very effectively by proper design.

The inherent ability of concrete to resist high compression efficiently makes it especially suited to deep draft vessels, such as the semi-submersibles for tension-leg platforms (TLP) where hydrostatic forces dominate.

Prestressed concrete floating structures will have a self-weight draft of two to four times that of steel. Hence, when draft is critical, the prestressed concrete may be combined with steel in a variety of ways, to be discussed later in this chapter.

Prestressed concrete offers both direct and indirect economies. The direct savings are the lower cost of the basic materials, and the more efficient use of the materials, especially when subjected to multi-axial loading conditions.

High-strength concrete is extremely efficient and economical as a compression-carrying member, whether the compression be from external loads or internal prestress. High-strength steel, typical of prestressing tendons, is likewise economical and efficient if it can be utilized in tension and protected from corrosion.

The recent development of high-performance lightweight concrete is especially meaningful for floating structures, reducing draft and enhancing the endurance against fatigue.

The indirect advantages of concrete floating structures include inherent durability, thus requiring little or no additional corrosion protection. Local materials and labor are utilized to a high degree, enabling high-quality construction in developing as well as developed countries. The construction sites are free from many of the restrictions that affect conventional shipyards.

Actually, the first application of reinforced concrete in the 19th century was by the French landscape architect Lambot, who constructed reinforced concrete boats. Some of them are still afloat over 100 years later. Prestressing, relatively new to such applications, improves the structural performance, crack resistance, economy, and durability of these structures.

When this chapter was written in the first edition (1971), the actual experience in prestressed concrete floating structures was extremely limited, while that of concrete floating structures in general was largely associated with the emergency construction of concrete ships and floating oil storage in World Wars I and II (Fig. 15.1).

Today, 20 years later, we can add a wide list of successful applications:

- A floating LPG (liquefied petroleum gas) storage vessel with 15 years successful experience in the Java SEa; eight times the displacement of any previous concrete vessel.
- Cargo and oil barges in the Philippines.
- A floating container terminal in Alaska.
- A floating phosphate extraction and process plant in Mexico.
- Floating breakwaters for small boat harbors.
- Floating pontoons for construction, plant, and equipment.
- Floating ferry terminals.
- Many floating marina docks.

Figure 15.1. Innovative concept for reinforced concrete ship of World War I.

- Prestressed concrete floating bridges in Washington and British Columbia, with a new one under construction in Norway, and one in design in Greece.
- A tension-leg platform under final design in Norway.
- Floating drydocks.
- Floating guide walls for locks on river systems.

The successful performance of the above structures, along with their economy and low maintenance, has led to the conceptual design of many other applications, ranging from floating airports to target submarines. Meanwhile, a substantial number of huge offshore platforms, underwater tunnel segments, and coastal terminals of prestressed concrete have been constructed, which must float during fabrication, transport, and installation. The principle of floating prestressed concrete is now well established.

15.2 PROBLEM AREAS AND SPECIAL TECHNIQUES

15.2.1 Durability

Earlier the statement was made that one of the great assets of high-quality prestressed concrete was its durability. Properly designed and constructed prestressed concrete is undoubtedly the most durable construction of all the practicable structural materials available for seawater environments.

However, it must be emphasized that this durability is not automatically obtained. Careless or improper practice may lead to disastrous results, most commonly due to corrosion of the reinforcing steel. The frequency of such adverse results of concrete sea structures is remarkably low, considering all the variables involved. However, when trouble has occurred, it has usually been general and progressive.

The exposure for floating concrete structures is of three types. The first is deep submergence with air on the inside; e.g., the hull of a deep draft vessel. In this case, the salt water tends to move through the concrete and evaporate from the inner face. The chlorides penetrate both in aqueous solution, through permeation and capillarity, and also through ion diffusion. On the inside face, salt crystals are left by evaporation and may set up electrolytic salt cells. Since oxygen is readily available, corrosion can take place. The greater the permeability and diffusibility of the concrete, the more serious the corrosion will be, particularly since carbon dioxide is available from the air to reduce the pH of the cement paste.

For such structures, therefore, the concrete should be as dense and impermeable as possible, adequate cover should be provided over the steel, and consideration should be given to an impervious coating on the outside.

The second exposure is in the splash zone, especially where the concrete penetrates the air-sea interface and is subjected to alternate wetting and drying, with adequate oxygen to drive the corrosion process. In some environments, freeze-thaw action and ice abrasion must also be considered. Use of a maximum air entrainment and highly impermeable concrete appears essential in a freeze-thaw saltwater environment.

The third exposure arises when concrete structures are used for storage of fluids such as oil and petrochemicals. Specific precautions relating to these chemicals must be taken. The general rules of a dense, impermeable concrete, low W/C ration, high cement factor, and adequate cover are helpful in all cases, but special additives (e.g., air entrainment, pozzolans, including condensed silica fume) or special coatings (such as polyurethanes or epoxies) may be of value.

Thermal strains occurring during manufacture may cause cracking of the external bulkheads, especially when concreting takes place during a period of low temperatures and winds. Insulation of the forms and blanketing of the bulkheads after form removal may be the optimum methods for controlling this.

Other means of reducing thermal strains are the replacement of a portion of the cement, say 15%, by PFA, and pre-cooling of the mix by batching ice or injecting liquid nitrogen.

As noted in Section 1.2.5, floating structures are subject to fatigue, due to the many hundreds of thousands and even millions of cycles of bending under dynamic wave action. Especially damaging are those cycles which cause the concrete hull to go into tension as this can lead to cracking even though the magnitude of tensile stress may be less than half the static strength in direct tension. In the typical floating structure, the bottom and deck of the hull will have membrane stresses, more or less uniform throughout their thickness. Hence any cracks will be through cracks subject to leakage. Continued cycling will widen the cracks and lead to bond slip of the reinforcing steel and eventually to failure.

Prestressing is generally considered essential in order to keep the bottom and deck of the hull always in compression.

15.2.2 Unit Weight

The weight of prestressed concrete may be in some cases a detriment to its use. Despite the value of the concrete's increased thickness, providing structural rigidity and resistance to local deformations, the strength/weight ratio is not as favorable as steel hulls of similar capability. Thus, for operations in shallow water, methods will have to be adopted to minimize the draft.

Since the wall thickness of concrete barges, ships, and hulls is usually fixed by parameters other than strength alone, the use of structural lightweight aggregate concrete will often show a substantial reduction in total weight and draft. The unit weight of high-quality, impermeable, structural lightweight aggregate concrete is approximately 75% that of conventional concrete, having a specific gravity of about 1.8. French research engineers have reportedly developed a polymer concrete with a specific gravity approaching 1.0 while developing a compressive strength of 30 to 35 MPa.

Another method of reducing weight is to reduce the wall thickness itself. Frequently wall thickness is determined by the need to fit in reinforcing steel and prestressing ducts. The use of pretensioning, instead of post-tensioned tendons, with their sizable ducts, is sometimes feasible. This method was employed for 20 cargo barges in the Philippines.

Posttensioned tendons and their ducts may be reduced in cross-sectional area by use of special strand and fewer strands per tendon, etc. Internal ribs may be provided paralleling the post-tensioned tendon path, so as to provide the necessary cover at this location, without increasing the overall thickness of the wall.

Although not yet utilized in the original construction of floating structures, the use of external tendons, a technique from state-of-the-art bridge construction, has been effectively utilized to strengthen existing floating bridges and has potentially interesting advantages for new structures as well.

External tendons would be located inside the pontoons. They would be encased in a polyethylene sheath and grouted after stressing, so that they would be fully protected from corrosion. This technique, which has been widely utilized in bridges in Europe, permits a reduction in wall thickness of internal bulkheads. It also permits future replacement of tendons.

As in bridges, the detailed design of anchorages and deviator blocks is very important.

Reference should be made to Sections 2.3.6 and 12.18 for guidance on construction with external tendons.

Where the determining factor for wall thickness is local strength and rigidity, ribs may be used to provide increased stiffness and flexural strength to the sides, decks, hull, and internal bulkheads.

Wall thickness is often controlled by considerations of shear.

Cellular construction (hollow-core) for walls has been proposed and is certainly worth careful consideration on very large vessels such as storage vessels. The cellular construction in effect gives a double wall with diaphragm connectors, which is a very efficient cross section in bending. The individual wall thicknesses (minimum thicknesses) must still be selected to give adequate cover over the reinforcement and space for proper concrete placement.

15.2.3 Construction

Floating structures are very sensitive to tolerances, especially in wall thickness. The walls and slabs typically have reinforcement on both faces, both directions, as well as central prestressing ducts. This makes it very difficult to place and vibrate the concrete so as to thoroughly consolidate it without honeycomb and laminations.

Precast concrete slabs have been used on a number of floating structures including the floating container terminal at Valdez, Alaska. For these, precast slabs were stood up on steel "chairs" in the construction basin, so as to form the longitudinal bulkheads and sides as well as the transverse bulkheads. Then the bottom slab was placed, making sure to drive out all air by vibration from under the support of the wall slab.

Timber ledges were bolted to the top of the walls, on which precast deck slabs were then set. Joints were of fine concrete, well vibrated.

Other systems of fabrication have involved precast match-cast hull units, each being then concreted against its mate so as to fit tightly against it when later assembled. Two adjoining units should be steam cured together (Fig. 4.15).

Other techniques for concreting deep, thin walls have included pumping up from the bottom, and placement by tremie from the top.

During assembly of slabs or units in a construction basin or dock, differential drying out may occur. Later, on flooding, differential absorption may take place. Some of these strains are of the same order as those from prestressing. They do not significantly affect the ultimate strength but may affect cracking. Judicious use of sealants during construction can minimize this problem.

Especially critical zones occur in the bottom hull and deck opposite the supporting bulkhead walls. Negative moment is a maximum, which can be especially critical on the bottom slab, where it is also accompanied by high shear due to hydrostatic forces.

The rules of classification societies generally require that no tensile stress exist under the 20-year return period wave, a condition difficult to meet.

Haunching of the bottom slab adjacent to the walls will usually be found to be efficacious in meeting this requirement.

Stirrups must often be employed to accommodate the shears, to tie in any changes in curvature of the tendons, and to confine zones where longitudinal reinforcing bars are lap spliced.

It is important to ensure that water flooding into the dock or basin have free penetration under the bottom, so as to overcome any suction. The hull concrete will generally not be prestressed until the hull is waterborne, since the post-tensioning will be resisted by the friction on the bottom.

15.2.4 Joints

Prestressed concrete structures of the sizes contemplated usually must incorporate some joints, whether they be construction joints with cast-in-place concrete or joints between precast segments. This is no different from other large structures, except that joint details are more critical, and the consequences of poor joints more serious. Poorly made joints may leak or weep. They may provide paths for electrolytic cell formation and corrosion. They may be susceptible to fatigue. Poor joints are usually caused by lack of consolidation of concrete in the joint due to congested space, by shrinkage in the joint, or by lack of bond with the adjoining concrete. Structural impairment may be caused by lack of shear transfer from one face of the joint concrete to the other face. Where reinforcement must be spliced in joints, the highly congested area may cause numerous difficulties, such as spalling from welding heat, eccentric stress transfer at bar laps, inability to place concrete around the bars, and lack of proper cover where two bars overlap.

The faces of the joints of concrete segments must be properly prepared, by casting of keys and high pressure water jetting or other methods. Even with match-cast joints (dry or epoxy joints), a light jetting may be desirable to remove the skin of cement paste from the faces.

Use of an epoxy bonding agent, for bonding of the segment to the concrete in the joint, or from segment to segment, will prevent cracks from forming at the boundary.

If the joint is to be concreted, i.e., if it will be a cast-in-place joint, best results are obtained when the concrete is poured and internally vibrated. This means a joint width of 75 mm (3 inches) as a minimum, preferably 100 mm (4 inches), and the use of pea-gravel (8 to 10 mm) ($\frac{3}{8}$ inch) as the coarse aggregate in the joint concrete mix.

Shrinkage can be minimized by the use of a low W/C ratio in the joint concrete. Since workability is essential, a water-reducing admixture is desirable. Prestressing across the joint of external hull elements is essential.

The steel details at the joint require extremely careful design. An excellent method is to prepare a full-scale drawing to insure that the bars and tendons, etc., can be physically fitted into the space provided. Eccentricity should be avoided in the splices of bars carrying principal stresses; this can usually best be accomplished by mechanical splices, butt welding, or use of an angle iron splice bar. For these reasons, use of numerous small-diameter bars will generally be preferable to a few large bars. Sleeve details for connecting ducts must be provided, as well as means of sealing the splice grout-tight, as by screwed coupling or sleeved wedge fitting.

Proper attention to detailing and proper execution in the field can assure that full continuity will be obtained across joints and that the joints will be crack and trouble-free.

15.2.5 Ferro-cement

Ferro-cement techniques have been successfully applied for many years to small boat construction, starting generically with Lambot's boats in 1855, resurrected by Nervi during the 1940's, and becoming an established method of boat building in the 1960's, although now largely replaced by composites and plastics. Prestressing and ferro-cement techniques are compatible. The ability, through prestressing, to keep the concrete under compression and to make the structure act as a whole, is important as is the structural strength attainable through prestressed ribs and beams.

In the other direction, however, prestressed concrete floating and submerged structures have much to gain from ferro-cement techniques. The use of closely spaced wire mesh near the concrete surface holds the concrete together under impact and collision and restricts and limits flexural and shrinkage cracks. In some cases, the use of the multi-layers of wire mesh may provide a needed resilience and flexibility. Finally, the advantages are shown of a high steel proportion, in the form of finely divided steel, whether in the form of mesh, tendons, or small reinforcing bars.

A prestressed ferro-cement soil boat won the ocean race from Sydney, Australia to Tasmania, a demonstration of the seaworthiness potential of this system.

15.3 DESIGN

Recommended design practices for prestressed concrete floating structures have now been standardized and documented. A number of these are listed in the references at the end of this chapter.

The design for floating and submerged structures should follow good prestressed concrete design practice with special attention to the following:

1. Durability in seawater environment.
2. Reversals of stresses due to waves (hog, sag, and torsional moments and shears). Fatigue endurance.
3. Pressure differentials due to potential hydrostatic heads in operating and accidental conditions.
4. Accident conditions: explosion, internal or external; collision, etc.
5. Detailing of joints and connections.
6. Abrasion from ice, sand, and mooring lines. Impact from other vessels and docks.
7. Marine growth.

8. Manufacturing and construction aspects; stages of construction including draft, stability, and connections. Special attention must be paid to tolerances in unit weight and dimension, and their effect on draft and stability.
9. Operational aspects: draft, stability, mass.

Structural concepts and techniques which can be beneficially exploited in floating vessels are:

1. Haunching of slabs near intersections, so as to minimize flexural stresses in negative moment areas and resist shears.
2. Shell configuration, so as to place concrete under uniaxial or biaxial compression under hydrostatic loading.
3. Use of shell configuration for internal bulkheads to give desirable deformations under hydrostatic loading of hull and differential thermal strains.
4. Use of internal trusses of prestressed concrete or steel tubulars, to provide internal global shear resistance.
5. Use of high-strength concrete.
6. Use of high-performance lightweight concrete.
7. Special reinforcing details for transverse shear and confinement.
8. Use of mesh and fibers in appropriate locations.
9. Cellular and hollow-core slabs and shells.
10. Unbonded tendons (plastic-sheathed).
11. External prestressing (inside the hull).
12. Mechanical splicing of conventional reinforcing steel bars.

15.4 LAUNCHING

15.4.1 General

Almost all structures designed for submerged or floating service have a common characteristic: they are extremely large and massive. This poses serious problems for manufacture and construction. Just as with long-span bridges, the methods of manufacture and construction must be considered as an integral part of the design process. Constructors have evolved some extremely ingenious methods for manufacture, launch, assembly, and final installation. A brief review of a number of these methods follows, to serve as a guide to past practice and a stimulus to the development of new and better methods. Certainly this is a field where ingenuity has wide scope for further advance.

15.4.2 Launching from Ways

Many concrete barges and caissons have been constructed on launching ways, and slid down the ways to flotation in the water. The ways must be built to withstand the loads involved and provide necessary structural support to the barge or caisson during both manufacture and construction. The caisson or barge must have the structural strength needed to withstand the moments and shears during launching. Side launching was used to launch a prestressed concrete phosphate-processing barge in Singapore (Fig. 15.2a)

It is usually easier to construct a barge on the level rather than in an inclined position. One solution is to employ a launching cradle, which rides down the inclined slope, while keeping the barge or caisson level. Another solution is to build the barge or caisson on the level at the head of the slope and then rotate it, by jacking beams, to the inclination of the ways.

Most concrete sections have been launched by gradual lowering down the ways

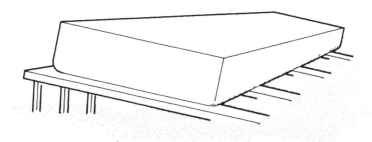

(a) *Side launching from ways*

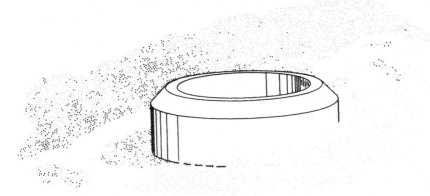

(b) *tidal launch*

Figure 15.2. *Launching (float-out) of prestressed concrete hulls from building site to water-borne condition has used many methods.*

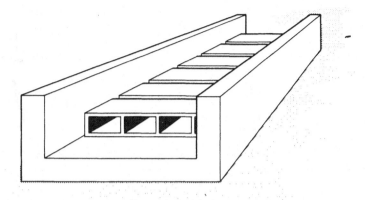

(c) *assembly in drydock or graving dock*

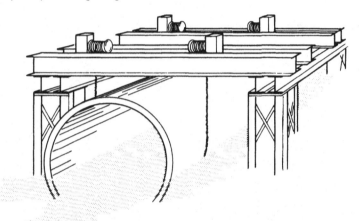

(d) *lowering down*

Figure 15.2. (Continued)

by hoists, or, if moving freely under gravity, with the ways extended well out into the water. However, the side launch (free-launch) technique, so successfully used with steel ship hulls, was adopted for the barge in Singapore.

15.4.3 Direct Lift

Floating derricks and shear legs with capacities from 600 to 3000 tons (and more) are now available for direct lift from barge or bulkhead, and for lowering into place. Direct lift, while afloat, was utilized in the assembly of the CIDS offshore platform in Japan. Syncrolift facilities in a shipyard may be similarly used.

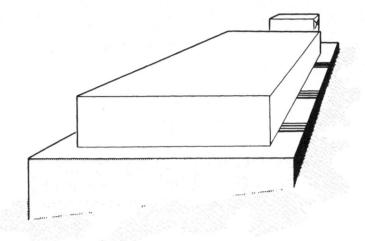

(e) barge launching

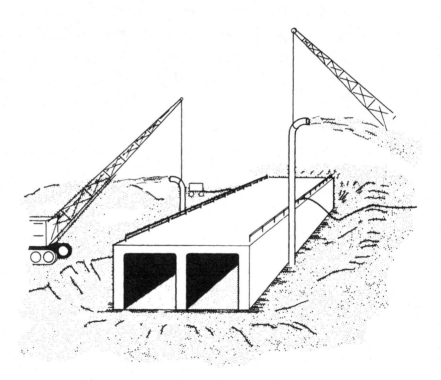

(f) sand jack

Figure 15.2. (Continued)

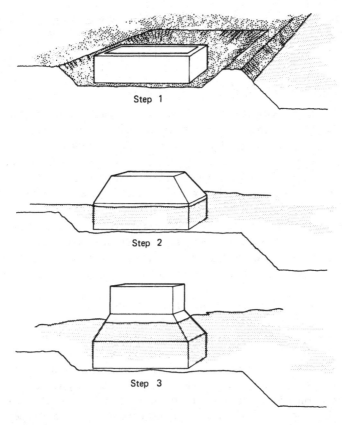

Step 1

Step 2

Step 3

(g) successive basins.

Figure 15.2. (Continued)

15.4.4 Tidal Launching

A large number of concrete structures have been launched by making use of the tidal rise, either alone or in conjunction with other methods (Fig. 15.2b). Temporary protection of the fresh concrete from seawater must be given during fabrication, e.g., in the form of a membrane or coatings.

15.4.5 Graving Dock/Drydock

Many concrete ships, tubes, etc., have been constructed in a floating drydock or graving dock (Fig. 15.2c). These have the advantages of providing stable support and dry working conditions during construction, and controlled flooding and launching. Charges for the use of drydocks are high, and the speed of construction of a concrete hull, for example, may be slow. A drydock is usually a very congested working area, and frequently presents some problems of accessibility.

Many of these problems can be minimized by manufacturing precast segments or sections elsewhere, floating or lifting them into the dock, and making the final connections in the dock. Means must be provided for alignment and positioning as each section or segment is set on the keel blocks. Guides can be used for this purpose, assisted by jacks.

15.4.6 Construction Basin

Many of the largest concrete structures have been cast in a basin, which is later flooded, the dike breached, and the structure floated out (Fig. 14.19).

In the Netherlands, for example, a bow levee is constructed, using a sand dike with sheet-pile core. After excavation, the basin is dewatered, and the structure constructed. Then the basin is flooded and the dike opened.

In other areas, basins may be excavated, kept dewatered during construction, then flooded and an access channel dug to the waterway.

A steel sheet-pile cut-off wall may be necessary to control water inflow, especially if the basin is to be used more than once. Prestressed concrete caissons have been used as gates where multiple use is contemplated. Well-point systems may be needed with French drains or under-drains to keep the basin dry during structure fabrication.

Basins should be designed with full consideration for access of materials and equipment. Because of long side slopes for deep basins, it may be necessary for the cranes, trucks, etc., to work in the bottom, or else for access trestles to be constructed.

It is very important that a good working surface be provided, such as a thick blanket of rock or even a concrete slab. Drainage must be provided.

Support during construction should be well thought out. Pile supports may be necessary in some soils. For many structures, concrete or timber sleepers on a rock base are adequate; they can be ballasted to grade, and they are sufficiently flexible to distribute the load more or less uniformly.

15.4.7 Lowering Down

A number of smaller floating structures have been launched by lowering down. A platform is built on or over the water, supported at the edges, or outside the final concrete structure. This platform may support the load partially by flotation (Fig. 15.2d).

At time of launching, overhead beams are placed and either a hoist, jack rods, or jack cables are used to lower the structure and platform to a point where the structure can float free. The platform is then raised for a second use. For reasonably sized structures, in which many uses are contemplated, this is a very efficient system.

The mechanized version of this method is "Syncrolift."

15.4.8 Barge Launching

Concrete barges, caissons, and similar structures are frequently constructed on a barge. Launching is then accomplished by flooding the supporting barge (Fig. 15.2e).

Once the deck of the supporting barge dips under water, the stability of the barge waterplane is lost. However the waterplane of the structure itself may still provide stability. If the supporting barge and its structure list, the caisson then slides sidewise. Such intentional launching during sinking is often aided by flooding the support barge so as to initiate the list in the desired location. However, side launching places very heavy concentrated loading at the edge of the barge. If supports are not properly designed, this can cause buckling of the support barge side or rupture of the concrete caisson.

The operation may be controlled by performing the launching in shallow water with a relatively flat bottom, so that the barge cannot tip more than an acceptable degree. Alternatively the support barge can be guided by spuds and hoists or jack cables. Additional stabilization can be accomplished with side columns or pontoons; the support barge then becomes a small floating drydock.

The support barge must be able to resist the external pressures at the maximum depth of launching without inward buckling. For safety, the maximum depth of submergence should be controlled. This is usually attained by launching in a predetermined depth of shallow water. When launching must be done at sea, control may be effected by winching down from adjoining barges. Once the structure floats off, the supporting barge, being completely submerged, has no waterplane stability and will need guides to prevent it from uncontrolled list as it rises up.

15.4.9 Construction Afloat

By this method, a base segment must first be built by one of the other methods described in this section. After it is launched, concrete construction is continued while the structure is supported by the buoyancy of the water.

Great care has to be exercised to ensure that during the construction, adequate freeboard is maintained against overtopping, adequate stability against excessive tilt, and especially adequate moment capacity. This latter is very demanding, since the walls being constructed have little capacity for compression under the sag condition. They can be given adequate tensile capacity in this stage by reinforcement and/or prestressing placed at the top of intermediate stages. It is usually desirable to maintain a slight hogging moment in the structure, since temporary additional tensile capacity is easier to provide than additional compression capacity.

The base raft may be a steel shell, which is designed to work in composite action with the concrete placed afloat.

15.4.10 Sand Jacking

An old method for lowering heavy structures is by the use of the sand jack. As applied to a large concrete floating structure, a slip would be dredged to a depth adequate to float the new structure. Jet and suction pipes are installed on the bottom. The slip will then be backfilled with clean sand to the ground line, at a working level which is well above water and accessible.

The structure is now constructed.

Then the sand is progressively excavated from under the structure's base slab by use of jets and suction pipes (Fig. 15.2*f*). Continuous soundings and deflection measurements are made during the sand dredging phase, until finally the structure floats free.

The slip may then be refilled with sand for a second operation.

The advantages of this method are the elimination of costly bulkheads, dewatering systems, and gates; also the greater accessibility for all construction operations.

15.4.11 Successive Basins

Previous sections described the use of tidal differences and basins for launching. A system used successfully for a number of very large concrete caissons and other floating structures has been as follows:

The first stage, including bottom plate and the lower walls, is constructed in a tidal basin, so that it may float off as a unit at high tide. Then it is moved to a prepared underwater basin, where it is sunk onto a level bed of gravel or sand at a depth just sufficient to expose the top of the walls at low tide. A second lift is poured, and, if necessary, the unit moved again. Thus, progressively, the unit is floated at high tide, sunk in a new basin at low tide, then its walls are constructed to a higher stage (Fig. 15.2*g*).

This method was used for construction of the concrete caisson for the Kish Bank Lighthouse in Ireland.

15.4.12 Summary

The above sections describe some of the methods which have been used in the past to construct and launch concrete barges and floating structures. Some of the more successful major installations have utilized two or more methods in sequence. The immense importance of this phase requires the integrated efforts of engineer and constructor. Extremely careful planning is essential.

Since the actual work is done by specific human beings, the entire crew must be thoroughly indoctrinated. Practice with a model in a model basin or swimming pool is invaluable to both workers and engineers. A dress rehearsal should be held so that every man understands his function and the correct sequence, communications, etc. For very large and complex installations, an emergency procedure should be developed, that is, prior planning of what steps to take if actual behavior differs from planned behavior.

Adequate communications and instrumentation are essential. The recent advances in ocean technology have made available a great many sophisticated instruments to measure depth, orientation, level, etc., as well as underwater television and diver reports. These instruments should be utilized to the fullest for all deep and important installations.

The launching of large floating structures of prestressed concrete can be one of the most challenging and demanding opportunities a construction engineer faces, for personal qualities of courage and decision are required as well as technical ability in the highest degree.

15.5 JOINTING AFLOAT

Large prestressed concrete floating structures can be built in sections and launched by one of the means described above. The sections can then be assembled and joined together by post-tensioning. The technique has been successfully employed on floating bridges and on a floating container terminal, and has been designed for the base portions of concrete Tension-Leg Platforms (TLP's).

The technique may be described as follows (Fig. 15.3):

In calm protected waters, two sections are floated into approximate position relative to each other. Using winches mounted on the decks, they are drawn into their mating position. This perfect fit is assured by mating dowels on the deck of one section being entered into female pipes on the other. If one dowel is made slightly longer than the other, it is easier to enter one, then the other.

The underwater portions of the two sections are rotated apart during this initial mating by ballasting so that the below-water decks mate first. Then after the doweling has assured the proper location and the decks are in contact, the ballast is shifted so as to rotate the decks into contact. Contact may be designed to compress neoprene seals previously attached to the bottom and sides. Then the space between the two mating bulkheads may be drained or pumped, assuring further compression. Railroad couplers temporarily fixed to the sides can be activated by divers so as to lock the joint.

Posttensioning ducts are sleeved and spliced. High-strength fine concrete or grout with an accelerating admixture is used to fill the intervening space and the units are posttensioned.

Variations on the above process include fixing of a portable cofferdam around the underwater portions of the joint, and the use of epoxy filler on closely fitting concrete contact zones to expedite the process of post-tensioning.

Lest this process sound complex, it should be borne in mind that steel barges have been mated afloat side by side, with full strength transfer, and that even steel supertankers have been joined afloat, end for end, with welded joints constructed in temporary cofferdams.

The above procedure was successfully used on the Hood Canal Replacement Floating Bridge in Washington.

15.6 APPLICATION AND UTILIZATION

Concrete has already been more widely utilized for floating and submerged structures than is generally realized. Performance has been generally excellent; economy has been somewhat erratic. A primary factor in the utilization or lack of utilization of prestressed concrete for ocean applications has been the lack of sufficiently strong and viable industry to develop, design, construct, and promote its use. Despite this handicap, the inherent advantages of prestressed concrete are leading to increased use in a wide variety of applications.

Among the applications for which concrete, both conventionally reinforced and

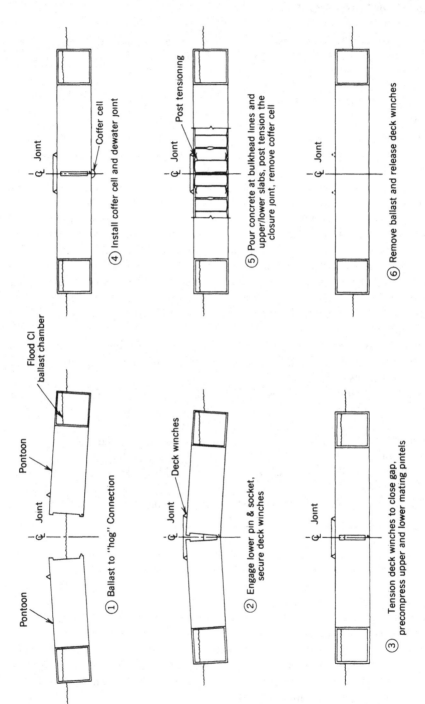

Figure 15.3. Method for jointing prestressed concrete structures afloat. FIP Notes. (Federation Internationale de la Precontrainte, London.)

prestressed, has been utilized to date are those listed in Sections 15.6.1 through 15.6.17.

15.6.1 Floating Oil Storage

In both world wars, an extensive concrete shipbuilding program was undertaken, amounting to a total of some 800,000 deadweight tons. The ships proved their seaworthiness in hurricanes, bombing attacks, collisions, and fire. They were watertight, vibration- and condensation-free. They were, however, very heavy and the reinforcing steel quantities were very great.

Both normal sand-and-gravel and lightweight aggregate concrete were used. Considerable difficulty was experienced during construction in fitting a double-curtain of reinforcing steel into wall thicknesses of 11 to 12.5 cm ($4\frac{1}{2}$ to 5 inches). Problems were also encountered with consolidation of concrete; vibration of reinforcement and form vibration were employed. Rich cement mixes were used.

Considerable variation in durability was experienced; with corrosion problems of the reinforcement mainly attributable to inadequate cover, lack of consolidation, and high W/C ratio. On the other hand, many of the vessels survive to the present day, in excellent condition, despite little or no maintenance. The ship Selma (6340 tons), built in 1919 of expanded shale lightweight aggregate and having minimum cover over the reinforcement, was beached near Galveston, Texas, in 1921. It is still in a remarkable state of preservation despite the semitropical saltwater and splash environment.

15.6.2 Floating Drydocks

Floating drydocks of reinforced concrete were similarly built during the wars and some are still in service today. They were generally built with adequate cover and with walls thick enough to enable proper consolidation of concrete. Those the author is familiar with have required a minimum of maintenance, and that which has been necessary has generally been directed at the metal fittings and appurtenances.

Prestressed concrete floating drydocks have been standard in the former Soviet Union for several decades because of their low maintenance.

A very large floating dock of prestressed concrete, in which the internals were a space frame of tubular steel, was designed to drydock large tankers in Geneva, Italy. Several sections were built. Unfortunately, the design did not recognize that the stiff concrete wing walls would have to carry heavy global in-plane shears as well as transverse bending and shear from the hydrostatic forces. Severe cracking occurred in the wing walls during trials, and the dock was never fully completed.

The durability criteria for floating drydocks are very severe because of the alternate wetting and drying, which tends to promote microcracks and hence greater permeability.

15.6.3 Ocean-Going and Inland Waterway Barges

A considerable fleet of ocean-going barges has operated for over 30 years in Southeast Asia. Pretensioning was extensively employed to provide a crack-free, durable hull. Steel frames have been incorporated in some of these barges to span cargo hatch openings. The design was by Alfred A. Yee.

These barges were built on a ways, with heavy anchorage abutments at each end. The hull was pretensioned longitudinally: bottom, sides, longitudinal bulkheads, and deck. Conventional reinforcement was used transversely.

These barges were launched in the traditional manner of steel barges, and have been used in a variety of services, such as general cargo and fuel oil transfer.

Despite the increased draft of these cargo barges, they reportedly tow at about the same speed as comparable steel barges, due to their lesser tendency to yaw.

They have required very little maintenance, since even marine growth has been less than that on steel barges. Most dramatic has been their self-limitation of damage, due to the high redundancy of reinforcement, and the ease of repair after accidents. These incidents include grounding on a breakwater during a typhoon, when companion steel barges were total losses, and external explosion of mines during service in Vietnam. In all cases, repairs have been readily effected by eccentric ballasting so that the damage zone was rotated above water level, then chipping away the damaged concrete with hand tools, replacing the few severed strands with conventional reinforcing bars, and applying shotcrete.

Inland waterway barges of concrete, both conventionally reinforced and prestressed, have been used for a number of years in rivers and harbors around the world. Such barges have served as cargo carriers, bulk product carriers, and liquid carriers. However, their use has been scattered and generally of limited extent. The greater draft of the concrete barges has been a deterrent in many cases. Construction methods have only recently begun to be developed to provide quality and economy. These methods include systems employing precast pretensioned segments, joined and post-tensioned transversely.

A significant number of heavy concrete barges have been built in Europe to serve as floating docks and to support floating concrete plants and floating sheer-leg crane barges. For these uses, mobility is not a prime requirement, but elimination of drydocking for maintenance is. Their use has proven so satisfactory as to become routine.

In Japan, prestressed concrete barge hulls have been built of prestressed concrete, with the internals of conventional steel, joined to the concrete with studs so as to give fully integrated behavior. The objective was to achieve low maintenance, especially where such barges were installed permanently for use as floating docks.

15.6.4 Prestressed Concrete Ship Terminals

At Valdez, Alaska, the steeply sloping and unstable sea floor from land to deep water led to the decision to use a floating container terminal instead of the conventional pile-supported wharf. This structure was constructed in Tacoma, Washington, in two sections, in order to minimize bending moments during the open ocean tow of almost 1000 km (600 miles). After arrival at the site and mooring, the sections were mated afloat by the construction of a grouted and post-tensioned joint (Fig. 15.4).

The dimensions of the dock, 200 meters long (650 feet) and of adequate width

Figure 15.4. *Floating container terminal of prestressed concrete at Valdez, Alaska. (Courtesy Berger/ABAM Engineers Inc.)*

and depth, were selected to ensure minimum list and trim during operation of the container crane. Hinged ramp bridges connect the wharf to shore and help position the dock during the 6 meter rise and fall of tide.

Each 100-meter section of dock was fabricated in a construction basin. Precast side and bulkhead segments were set by crane, and the floor concrete placed. Then precast half-depth segments were set to form the deck and a cast-in-place placement joined all segments for monolithic action. The structure was then post-tensioned.

15.6.5 Floating Piers and Docks

An extensive number of prestressed concrete pontoons have been constructed, of various shapes and sizes, to provide floating docking facilities for small boats, sea planes, ferry boats, recreation, and, in some cases, ship cargo wharves. Some were constructed in Germany during World War II and are still in service. Concrete was selected for the pontoons for seaplane docks to eliminate the electrolytic corrosion of the magnesium hulls from steel pontoons. Concrete has been selected for the other docks primarily for economy and low maintenance and where steel was unavailable (Fig. 15.5).

15.6.6 Concrete Barges for Compressor Stations, Petroleum Production Facilities, Remote Refineries

The oil and gas industry has made extensive use of concrete barge hulls on which have been mounted production and processing equipment of various sorts. More than 1000 such barges have been constructed for the Gulf of Mexico alone. These

Figure 15.5. Floating ferry terminal for Puget Sound, Washington.

barges have been towed to the site, as far as from Louisiana to Nigeria (Fig. 15.6) and from Antwerp to Libya, and either moored afloat or sunk in shallow water.

Many of these barges are designed so that the barge hulls will be underwater in the final installation, with only columns extending through the wave zone, and the deck with its equipment well above the storm-wave height.

The submerged hulls have been utilized to a limited extent for storage of oil and fresh water.

These barges have made extensive use of precast concrete segments. They have been set in place to form the bulkhead and sides, then shotcrete has been used to construct the joints. On many of these, shotcrete has also been used as an overlay to provide continuity of the precast segments.

These have been typically built on a barge. When the deck has been completed, the barge is moved to a prepared shallow water site and ballasted down. The new concrete barge helps to provide stability until the two are seated on the sand bed of the sea floor. Deballasting of the new concrete barge enables it to float off. Later the construction barge is deballasted and prepared for another construction cycle.

Stability of the construction barge during this recovery phase is provided by columns at the ends of the construction barge.

For the typical barge, concrete columns are cast and precast deck units placed so as to form a posted barge. The equipment is then mounted and the barge is towed to its destination, where it is ballasted down on to the mud bottom.

The performance of such barges in a saline environment has been excellent, many being 30 to 40 years of age, with no below-water maintenance.

Figure 15.6. *Prestressed concrete production facility under tow from Louisiana to Nigeria, where it will be seated in shallow water.*

The barge that was constructed in Antwerp and towed to Libya encountered a major storm in the Bay of Biscay. It was supporting a complete refinery, and when the towboat had to cut loose the tow because of the heavy seas, there was fear of damage and perhaps loss of the refinery. In actuality, both the barge and process equipment came through without serious damage, which was attributed largely to the stiffness of the concrete hull.

15.6.7 Pontoon Bridges

A number of notable pontoon bridges have been built of concrete, initially of conventionally reinforced concrete and more recently of prestressed concrete. These include the several bridges across Lake Washington at Seattle, the Hood Canal Bridge, also in the State of Washington, Lake Okanagan in British Columbia, and the Derwent River Bridge at Hobart, Tasmania. Both the initial Hood Canal and the Derwent Bridges suffered extensive damage from wave action. Although attempts were made to reinforce the Derwent River Bridge by cables and new expansion joint material, damage continued to be excessive. The bridge was eventually replaced by a high-level fixed bridge of prestressed concrete.

The Hood Canal Bridge was reinforced by longitudinal posttensioning together of pontoons in groups of four, with epoxy joints. The posttensioning tendons were "external tendons," run through the open cells of the pontoons and through holes drilled in the bulkheads.

This modification solved the previous problems of joint damage during storms for many years and, in fact, the Eastern half of the bridge has suffered no further damage and is still in use.

The Western half of the bridge survived for 20 years, but then was destroyed in a storm having a return period greater than 100 years. Many factors contributed to the loss: resonant "galloping" of the bridge along its length due to the waves impinging at a slight angle, so that the effective wave was long; dragging of anchors; impounding of water on the deck; and open deck manholes. After several hours, the pontoons broke apart, turned upside down, and sank to the bottom in more than 100 meters of water.

The Hood Canal Replacement Bridge was constructed in sections using precast segments and cast-in-place joints. They were then post-tensioned both longitudinally and transversely. This fabrication was carried out in a dewatered construction basin. After completion and flooding of the basin, the sections were joined into lengths of 400 meters by mating, dewatering of the joint, concreting, and post-tensioning. The replacement bridge, which includes a retractable drawspan, has functioned for the subsequent decade with no difficulties (Figs. 15.7, 15.8).

The more recently built bridges in Washington have used normal sand-and-gravel for the bottom and sides of the hull, and structural lightweight concrete for the decks. They are post-tensioned longitudinally. The box-section pontoons have been constructed in shallow graving docks, and floated out at high tide.

These floating bridges across Lake Washington have performed exceptionally well, carrying very heavy traffic loads as part of the Interstate Highway System. Some minor problems were originally encountered with heavy spray during intense storms. The vertically sided barges created a standing wave; the wind then blew the top of this wave as spray across the roadway. Splash walls appear to have mitigated this problem satisfactorily.

The original reinforced concrete floating bridge, across Lake Washington, constructed in 1939, with no prestressing, was under reconstruction during the winter of 1990. During and immediately after a severe storm, in which water flooded into the pontoons through temporary construction openings, the dynamic response is believed to have opened wide cracks in the hull bottom, leading to progressive

Figure 15.7. Erecting precast segments in construction basin for replacement bridge at Hood Canal, Washington.

Figure 15.8. *Precast concrete segments will be joined by cast-in-place concrete and post-tensioning to form pontoon bridge.*

sinking of the bridge. The failure scenario was apparently initiated by bond slip at an unconfined tension splice as a result of millions of cycles of fully reversed loading in the submerged concrete structure. The damage was aggravated by torsional moments and shears.

The replacement bridge, currently under construction, is heavily prestressed and will have much more subdivision into watertight compartments.

A floating bridge now under construction across a deep fjord at Salhus on the West Coast of Norway will be supported by prestressed concrete submerged pontoons. Indeed, the concept of isolating the bridge spans from the wave excitation by use of the semi-submersible concept appears to be a logical next step for floating bridges subjected to waves and seasonal ice.

The current design for a floating bridge between Rion and Antirion in Greece employs independent semi-submersible pontoons, with both vertical tethers and lateral moorings.

Proper subdivision of these pontoons for floating bridges must be constructed, using the principles of naval architecture in order to ensure safety in the event of accidental holing.

One of the concepts being considered for the crossing of the Strait of Gibraltar is a floating bridge supported on semi-submersible pontoons of prestressed concrete spaced at relatively short distances of about 300 meters (1000 feet). These pontoons would be submerged to a depth of about 60 meters (200 feet) where they would have minimum response to the waves.

To resist the strong currents, they would require both lateral and vertical moorings.

Design of the pontoons would consider safety under impact, e.g., from a ship or submarine, and would hence incorporate heavy in-plane and out-of-plane reinforcement in the walls. Elsewhere, closely spaced reinforcement would control cracking and provide redundancy. Concrete mixes would be designed for maximum durability and leak-tightness, and would require thorough consolidation by vibration during placement.

15.6.8 Prestressed Concrete Marina Pontoons

Concrete pontoons have been extensively utilized for marinas. They are typically made of lightweight concrete, and have very thin walls, reinforced with wire mesh.

Prestressing has been employed on a number of these, using pretensioning or else posttensioning with greased-and-plastic sheathed single strand tendons.

The use of external tendons, inside the pontoon, may be worth consideration.

The main deterrent to the wider use of prestressing is the protection of the anchorages against corrosion. Recessing and use of epoxy mortar appears to be the best solution.

15.6.9 Caisson Gates for Graving Docks

In the former USSR and in West Germany, the caisson gates for large drydocks have been constructed of prestressed concrete. This is an excellent selection, because the moldability of concrete can be fully utilized in the complex shape and varying cross section required. Weight is an advantage; steel caisson gates must usually be partially filled with concrete for stability.

The security of a favorable mode of failure, i.e., restricted zone of damage, is provided by prestressed concrete, and is again of great importance. A drydock is usually located in busy shipyards, where the caisson is always subject to the possibility of collision or impact, and must still protect the exposed ship and workers behind it. Concrete gates minimize the internal stiffening required and reduce maintenance costs.

Based on this successful experience, and driven by the demand for larger graving docks to serve as construction basins for steel and concrete offshore platforms, a number of large prestressed concrete gate structures have been built in such antipodal locations as Scotland and Australia. They have proven extremely efficient in multiple operations.

Similar "caissons" of prestressed concrete are also being currently designed to serve as floating guide walls for approaching tows of barges in the deep reservoirs upstream of dams of the Columbia River.

15.6.10 Temporarily Floating Structures

In the first edition of this book, floating and submerged structures were treated as one subject and contained in one chapter. Since publication of that edition in 1971, a great many bottom-founded or gravity-base structures have been constructed, to serve as offshore platforms, marine terminals, power plants, breakwaters, pipelines, and subaqueous tunnels. These all are floating structures for the final construction, deployment, and installation phases.

However, because of the different and specialized criteria involved, they have been treated in a separate chapter (Chapter 14) in this second edition.

However, it must be emphasized that during their floating stages of life, they are subject to all of the same construction and naval architectural principles as the vessels of prestressed concrete which are destined to float their entire lives.

15.6.11 Buoyant Subaqueous Tunnels

In recent years there has been a strong interest in the use of permanently buoyant prestressed concrete tunnels which are kept submerged by being tethered to anchors on the sea floor.

Such a system has been proposed for crossings of fjords in Norway, of deep channels in Greece, and of the Strait of Messina in Italy. Possibly the first such tunnel proposed was for a crossing of Puget Sound between Seattle and Bremerton in the 1930's.

The most advanced engineering studies have been carried out for crossings in Norway, where the combination of very deep water, sea ice, and shipping makes the submerged buoyant tunnel concept attractive.

Model studies appear to indicate the desirability of concentrating the anchoring tethers at the third points of the crossing.

Other concepts employ the concept of a horizontal arch, tethered by prestressing tendons to the shores, so as to act as an arch in compression for normal loads and as a catenary in tension for loads from the opposite direction. This was the scheme employed on the Derwent River Bridge in Tasmania, referred to earlier.

While the concrete tube is ideal for resisting the hydrostatic loads in ring compression, consideration has to be given in design to such events as a dropped or dragging anchor, internal explosion, and even a sinking boat or ship. While the probabilities of such events are low, the consequences are catastrophic. Various concepts have been proposed, therefore, which embody redundancy in the form of buoyant blisters, and so forth.

There are construction problems in assembly and tethering. The basic approach proposed is similar to the jointing of subaqueous bottom-founded tunnels, with the addition of longitudinal prestressing of the joints after connection and dewatering.

Tethering down to the desired profile is proposed to be accomplished by a combination of linear jacks from within the tunnel, water or dry ballasting, and restraint by spar buoys.

15.6.12 Floating Oil Storage

Where floating production systems are used to drill and produce oil, the use of oil storage vessels or facilities may be required. In some current cases this is provided by moored tankers or dumb hulls, but in very deep water, remote areas, and areas where there may be sea ice, storage in moored vessels of prestressed concrete is being considered.

For surface floating structures, the problem of mooring in multi-directional winds and seas can be solved by a simple-point mooring, in which case the concrete vessel will resemble a conventional steel tanker, or by a fixed mooring of a large round vessel, compartmented by inner rings and radial walls. Such a configuration gives great stability and damage control in the event of ship collision.

A different approach is to submerge the storage vessel, using a column extending up through the water plane to provide access and stability. In this case, the different weights of oil and saltwater ballasts must be compensated for by a separate tank, which is emptied progressively as saltwater ballast replaces oil during discharge, and later filled progressively as oil is produced to storage.

The main attraction of prestressed concrete for storage vessels is their ability to stay more or less permanently on station, without periodic docking. This and the flexibility in configuration and the favorable shell behavior of the concrete tanks under hydrostatic external pressure all favor the use of prestressed concrete.

Design of such oil storage vessels has to take into account temperature differentials which create tensile strains in the outer face. While the overall response of a cylindrical storage vessel to external hydrostatic loads is in circumferential compression, for which concrete is well suited, the typical end closures will develop shear and moment at their connections to the cylindrical shell, which will require both through-thickness and around-corner reinforcing. Adding to the steel congestion in these areas will be reinforcement to resist tension due to accidental overpressure.

Therefore, placement of the reinforcing steel in this area has to be done with great care to ensure accuracy, since even minor deviations have adverse significance. Further, the concreting and vibration will be difficult.

Internal bulkheads for compartmentalization produce plate bending stresses in the cylindrical wall. Various schemes have been devised to minimize these, such as haunching of the wall.

15.6.13 Floating Breakwaters

A number of floating breakwaters have been built to protect marinas and small boat harbors. The walls are typically made relatively thin (150 to 200 mm) (6 to 8 inches).

While longitudinal prestressing provides the moment resistance for the spans whose length is determined by the incoming wave, shear against impact from boats and debris is usually critical. The walls are generally too thin for transverse steel, except in locations near the internal bulkheads. Additional mild steel may also be used over the supports.

To prevent flooding and loss in the event of accidental holing by boat impact, transverse bulkheads are typically provided at relatively close spacing. Smaller floating breakwaters may be filled with foamed polyurethane, which will prevent flooding even in the event of holing.

Much larger floating breakwaters have been designed for the protection of floating airports and exposed harbors. Several systems have been proposed, using assemblies of floating pontoons, where the width of the breakwater is related to the design storm wave length. Various schemes for dissipation of wave energy have been developed, including multiple holes and baffles. An experimental breakwater made up of prestressed web members of hollow concrete was deployed at Ardyne Point in Scotland and was reasonably successful. The problem, however, with most of these larger breakwater concepts arises in major storms, when peaked wave energy causes erratic structural behavior, leading to local and then overall failure.

Caissons for breakwaters which are founded on the sea floor have been extensively employed in the United States, Canada, South Africa, and Australia. In all cases, the caissons have been constructed at a building site, usually a construction basin or a slipway, then towed to their final location and ballasted down on the seafloor. Both conventionally reinforced and prestressed concrete have been employed. The most spectacular of these is the Ekofisk Barrier (Fig. 14.2).

Some of the caissons have been towed in the self-floating mode even in the open ocean; others have been loaded onto a submersible barge for the ocean tow, then launched at the site (Fig. 15.9).

15.6.14 Floating Storage Vessels for LPG and LNG

A major breakthrough in the use of prestressed concrete for floating storage came with the design, regulatory approval, construction, deployment, and highly successful service performance of the Seki Arjuna, a storage, refrigeration, and transfer vessel for LPG in the Java Sea. This is the largest concrete vessel built to date, displacing 66000 deadweight tons (DWT).

This vessel was built of prestressed concrete in Tacoma, Washington, where it was completely outfitted with storage tanks, refrigeration equipment, and transfer facilities. It was then towed almost 10,000 miles across the Pacific Ocean to the Java Sea, where it is permanently moored (Figs. 15.10, 15.11).

It has now had 17 years successful experience in a tropical marine environment. The only maintenance required has been to the post-tensioning anchorage zones, especially those in the deck where seawater can stand and evaporate. Some minor corrosion has occurred to the aluminum and steel anchorage assemblies there. Other minor local damage has been experienced in the hull sides where impact has occurred.

Prestressed lightweight concrete is an ideal material for floating storage of

Figure 15.9. The four caissons for Tarsuit Caisson-Retained Island, in the Beaufort Sea of the Arctic Ocean, were made of prestressed lightweight concrete. They were towed 2500 miles on a large submersible carrier vessel.

Figure 15.10. Launching the prestressed concrete LPG floating terminal at Tacoma, Washington. (Photo courtesy of BERGER/ABAM Engineers, Inc)

Figure 15.11. *The 66,000 ton displacement Seki Ardjuna sets out on a 10,000 mile voyage to the Java Sea. (Photo courtesy of BERGER/ABAM Engineers, Inc.)*

cryogenic materials because of the favorable behavior of prestressing steels and lightweight concrete at very low temperatures.

In addition, prestressed lightweight concrete has high energy absorption and impact resistance due to its redundancy; that is, the multiple tendons and reinforcing bars go into catenary behavior after the concrete crushes, hence limiting the damage.

Periodic hull inspection can be carried out on station. Since painting is not required, periodic drydocking is eliminated.

Most important of all, it is feasible to design the LNG containment, the insulation, the test spaces, and the hull as an overall system that eliminates or at least extends the interval for periodic shutdown for inspection for leakage. Since such shutdown requires warming up, and subsequent cooling down, both over considerable periods, the longer interval possible with prestressed concrete has major economic advantages.

The US Coast Guard has indicated that it will consider that a prestressed concrete hull has adequate collision resistance if it has the same energy absorption as that developed during the specified 1.5-meter penetration of a steel hull. This can be obtained in prestressed concrete hulls by providing a high percentage of through-thickness (transverse) reinforcement.

The major difficulty for all LNG floating storage and transport, both prestressed concrete and steel, is with the transverse bulkheads. Although these can be insulated, as are the sides, they cannot be prevented from eventual cooling down and contraction, unless an internal source of heat is provided to the bulkhead.

Active heating is used in some existing steel LNG ships but passive control of the thermal regime is much preferred by regulatory authorities. Various schemes of transferring heat from the seawater are under development but are not yet operational.

A concept proposed by a German engineering firm is to use curved bulkheads, with intentionally thinned sections at the connections to the longitudinal structural members, allowing the concrete to distort sufficiently to accommodate the change in dimensions upon cooling. Alternatively, stainless steel connection strips can be similarly designed to bend to accommodate the change.

15.6.15 Floating Airfields and Offshore Bases

Most grandiose of all proposed uses of prestressed concrete structures afloat have been the periodic revival of concepts for floating airfields and Mobile Offshore Naval Bases (MOBS). Some have been designed for use in protected waters, others for use in the open ocean.

Prestressed concrete appears uniquely qualified for the deeply submerged pontoon hulls for the semi-submersible concepts which are generally believed to be required for open-ocean deployment.

For more protected environments, e.g., the Thames Estuary, multiple precast concrete pontoons are proposed to be connected by multi-directional post-tensioning into a large flat plate (Fig. 15.12).

The problems with the design and construction of such mammoth structures are those of naval architecture rather than prestressing. Waves acting obliquely to the airfield or base can produce very high bending moments, especially in the end sections.

Mating of huge sub-assemblies in the open ocean presents serious hydrodynamic problems of heave, even under very low sea states, due to the large inertial forces. It is probable that solutions will be found, possibly by use of the large equalizing jacks employed for installation of the tethers of tension-leg platforms (TLP), where similar forces are encountered.

Figure 15.12. *Proposed floating airport for Thames Estuary UK.*

An alternative method being currently explored has been directed to progressive assembly of relatively small segments, each made to conform to the movements of the large structure.

15.6.16 Floating Industrial and Power Plants

A prestressed concrete barge was constructed in Singapore, with internal configuration specially designed for the barge's future service as a floating phosphate production and concentration plant. It was towed to Hawaii where it was outfitted, thence to the West Coast of Mexico where it continuously dredges concentrates and loads phosphates for shipment and use as fertilizer.

Concepts have been developed for other complete plants for support of coal-burning power production. The ability to stay in service continually, utilizing the easy transfer of powdered coal, the unlimited source of cold water, and the dissipation of heat (thermal pollution) into the vast sink of the open ocean, makes this concept extremely attractive, especially now with the development of high capacity cables for subsea power transmission.

Most exciting of all for the future of prestressed concrete were the several ocean thermal energy concepts (OTEC), with a major floating power plant from which would hang 2000 to 3000 meters (6000 to 10000 feet) of cold water pipe. Many concepts were considered for the large diameter pipe, including prestressed lightweight concrete. Installation of the pipe, in successive segments, or by progressive upending, presents a major practical problem.

Almost all concepts employed a hull of reinforced and prestressed concrete, in order to minimize both first costs and long-term maintenance.

To date, only small test OTEC structures have been built, but they have proven successful in producing power. The current problem is largely that of economics since the efficiency of power development with low-head differentials is inherently low.

Although the previous systems were based on a closed-cycle process, more recently a fresh look is being taken at the open-cycle. Similarly, it is likely that the first prototype plants will be shelf-mounted, that is, installed on the coast of a volcanic island, where power from other sources is very expensive and where the seafloor dips steeply down for 2000 meters (6000 feet) or more, from where cold water may be drawn.

The possibility of combining a fossil-fueled floating plant with a cold water cooling system using the very cold water from the depths to gain another 20°C (36°F) of coolant increases the long-term potential of the OTEC concept, in which prestressed concrete appears destined to play a major role.

Floating nuclear plants were proposed during the 1970's in order to permit standardized plant design and approvals and to isolate the plants from the lateral components of seismic events. Although the initial designs were based on a combination of steel hulls with concrete added to give composite action along with large precast concrete modules, the use of prestressing would appear to offer advantages in developing fully monolithic action.

Since it was proposed to install these floating nuclear plants in sheltered waters, fully protected against wave action, the engineering problems then become those of vertical impulse of the earthquake (seaquake) and of mooring in hurricanes and tornadoes.

15.6.17 Tension-Leg (TLP)

For the development of the Heidrun Field, in the far north of the Norwegian Sea, Conoco has selected a concrete TLP, which is now under construction. This concept is based on a prestressed concrete semi-submersible tethered to a prestressed concrete gravity anchor box on the seafloor. In-depth studies by Conoco and their engineer-contractor, Norwegian Contractors, have shown it to have advantages over steel hulls in minimizing motion in response to the waves and fatigue in the tethers. These favorable responses to the waves are primarily due to its deeper draft. Concrete is especially well-suited to resist the external hydrostatic head of deep submergence by its shell action.

Recent studies indicate that the prestressed concrete TLP may be the least costly and most efficient concept for the development of offshore oil production in the deeper tracts of the Gulf of Mexico. In-depth feasibility studies continue.

15.6.18 Very Large Crude Carriers and Commercial Submarines

With the current prices of steel hulls and the criteria which they are required to meet, there are not even any serious current studies for large crude carriers of prestressed concrete. However, there is also a current public demand for greater safety of the hulls, such as double hulls, to prevent release of oil in a collision or grounding.

Thus, prestressed concrete, properly reinforced in both the in-plane and out-of-plane directions, could fill a very important role in reducing the costs of large crude carriers.

There is no question but that a prestressed concrete hull will have greater weight, hence greater required displacement, and consume more fuel if the same speed is to be maintained. Thus, the savings in first cost will have to be significant.

On the other hand, published studies indicate greater safety and reliability of prestressed concrete vessels as compared with current single-skin steel vessels.

One service in which prestressed concrete hulls should show dramatic advantages is in arctic and subarctic service, where ice-strengthening is required. Not only is prestressed concrete admirably suited to resisting local impact due to ice, but its greater mass is an advantage in plowing through pressure ridges.

A similar situation exists in regard to submarines, at least those that will operate at moderate depths. The current costs of submarine pressure hulls are so great as to discourage consideration of commercial submarine transport, even though many studies have shown advantages. The "submarine" need not necessarily be totally cut off from the surface: some designs marry the submerged hull with surface-piercing shafts for air and pilotage.

An in-depth study of concrete as a pressure hull showed that current technology would enable operation at depths up to 500 to 1000 meters (1600 to 3000 feet). It also showed that ultimate safety against collapse can be enhanced.

Thus, there appears to be a long range potential for prestressed concrete even in the transport field.

15.7 SPECIAL CONSTRUCTION CONSIDERATIONS FOR FLOATING STRUCTURES

Weight control is essential for all floating structures, since self-weight determines draft, list and trim, and carrying capacity. Hence the mix must have a very carefully controlled weight for the fresh concrete. Entrained and entrapped air may affect this weight by several percent. Trial mixes are essential.

Where lightweight concrete is used, dried coarse aggregate can run 10% or more less dense than saturated aggregate. Most of this will be retained even after mixing; at most 3 to 4% will be reabsorbed from the fresh concrete.

Weight can also be affected by tolerances. Forms usually spread, especially with intense vibration. A plus tolerance of 12 mm ($\frac{1}{2}$ inch) on a wall thickness of 300 mm (12 inches) means, 2 to 4% greater draft.

Presumably the weight of the reinforcing steel, both passive and prestressed, has been accurately calculated by the designer, but splices, length tolerances, and auxiliary steel, can all add weight.

Thus careful checks must be maintained of the actual weight of concrete and reinforcement being placed.

Weight can increase due to absorption of seawater over extended periods. Such absorption is also detrimental to durability, since it provides a path for chloride ion diffusion and permeation as well as for oxygen.

Although most aggregates are dense and impermeable, this is not true of all limestone aggregates. The matrix itself can be rendered highly impermeable by inclusion of fly ash or other microsilica and by a low W/C ratio, obtained by use of HRWR agents.

For floating concrete structures, reinforcing steel is typically very dense; ratios typically run 200 to 300 Kg/m^3 (340–510 lbs/yd^3).

Concentrations will be more dense at corners and bulkheads. Therefore steel placement will be slow and difficult, since prestressing ducts must also be threaded through the double curtain of reinforcement.

It is essential that steel placement tolerances be closely adhered to, as otherwise concrete cover will be deficient, with danger for early corrosion.

Stirrups give the most trouble. In many installations, they can be beneficially replaced by mechanically headed T-bars, so as to lock behind the in-place reinforcement (See Figs. 3.1 and 3.2). These not only can be placed in a fraction of the time that stirrups take, but fewer are required, because of their greater efficiency. They accurately fix the spacing of the two curtains of steel, thus preventing spreading and loss of cover. Another alternative is to use U-stirrups, overlapping in the core concrete.

Epoxy coated reinforcing steel is being increasingly specified in order to give greater assurance against future corrosion. Epoxy coated bars have almost no adhesive bond, relying on the deformations to give mechanical bond. Thus splice and development length must be increased, as required by the designers, which again usually means greater congestion of steel at intersections, joints. etc.

Fusion-bonded epoxy coated bars are usually straight bars, then bent afterward. There is often minute cracking at the bends which requires very careful inspection and touch-up as indicated. The fluidized-bed method of hot-dip coating can be applied to prebent bars and fabricated cages.

Epoxy coated bars must be handled and placed with care to avoid abrasion. Fiber slings should be used.

Attachments and inserts must be placed with care and held against movement during concreting. Where permitted, they can be affixed to the reinforcing bars. In some cases, electrical isolation is required, in which case spacer "adobe blocks" of mortar or plastic chairs may be placed between the attachments and the bars, and tied with plastic ties.

Welding to attachments will almost always cause warping of the plate. A wood fillet around the plate prior to concreting can give space for expansion in the plane of the wall and thus prevent cracking. This space can later be filled with epoxy putty. Provision should be made for injection of epoxy behind large plates after the weld has cooled.

To ensure thorough consolidation of the mix, to prevent honeycomb voids and rock pockets, and to drive out the entrapped air, intensive internal vibration is required. This may be supplemented by external form vibration, but that is usually effective only on the cover concrete.

The problem is how to use internal vibration in thin walls full of congested reinforcement and prestressing ducts. A small diameter "pencil" vibrator is useful but cannot always find the place to penetrate through the steel.

Forms may be marked to show the best locations at which to insert the vibrator. An ingenious method places a tube of slightly larger diameter through the steel. The vibrator is inserted into the tube as the concrete is placed, and the tube withdrawn over the vibrator and hose.

Curing is critical for the concrete hull, bottom, sides, and deck. The concrete itself is relatively impermeable, hence drying shrinkage will be minimized. However, thermal shrinkage due to cooling after heat of hydration expansion may be serious, especially in members over 600 mm (24 inches) thick, since rich mixes will have been employed. The bulkheads will prevent contraction, thus leading to cracks.

Heat of hydration expansion may be controlled by substituting PFA for a portion of the cement, by selecting coarse ground cement, and by pre-cooling of the mix.

The external forms should be insulated, as by sprayed-on polyurethane, in order to prevent rapid loss of heat. This will minimize differential contraction across the walls.

After stripping forms, membrane curing compound should be sprayed on the surface and blankets draped to prevent rapid heat loss. Cold, drying winds in winter must be especially guarded against, but thermal cracking has occurred in the Middle East where the very hot day temperatures drop over 20°C (36°F) at night.

Piping penetrations present problems in preventing leakage and in ensuring against accidental rupture. Detailing is important but the actual construction may be even more so. Special effort should be made to ensure thorough consolidation of the concrete around the penetration. Revibration while the concrete is still workable will drive out the early bleed water that tends to collect under the penetration.

With large penetrations, a plastic tube or tubes can be inserted under the penetration, through which epoxy can be injected after the concrete has hardened and the bleed water has been reabsorbed.

When openings are left in the structure in order to provide access for mechanical installations during construction, the closures require special detailing and care in construction to ensure that they will behave monolithically with the basic structure. This is especially important if the opening will be subjected to alternating tension and compression or alternating shear. Couplers can reestablish reinforcing continuity. The joint surfaces can be configured to minimize bleed and the "window box" technique of concreting can be used to minimize the accumulation of bleed water on the underside. Steel waterstops can be employed to minimize direct leakage along the joint. The exterior surface of the concrete can be coated with epoxy.

Prestress tendons are typically deviated around these temporary openings rather than through them, but in the case of very large openings, provision should be made for splicing ducts and installing a rock anchor from the top, then stressing.

15.8 CONCLUSIONS

Much of the future exploitation of the ocean depends on the construction of large, fixed installations, either floating on the surface or beneath it, or sunk and anchored to the ocean floor.

This is a relatively new concept in naval architecture, since the overwhelming preponderance of man's concern with the oceans in the past has been as a means of transport. The outrigger canoe of the Polynesians, the oar-powered galleys of the Phoenicians, the wide-ranging boats of the Vikings, differ only in size from the modern supertanker or ore carrier. This difference in size, however, is more significant than would appear at first thought, because of its relation to the wave length. Therefore, as indicated earlier, prestressed concrete now deserves serious consideration for vessels in the transport industry, particularly specialized vessels.

Storage, operational, and support facilities, however, are of a completely different nature from those which serve the transport function. The criteria no longer require lightness of weight. Durability, security, economy, constructibility and rigidity become the fundamental parameters when selecting a structural material. Prestressed concrete is rapidly becoming recognized as a primary material for the emerging exploitation of our ocean resources.

REFERENCES

1. DNV "Guidelines for the Design, Construction and Classification of Concrete Floating Structures," Det Norske Veritas, Oslo Norway, 1979.
2. FIP "The Inspection, Maintenance, and Repair of Concrete Sea Structures," Federation Internationale de la Precontrainte, London, 1982.
3. Final Report of Committee V. 2, 10th International Ship and Offshore Structures Conference, "Concrete Marine Structures," 1990.
4. ABS Draft Report, "Proposed Rules for Building and Classing Concrete Vessels," American Bureau of Shipping, New York, 1981.
5. Conference Proceedings, *Concrete Ships and Floating Structures*, University of California Extension Division, Berkeley, California, 1975.
6. ACI 357-2R-88 "State of the Art Report on Barge-Like Concrete Structures" American Concrete Institute, Detroit, Michigan, 1988.
7. FIP Recommendations "Design and Construction of Concrete Ships," Structural Engineers Trading Organization, London, 1986.

Prestressed Concrete Tanks

16.1 GENERAL

One of the earliest uses of prestressed concrete was for water tanks. Since then, many thousand tanks of prestressed concrete have been built for water, oil, gas, sewage, granular and powdered dry storage (silos), process liquids and chemicals, slurries, agricultural waste, and, more recently, cryogenics. The nuclear reactor pressure and containment vessels are a special form of tank (see Chapter 17), as are some cooling towers. Prestressed concrete tanks are also proposed for storage of high-pressure high-temperature processes associated with coal gassification.

Elevated water tanks are widely used in Northern Europe, the Middle East, and the Midwest of the United States. Some are outstanding architecturally as well as functionally.

In many cases, the elevated tanks are constructed at ground level, then jacked up progressively as the column support is built. In other cases, the column or shaft is first constructed, then the tank is cast and stressed around it at ground level, and finally, the tank is jacked up and fixed at its final elevation.

The walls of tanks are subject to high membrane tension due to the internal hydrostatic pressure, as well as to vertical bending at the joints with the floor and roof. Circumferential prestressing is applied to prestress the concrete sufficiently so that it remains crack-free under normal working stresses with a suitable safety allowance selected on the basis of the probability of overload and the consequences of cracking. For gravity storage of water and other liquids, this usually requires zero tensile stress at design load. In some areas, seismic loading of tank and the dynamic response of the contents control the design loadings.

Most prestressed concrete tanks have been circular in cross section, with circumferential prestressing sufficient to eliminate tensile stresses at each horizon. The circumferential tendons may be continuous, applied by wire-winding under stress, continuous strand paid out under tension, or applied hand-tight and wedged or jacked out by a sequential operation to provide the necessary stress. Strips of high tensile steel can be placed in grooves in the walls, applied under tension by a winding machine. Circumferential tendons may also consist of internal post-tensioning tendons which are anchored at buttresses, where they overlap with the adjoining tendons. For example, there may be six buttresses, with each tendon spanning one-half of the circumference.

Summarizing, prestressing tendons may consist of one of the following systems:

1. High-tensile wire, in which the stress is obtained by drawing through a die.
2. High-tensile seven-wire strand, black or galvanized, or greased and plastic-encased. This strand may be applied by braked machines, reacting against a chain, so as to apply stress, or be laid out unstressed and then wedged or jacked out to create the desired stress.
3. Cold-drawn steel strip, wound under tension within a steel channel, which is then covered with a steel plate, and the channel filled with corrosion-inhibiting grease.
4. Multiple strands in a duct, posttensioned and grouted.
5. Bars of alloy steel, in ducts, which are posttensioned and then grouted.

Prestressed concrete tanks may also be prestressed biaxially by helical post-tensioning tendons crossing at 45°. This permits the use of relatively short tendons, such as bars, and minimizes friction. This permits posttensioning to be carried out from the upper edge beam.

Tanks may also utilize prestressing in the vertical direction. The prestressed tendons may be either pre- or posttensioned. Since the vertical bending moments are usually a maximum at the bottom and top, either mechanical anchors or supplemental reinforcing steel are required at the ends. When precast pretensioned slabs are used as staves of the tank walls, indented strand is useful in reducing the development length.

Square tanks may be required for industrial uses and may span either horizontally or vertically. The effect of deflections under load must be considered, especially since the load follows the deformations. Square (or hexagonal) tanks offer advantages for storage in congested urban and industrial sites because of more efficient space and land usage. For these, the walls will usually be designed to span between floor slab and roof. Vertical prestressing will usually dominate the prestressing in the walls, with post-tensioned anchors in floor slab and roof. The top and base slab may be conventionally reinforced or posttensioned with tendons crossing at 60° or 90° angles.

Circular tank walls may also be designed to span vertically between a top and bottom ring or frame. The rings may be prestressed circumferentially, and the walls vertically. Thin concrete shells, such as hyperbolic paraboloids, may be used to span vertically between the rings. For these, the matter of deformation at joints and its effect on load transfer and leakage, etc., must be considered in design. These joints will be subject to high local stresses, so that weld details, etc. will be very demanding.

Multiple tanks and silos have been constructed by utilizing interlocking polygons, such as hexagons, prestressed together to act as a unit with transverse and/or vertical tendons.

Elaborate doubly-curved shells have been built to store sewage and other active substances where both efficiency of storage and efficiency of operation can be facilitated by the shape (Fig. 16.1). Similarly, double-curved shells have been built for water tanks and cooling towers, to take advantage of the efficiency of shell-action of the concrete, combined with prestressing at the edges.

Tank designs have not always taken full advantage of the possibility of internal bracing and tying. With water and oil, etc., such ties offer an efficient structural solution. Obviously, corrosion problems must be considered in the light of the usage, and the necessary protection provided. External tendons, greased and encased in a polyethylene tube, with subsequent grouting, would meet this problem.

Prestressed concrete is essentially watertight but it is not gas-tight. Where vapor must be retained, a thin membrane liner of steel may be used. This has proven so successful that the metal liner concept is being increasingly employed. The liner may be fluted, to permit expansion and contraction under alternating filling-emptying conditions and under prestress, or it may be flat, with suitable anchors to the concrete. Such a liner may also be considered to provide a portion of the vertical reinforcement. The liner may be either external or internal to the structural concrete wall, depending on the product to be stored.

Composite construction permits the ductility and membrane characteristics of

Figure 16.1. Doubly curved shells of prestressed concrete serve as sludge digestors for sewage treatment facilities, Los Angeles.

steel to be combined with the economy, rigidity, and virtually unlimited tensile capacity of prestressed concrete. The thin liner is readily welded; thus tanks of larger diameter and height, or greater pressure resistance, can be obtained than are possible with steel alone. Alternatively, the membrane joints may be lock-seamed and sealed with polysulfide or epoxy glues.

This concept is especially important when special steel membranes are required such as nickel steel (or aluminum) for cryogenic storage.

Plastic liners may in some cases be used in lieu of steel, either as sheets or sprayed or troweled on after construction. For these, surface preparation must be meticulously performed. Epoxy and polyurethane coatings have recently been developed which have sufficient ductility to span cracks in the concrete up to 0.5 mm (0.02 inches) without reflecting the crack.

For underground or underwater tanks, prestressed concrete is able to economically resist both the design internal and external heads. Proper corrosion protection is important, with special attention paid to construction joints and boundaries.

The concrete for tank walls may be cast-in-place concrete, precast concrete panels, or shotcrete.

Cast-in-place concrete walls are often cast in alternate segments to the full height, to allow shrinkage to be dissipated. Panel forms are utilized, usually so framed that no ties pass through the walls. When an internal steel liner is used, form ties using welded studs may be used, with the outside end of the tie being later removed and patched. Form ties which incorporate integral waterstops may also be used.

At construction joints, both vertical and horizontal, the joint must be cut back by water jet or sandblasting so as to expose the aggregate. Thorough soaking before the next pour, or an epoxy-bonding compound, will aid bond. The epoxy must, of course, be stable and compatible with the material to be stored.

Slip-forms are widely employed for cast-in-place concrete construction of tanks and silos. In this case, the complete circumference is cast in a slowly rising sequence. This eliminates construction joints but shrinkage strains must be recognized and controlled. Careful control must be maintained of out-of-roundness and thickness tolerances.

Precast panels may consist of vertical "staves," horizontal "planks," diagonal slabs, geodesic or folded plates, or thin shells. Such precast panels, staves, etc., are often prestressed longitudinally with circumferential prestressing applied either internally or externally after erection (Fig. 16.2).

Vertical prestressing is very effective because in conjunction with the later circumferential prestressing, it creates a multi-axial prestress that, if properly designed and constructed, can eliminate all cracking.

The precast wall panels or staves are often posttensioned, using plastic encapsulation (PVC) with an anti-corrosive lubricant, which permits stressing after the concrete hardens. High-strength bars and standard posttensioning are also employed. With the latter, care must be taken to ensure complete filling of the vertical ducts by grout. Thixotropic grouts and appropriate procedures for vertical tendons are required in order to prevent voids near the top.

Ducts may be of semirigid or rigid tubing. The latter will often be found to be

Figure 16.2. Precast concrete panels are erected to form a dike wall around an LNG tank. After jointing, they will be circumferentially wrapped with wire under tension. (Courtesy of Preload, Inc.)

best because it requires less support. Ducts should have a plastic cap installed to prevent the entry of foreign materials.

Since the largest moments in the panels often occur at the ends, posttensioning offers the advantage of positive end anchorages.

With pretensioning, it will often be required to install supplemental mild steel near the ends to provide adequate tensile stress transfer in the bond development zone. Of course this conventional reinforcing may be used effectively to transfer stress to the roof or dome. In the case of fixed base-wall connections, it may also transfer moment around this joint. As noted earlier, use of indented strand will reduce the development length.

Precast wall panels which incorporate a diaphragm or membrane of steel are usually formed with a reentrant key that locks the panel to the concrete. Plastic liners may have knobs that perform a similar function.

The membrane liner can be placed on the soffit and the concrete cast on top. The soffit should be curved to the contour of the finished wall, but in some cases is constructed flat, so as to form a chord in a multi-segment circle.

Care must be taken in construction that no penetrations, punctures, or tears are made in the liner, and that the exposed strips on either side are free from concrete.

If the panel is being constructed as an arc, the screed should be properly contoured. In any event, internal ducts should be properly supported to a tolerance of plus or minus 6 mm ($\frac{1}{4}$ inch).

Posttensioning of the individual panels is carried out prior to lifting from the soffit form. To minimize friction, the soffit form may be covered with a polyethylene sheet.

Side forms and end forms must be true to ±3 mm ($\frac{1}{8}$ inch) vertically and horizontally and tank radius to ±18 mm ($\frac{3}{4}$ inch). The bottom of the wall panel is especially critical, for it must seat vertically on the wall base pads.

Panel inserts must be properly installed and care taken to ensure proper consolidation around them.

Special attention must be paid to the quality and uniformity of the concrete, and its proper consolidation and curing, since it will have to sustain a high compressive stress while still at an early age. For example, the wall may require prestress to $0.55f'_c$.

Stressing of panels should start at the center of each panel and proceed in alternating patterns towards the sides.

The surfaces of the longitudinal joints should be sandblasted or jetted so as to remove laitance, in order to provide a proper bond with the jointing concrete.

Prior to erection of precast wall panels, the wall bearing pads will have been placed on the circumferential edge of the floor slab. These bearing pads are intended to allow radial deformation of the walls as they are prestressed and, later, as they are extended by hydrostatic pressure.

Radial shear keys and unstressed strand ties may be provided to transfer shear from the walls to the base slab in the event of earthquake or tank rupture.

In lifting the panels or staves, care should be taken not to damage the bottom corner by crushing. A softener of wood can be used or the panel can be turned in the air by appropriate rigging.

Erection of the staves requires that they be accurately aligned in a true circle and held rigidly.

Tolerance for setting of the panel is normally ±10 mm ($\frac{3}{8}$ inch) radially and ±12 mm ($\frac{1}{2}$ inch) circumferentially. Verticality should not vary from a true line by more than ±10 mm ($\frac{3}{8}$ inch) radially or ±30 mm ($1\frac{1}{4}$ inch) circumferentially.

After erection, the panels must be rigidly braced, either against scaffolding or by standard tilt-up braces.

The jointing is extremely critical, especially with the current tendency to use high-strength concrete in the staves. The joint concrete must develop strength just as high as that of the panels themselves.

After erection, if a membrane or diaphragm of steel or plastic has been used, the joints must be made. Often these are by welding, although care must be taken to prevent excessive heat from spalling the adjacent concrete. Perhaps most typical is the use of glued joints, with polysulfide or epoxy glue, held in place temporarily by screws or bolts at close spacing.

Polysulfide should not be used with sewage or crude oil storage since anaerobic bacteria may attack it.

If the circumferential prestressing is to be by internal post-tensioning in ducts, then these must be connected (spliced) across the joints. Sleeved couplers, taped with heat shrink tape, appear to be the most reliable.

Prior to concreting, the joint surfaces should be wetted, then allowed to become surface dry. Epoxy bonding compound is not recommended for this application, due to time control constraints.

Gaps for joints vary in practice from 75 to 600 mm (3 to 24 inches) but 300 mm (12 inches) seems best.

The concrete joint may best be constructed by placing a high-strength concrete mix whose maximum coarse aggregate is 10 mm ($\frac{3}{8}$ inch). This concrete may be placed at the top and permitted to free fall, unless there is a heavy congestion of reinforcing steel in the joint, in which case a tremie pipe may prove necessary to prevent segregation. However experience in the free-fall technique shows that it can produce satisfactory concrete after a drop of as much as 8 meters, even with a moderate amount of joint reinforcement.

A trial placement is indicated in doubtful cases, since the final quality of the concrete in the joint is so important.

In many recent projects, a highly workable concrete has been pumped into the joint from the bottom. The forms must be designed to accommodate the full fluid head of concrete. Narrow joints, 50 to 75 mm (3 inches) in width, can be grouted, using a grout mix designed for high strength.

Another means of concreting the joint is by shotcrete, in which case special care must be taken to prevent trapping of rebound. The sides of the precast staves should be specially configured to facilitate shotcrete.

Curing of joint concrete must be thorough. The author prefers a heavy coat of membrane curing compound, applied twice at 8- to 12-hour intervals in warm climates.

An inside base curb may then be concreted to provide for a waterstop connection to the concrete floor of the tank. Care must be taken that the concrete is consolidated on both sides of the waterstop.

This curb may also embed the seismic restraint cables, if employed.

Typically the curb is connected to the wall panels by reinforcement steel and cast against it so as to bond to the panels.

Once the concrete in the joints has attained sufficient strength, the circumferential stressing is applied. Most tanks employ an external system, either by wrapping wire, with the tension imparted by drawing the wire through a die, or by placing strands, with a braked drum used so as to maintain the designed tension. Tolerances in tension of ±5 to 7% are usual.

As the tension is applied, the internal temporary braces must be released.

Wrapping usually involves several laps. After each wrap, shotcrete is applied so as to cover the tendons to about one diameter. Pre-placed grade wires help to control the thickness.

When wrapping requires more than one coil of wire or strand, the ends should be clamped to an adjacent turn so as to maintain the stress, and to have an ultimate capacity equal to that of the tendon.

After each layer is coated, another wrap or layer may be placed.

The final coat of shotcrete, the finished coat, usually contains finer sand.

Some practitioners, in applying a coat over any individual layer, first apply a flash coat of pure cement and water, then follow immediately with the shotcrete before the flash coat has set. Others feel that the tendency for the flash coat to dry quickly and crack before the shotcrete negates the value of this procedure and prefer a rich mix for the shotcrete.

Both wet and dry shotcrete systems are employed. The author prefers the wet system because of less rebound and hence a lesser tendency to trap rebound.

For underground tanks and those in aggressive environments where durability is a concern, external coatings of the finished structure are usually advisable. These may be coal-tar epoxy coatings or bitumastic or polyurethane. Special attention has to be paid to the joint at the top between the roof and the wall, since water may seep down, leading to corrosion of tendons and delamination or spalling.

The coatings should be applied when the tank is filled with its test filling of water.

When internal posttensioning is employed, once the walls have been constructed and jointed, the tendons are pushed and/or pulled through the ducts. Tendons are then stressed individually around the circumference, being careful to progressively proceed around the tank so as not to create an adverse condition of high tension behind the anchorage.

The initial stressing of the first few layers of tendons must consider the losses which will occur due to elastic shortening of the concrete. For that reason, it is best to stress this initial group to 50% of final value as a first stage, proceeding around the circumference. Once two or three levels have been stressed to 50%, the stressing can be raised to 100% and the earlier levels restressed to their final values.

Center-stressing anchorages permit stressing of overlapping tendons at the buttresses or in block-outs in special stressing panels.

There are many issues of constructibility that apply to tanks and that affect their performance. A few of these have been presented in the previous sections; others follow.

Construction tolerances, both in wall thickness and out-of-round (or other shape), affect both the final design conditions and the stability of the structure during prestressing.

Shrinkage is particularly critical for tanks because of the thin sections and large exposed areas. Adequate curing is extremely important for durability and the prevention of surface cracks and crazing. Water cure may be provided by soakers or spray, with burlap. However, high walls exposed to strong winds may require more positive protection, e.g., membrane curing.

Nonprestressed reinforcement may consist of welded wire mesh or conventional mild-steel bars. A liner plate may also serve to provide a portion of the reinforcing requirements.

When buttresses are employed, they must be detailed so as to provide adequate clearance for the anchorages and the stressing jacks. The buttress is usually a highly congested area, and details should be laid out full scale to ensure adequate room for concreting. Particular attention should be paid to vibration in these areas and to selection of a mix that can be properly placed. Anchorages must be rigidly held to prevent displacement during concreting.

Since the tendons often have relatively sharp curvature into the buttress, radial shear forces during prestressing must be resisted by stirrups, to prevent pull-out of the tendon.

As usually designed, the wall panels must be free to contract under circumferential prestressing. Hence they are usually set on neoprene pads or low friction shims so that they can move radially inward as they are stressed. Later, the gap will be grouted and sealed, e.g., with a polyurethane sealant.

Roof joint details must not only provide the desired condition of restraint or freedom but must also provide a positive seal to prevent moisture penetration between steel and concrete, or between concrete and shotcrete, etc., which would lead to corrosion of the prestressing tendons. If the roof is constructed before the tank is post-tensioned, the roof-joint detail must permit free movement inward of the tank walls as they are stressed.

Joints between precast panels must be designed to transmit the shear and local bending stresses and deflections. This is of particular significance with shell-type units such as geodesic plates, folded plates, hyperbolic-paraboloids, etc. Welding details must consider the effect of heat and the possibility of spalling of adjacent concrete. Concreted or grouted joints must be detailed to ensure that the joint will possess the required strength. Fully concreted joints, 150 mm (6 inches) or more in thickness, are preferable to thin grouted joints. Epoxy-bonding compound applied to the joint surfaces will help to prevent shrinkage cracks. Joints should be well-cured.

The installation of the prestressing wires and the stressing of the tendons will produce temporary wall bending stresses. The sequence and stages of stressing must be set forth so as to maintain wall-bending stresses within allowable limits.

Openings are generally accommodated by deflecting the strands above and below, in bands. Individual tendons may have to be spaced so as to prevent excessive concentration of force such as might occur with bundling. Additional mild steel may be placed at right angles to the prestressing tendons to contain bursting and radial forces.

Brackets introduce localized structural loads which must be distributed. Any change in concrete cross-sectional area should be gradual (tapered) to avoid stress concentrations.

A long delay in filling of a tank may permit excessive shrinkage and creep.

Mild-steel anchor bars from floor to wall and wall to roof must have adequate embedment and anchorage to prevent pull-out under dynamic loading.

Earthquake cables, consisting of unstressed galvanized strands, are frequently used to anchor the walls to the floor slab when the joint details permit rotation. Sleeves of rubber or similar material may be placed around the strands so as to permit radial movement of the wall. The portion of cable to be encased in the sleeve should be given a protective coat against corrosion.

With dome roofs, two layers of steel should be used near the edge, in the meridional direction, to resist edge-bending moments. The dome may require thickening in this edge region. These moments can be minimized by proper detailing of the joint.

To prevent shrinkage and temperature cracking in the dome roof prior to prestressing, mild steel, such as wire mesh, should be placed and water curing continued until the prestressing is completed.

Positive keys or stops or anchor cables must be provided with unrestrained joints so as to prevent displacement of the roof relative to the walls.

Floors must be designed to resist hydrostatic uplift when empty. They may be anchored to the subsoil by stressed or unstressed rods. With proper gravel underdrains, floor pressure relief valves may be a solution. The gravel should be protected by a sheet of polyethylene from contamination and clogging by soil.

If sand is used under the floor slab, the edges should be protected from erosion or wash out. A low concrete wall or bitumastic seal coat may be used to retain the sand.

When steel liners are used, vertical ribbing or fluting is desirable, in order to permit shortening under prestress. Sheets should be seal-welded, brazed, or sealed with elastomeric caulking and sealing compound to form a permanent, flexible seal. Polysulfide liquid polymers may be used, applied with a caulking gun (but not for sewage or oil storage). Epoxies are also used, applied by pumping up the joints between the sheets and the wall.

Waterproofing of the exterior is often required, especially when the tanks will be backfilled. The thin concrete cover over highly stressed tendons makes such additional steps desirable to ensure durability despite any latent errors in concreting or shotcreting, or any porosity, etc. External paint may be epoxy, bitumastic, polyurethane, or one of the paints listed in the next paragraph.

Internal paints are employed when the tanks will contain acids, sewage, gasoline, etc. Rubber-base paint, polychloride vinyl-latex, polymeric vinyl-acrylic, polyurethanes, and epoxy paints are used, selected for their durability in continuous contact with the stored material.

Waterstops are usually of rubber, or plastic such as extruded polyvinyl chloride. Butted joints should be fused together to ensure water-tightness. Waterstops should be placed in joints in floors and wall footings. Waterstops are usually used in wall joints unless other positive means are taken, such as liners, or the joint surfaces are specially treated, e.g., by exposing the coarse aggregate by water jet and use of bonding epoxy.

Bearing pads are normally of neoprene or natural rubber. A combined bearing pad—waterstop of extruded virgin polyvinyl chloride may be used.

Sponge filler should be closed-cell neoprene or rubber.

Concreting of a dome roof should be performed in circumferential strips, working from the exterior edge towards the center. Circumferential and tangential screeds should be placed to ensure uniform thickness. The thickness can also be checked by probing the fresh concrete. The dome roof should be supported by scaffolding or posts until prestressed.

16.2 CIRCUMFERENTIAL PRESTRESSING

16.2.1 Internal Prestressing

Many prestressed concrete tanks are prestressed with internal post-tensioning tendons running through ducts in the walls. Because of friction due to curvature, it is usual to limit the arc over which any individual tendon runs. For example, tendons extending over 120° or 180° of arc are often employed, staggered so that one-third are stressed at each 60°.

To minimize frictional losses, ducts should be sufficiently rigid so as to prevent excessive wobble. Steel ducts are most common. Plastic ducts may be used provided it can be demonstrated that stressing of the strand will not cut the wall of the duct.

When precast panels are used, the ducts must be continued through the joints, without grout leakage and without abrupt changes in alignment.

The stressing points, e.g., at 60° locations, may be enlarged buttresses or may be special panels with blockouts. The anchorages are usually located externally, rather than internally, so as to minimize the curvature.

Stressing may be by conventional jacks or by special center-stressing jacks. Stressing both ends simultaneously will reduce prestress losses.

16.2.2 External Wrapping

Wire or strand wrapping under tension is extensively employed for circular tanks. Specifications commonly require a tolerance in tension of ±7%. As the wire is wrapped, temporary anchoring clamps should be attached to prevent loss of prestress if a wire should break. Anchoring clamps may be removed as the cover coat is applied. Terminals of individual wire coils may be spliced with splices capable of developing the full tensile strength.

The wrapping may be kept under tension by drawing through a die, or the tension may be provided by a braked drum.

A calibrated stress recording device should be employed continuously to record the tension levels during the wrapping process. In smaller tanks, intermittent readings are customarily taken; however, continuous readings are obviously preferable. Readings should be identified with height and layer of wrap.

If wire stress falls below the specified tolerances, additional wraps can be placed to compensate.

Wire spacing in a layer should normally be 8 mm ($\frac{5}{16}$ inch).

When strand or bar tendons are used externally, they may be stressed in arcs of 120° or even 180°. Stressing points are usually staggered 60°. Small saddles of steel are usually set at frequent intervals to minimize friction. These saddles should be staggered, or placed on a diagonal, so as to distribute the concentrated radial stress. Such saddles may be part of a wedge stressing device.

A ratchet-type jack has been developed that can stress bars at a coupler, pulling them together and anchoring. Other jacks permit tensioning on a slight outward curve, with minimum deviation from the circular path.

When tendons are placed, it is essential that corrosion protection be applied within as short a period as possible, e.g., 24 hours or less. It is especially important that no salt crystals be deposited by salt spray or salt fog. Prior to coating, the tendons should be thoroughly washed with fresh water under pressure.

Tendons likewise should not be exposed to a combination of moisture and sulfides, such as can occur near a refinery, especially during a fog. Tendons inside ducts can be protected by VPI powder, dusted in and sealed. External tendons should be washed as above and coated as soon as possible.

Under conditions of severe exposure, underlying concrete (or flash coats) should be jetted with fresh water both before wrapping and before coating.

Wire-wrapping machines impose a vertical bending movement on the walls. The walls should be suitably reinforced for this condition, since wire winding usually is performed prior to vertical prestressing. When wire is extruded through a die, means must be provided to cool the wire so that the heat generated does not affect the wire's properties. In any event, temperatures should be kept well below the 150°C (300°F) level.

16.3 PROTECTION OF EXTERNAL TENDONS

Protective coatings over prestressing tendons may consist of shotcrete or cast-in-place concrete. Shotcrete is most commonly employed. It must be emphasized that the placement and control of shotcrete is a skilled art. Failure to properly apply shotcrete may result in trapping rebound behind the tendons, leading to porosity and susceptibility to electrolytic corrosion.

Shotcrete should conform generally to the provisions of ACI 506. Calcium chloride must never be used in the mix. Shotcrete should be applied with a very dry mix, so that frequent slightly dry spots appear on the regular glossy surface. The wet process of shotcreting is preferable to the dry process, because it minimizes the entrapment of rebound. Joints should be cleaned with an air-and-water blast. They should be as nearly at right angles to the surface as possible.

Walls should be built up of individual layers 50 mm (2 inches) or less in thickness, using tensioned vertical ground wires not more than 1 meter (3 feet) apart to control thickness.

Each layer should be initially protected with either a cement slurry coat or a flash coat. The cement slurry is desirable but difficult to apply; the flash coat usually achieves satisfactory results. The outer layer should be protected by a flash coat, body coat, and finish coat. A suitable paint sealer may be provided over the final coat to ensure watertightness, and to prevent shrinkage cracks. The cost of painting is often justified because the cost is small in relation to the additional assurance achieved. Painting or coating is strongly recommended in coastal and arid regions, and near industrial areas.

The flash coat provides the initial protection to the steel, ensuring intimate contact and passivation. Care must be exercised to ensure no voids, trapped rebound, etc. The flash coat should be a 1:3 mix, wet but not dripping, and should provide 3 mm ($\frac{1}{8}$ inch) cover over inner wire layers and 10 mm ($\frac{3}{8}$ inch) over the outside layer.

The nozzle should be held at a 5° upward angle, moved constantly and regularly, pointing always at right angles to the surface (i.e., towards the center of a circular tank). The nozzle should be held at such a distance that the shotcrete does not build up or cover the front face of the wires until the spaces between the wires are filled.

Immediately after placing the flash coat, a visual inspection is made. If wire patterns show up as continuous horizontal ridges, the shotcrete has not been driven behind the wire properly. If the surface is substantially flat, with no particular showing of the wire pattern, then the space behind the wire is essentially filled.

The flash coat should be kept moist until succeeding layers and/or coatings are placed.

The body coat is usually 10 mm ($\frac{3}{8}$ inch) of 1:4 mix. It should be screeded prior to final set, and damp-cured.

A finish coat of 6 mm ($\frac{1}{4}$ inch) is frequently applied. Use a 1:4 mix.

Total coating thickness over the tendons should not be less than 25 mm (1 inch). The completed shotcrete coating should be continuously damp-cured for 7 days. Alternatively, especially in windy areas, a membrane curing compound may be used, applying two coats 8 to 12 hours apart, or an epoxy or similar membrane.

The tank should be filled with water prior to the application of the body coat, so as to prevent expansion and cracking. Where this is impracticable, a membrane coating after first filling is recommended. Flexible membranes, capable of spanning over cracks, are currently being developed.

16.4 TOLERANCES

A maximum out-of-round tolerance of plus or minus 75 mm (3 inches) per 30 meters (100 feet) of diameter should be maintained for circular tanks. Wall thickness should be kept to a tolerance of ±6 mm ($\frac{1}{4}$ inch). All transitions should be gradual. Walls should be vertical with a tolerance of ±10 mm ($\frac{3}{8}$ inch) per 3 meters (10 feet) of height.

Elastomeric bearing pads should be attached to the concrete with adhesive to prevent displacement during concreting. Any voids or cavities occurring between butted ends of pads and waterstops, pads and sleeves, or sleeves and waterstops should be filled with a soft nonpetroleum-base-mastic compatible with the pad, sleeve, and waterstop materials, and with the material to be stored in the tank.

Sponge-rubber fillers should be ordered 12 mm ($\frac{1}{2}$ inch) wider than the gap, in order to facilitate placing and reduce possibility of voids between sponge rubber, bearing pads, and waterstops. Sponge rubber should be secured to the concrete with an adhesive.

When wire mesh is employed in thin shells, "dobe" blocks or plastic spacers should be employed to keep the mesh at the proper distance from one face. For such thin shell applications, galvanized mesh is desirable.

16.5 STORAGE TANKS FOR SPECIFIC PRODUCTS

16.5.1 General

The previous sections have dealt with the general aspects of prestressed concrete tanks and because of the dominance, numberwise, of water and sewage storage tanks, has dealt primarily with them. In the succeeding sections, the special problems which arise with specific stored products are discussed.

16.5.2 Water Storage

In hot climates, thermal strains need special attention. These are due to the radiant heat of the sun, which affects one sector more than the others, and to the rapid cooling at night after the walls have been heated by the sun and the air. The water inside will of course have a relatively uniform temperature. The inside will be at almost 100% relative humidity, whereas the outside may be dry, which will induce additional strains due to moisture differentials.

Conversely, in climates exposed to freezing, an even more serious situation may arise, that of freeze-thaw damage internally. The tank walls are fully saturated, the inside portion of the walls is normally at or above freezing, whereas the external portion of the walls is below freezing. Thus there is a freeze front internally, which moves back and forth as the external temperature and sun exposure change. Individual crystals of concrete are thus subjected to many cycles of freeze-thaw in a condition of critical moisture. Even large quantities of air entrainment will not suffice.

The recommended solution for this phenomenon is to install either a steel membrane or plastic liner, so as to allow the concrete to remain dry.

Of course, if the below-zero temperature continues long enough and is deep enough, the surface of the water inside will freeze and exert pressure trying to expand the walls. The age-old preventive of floating a spar of softwood can be improved upon today by vertical blocks of styrofoam affixed to the walls.

The partial water saturation of the concrete walls also makes them vulnerable to corrosion. Above-ground tanks near the sea coast are exposed to chloride corrosion. Carbonation from the air may lower the pH and render the steel depassivated.

Underground tanks may be exposed to chemical attack from the soils, e.g., sulfate attack on the concrete or chloride attack on the steel.

The prevention is to ensure that the concrete is as impermeable as possible by taking special pains with the cover coat and finish coat of shotcrete to prevent trapping of rebound. External coatings of epoxy, bitumastic, or polyurethane are indicated for all underground tanks and for those above-ground tanks with particularly aggressive exposure. An internal membrane will, of course, eliminate the saturation from the stored water.

16.5.3 Sewage Digester and Storage Tanks

Prestressed concrete has been widely used for sewage digester and storage tanks (Fig. 16.3). Some digester tanks are shaped as eggs in order to facilitate the processing: others are simply cylindrical tanks.

Because sewage products are corrosive, the basic concrete tank must be free of cracks prior to prestressing. Such cracks may result from shrinkage or thermal

Figure 16.3. Sewage treatment facility in San Antonio, Texas, is comprised of tanks constructed with precast panels incorporating a steel membrane (Courtesy of Preload, Inc.)

strains. Provision of mild-steel circumferential reinforcement may be an effective means of resisting and controlling strains prior to prestressing.

With externally wrapped tanks, special care must be taken at the upper joint between the basic concrete and the wrapping layers. The atmosphere in the vicinity will be corrosive and any moisture entry will result in corrosion of the tendons. Sheathing and coating are therefore essential.

Similarly, in view of the inherently corrosive atmosphere, external coating of epoxy or polyurethane appears appropriate.

Attacks on high-quality concrete by the products, such as H_2SO_4, generated by anaerobic bacteria are only of concern if air will be trapped at the top. In that case, epoxy coating or plastic sheets will be advisable in that air zone.

16.5.4 Oil and Liquid Petroleum Products Such as Diesel and Gasoline

Concrete, especially when high quality, with adequate content of cementitious materials and low W/C ratio, is relatively immune to chemical attack from crude oil, even at elevated temperatures such as 60° (140°F). Of course thermal aspects need to be considered, but these may be minimized by internal convection currents and paraffin coatings that form on the walls.

On the other hand, refined products, such as gasoline, may be aggressive. Hence steel or plastic liners are used. As with all liners, it is necessary to anchor them to the walls, either by lugs, as usually used with steel liners, or by adhesion in the case of plastics.

Crude oil often contains anaerobic bacteria which generate sulfides upon contact with seawater. If oxygen is available, they convert to weak sulfuric acid, which can attack weak, porous concrete, although studies indicate that this is not a real problem for the grades of concrete normally used with prestressing. Some recent experiences indicate that if air will be present over the crude oil, epoxy coating is warranted in the above-oil zones. The H_2S and H_2SO_4 which form may, however, cause some corrosion of steel piping and embedments. Epoxy coatings are generally adequate to prevent this.

16.5.5 Heated Storage

A number of storage vessels of prestressed concrete have been designed and constructed for storage of fluids at elevated temperatures. Here the design must consider not only the global expansions and contractions but also those thermal variations through the wall and at the level of partially filled storage.

Provision will have been made for either displacement or rotation at the junctures of the floor and roof with the walls.

Both vertical and circumferential prestress will normally be applied. The contractor will have to install ducts with great accuracy to prevent the introduction of unwanted secondary and tertiary stresses. Use of heavy wall duct, pre-bent to the required radius, will be appropriate in the more demanding installations.

Wire mesh in the exterior face may inhibit thermal cracking.

16.5.6 High-Pressure Storage Vessels

Proposals have been made to store superheated steam, at very high pressures, in prestressed concrete tanks. Appropriate liners will be included in the design, to ensure against leakage.

Because of the high pressures envisaged, thick walls of high-strength concrete and very heavy prestress will be required. One system applies the prestress in grooves in the walls, in order to permit higher compressive stress in the concrete.

During construction, the heat of hydration and subsequent cooling of the walls will create a potential problem of thermal strains and potential cracks. The following methods would appear appropriate.

1. Design of the mix for low heat, by use of PFA or BFS-cement in appropriate percentages.
2. Use of coarse ground cement of moderate or low heat, but adequate to cause the reaction in the PFA or BFS.
3. Pre-cooling by liquid nitrogen.

4. Provision of mild reinforcing steel in both directions, both faces, to resist, disperse, and later close cracks.

5. Insulation of forms.

6. After forms are stripped, insulation of surface.

7. Appropriate curing.

16.5.7 LNG and LPG Tanks

Prestressed concrete has been shown to have excellent cryogenic properties at temperatures as low as −173°C (liquid nitrogen). This is true of both normal-weight and lightweight concrete. Thus the sudden impact of liquefied gas from a ruptured inner tank does not lead to failure. However, it of course does lead to cooling and contraction of the cold face, which will produce cracking of the concrete unless countered by careful design details.

In the typical LNG storage tank, there is an inner tank, with a 9% nickel steel membrane, with an annular space between it and the prestressed concrete outer tank (Fig. 16.4). The sudden surge of liquid as it flows around this annulus will produce ovalling and dynamic pressures that may rupture the tank. Thus the structural design must include this effect as a design criterion.

Other tank systems attach the nickel-steel membrane to the inside of the pre-

Figure 16.4. Two LNG storage tanks under construction at Staten Island, New York. The walls were constructed of precast panels and, after jointing, were circumferentially wire wrapped. (Courtesy of Preload, Inc.)

stressed concrete wall, with ductile anchor and/or pleats in the membrane to accommodate the sudden contraction forces from rupture of the tank.

Insulation systems have to be provided integrally with the tanks. The sequence of wall and roof construction becomes extremely critical, so as to ensure that all needed work can be properly performed and inspected.

Many such tanks are designed with external fire-resistant insulation as well.

To meet the extreme conditions imposed by thermal extremes, concrete of low modulus, e.g., structural lightweight concrete, may be selected for the walls and roof. Control of the mix and proper placement will be very exacting because of the demands and tolerances for compressive strength, tensile strength, and modulus. These can often be met by the use of precast concrete "staves," pretensioned in the longitudinal direction. After erection and jointing, they are posttensioned circumferentially by either internal grouted tendons or wire-wrapping. A cover of concrete or shotcrete is then placed.

16.5.8 Pulp Treatment Tanks

The pulp and paper industry is now increasingly required to treat the effluent prior to discharge. Prestressed concrete tanks are increasingly employed in this service. They have the additional requirement that they must contain oxygen in the upper level of the tank.

Very impermeable concrete is achieved by the incorporation of microsilica, and by caulking of the joints from the underside with a polyurethane sealant.

Internal coating with epoxy or polyurethane may be warranted.

16.5.9 Storage of Agricultural Waste

Throughout much of Europe, farmers are now required to store agricultural waste and manure so that it can be partially digested, with the residue discharged only during the rainy season. This practice can be expected to spread.

Because of the great number of moderately sized tanks required and the need for extreme economy, segmental precast tanks have been developed which can be readily shipped and erected. Most of the solutions include circumferential stressing by means of bars which are coupled. Stressing is carried out by a scissors jack.

16.6 EVALUATION

Prestressed concrete tanks have a long history of use for water storage and, to a lesser extent, for storage of other materials. Unfortunately, a few cases of corrosion and failure have occurred. Most of these can be traced to the following causes, aggravated by the continuous presence of water as an electrolyte and of oxygen over the large external area exposed to the atmosphere.

1. Sewage or other aggressive materials penetrating to the tendons through large open shrinkage cracks in the primary wall.

2. Salt-cell electrolytic corrosion in coastal regions due to porosity of shotcrete.

3. Use of calcium-chloride in the shotcrete mix.

4. Tendons left stressed for long periods prior to encasement, in an industrial area exposed to fog.

5. Separation of the joints between wall and cover coat, allowing moisture to penetrate to the tendons from the top.

6. Corrosion of sleeves and valves at penetrations through tanks.

7. Inadequate sealing and electrolytic corrosion of fittings.

Any tank used to store salts, particularly those containing chlorides, should be waterproofed on the inside to prevent salt-cell electrolysis.

The use of a steel or plastic liner, external painting or coating, and sealing of the top of the wall (juncture between roof and wall) will generally solve even the extreme durability problems. Adequate anchoring of liners must be ensured.

Thermal strains need to be addressed with care during construction to eliminate macro (visible) cracking.

Prestressed concrete tanks can be reliably and economically built to perform a wide range of services, from underwater oil storage, to buried fuel tanks, to elevated water storage, to silos for granular materials. Both hot liquids and cryogenic materials are safely and practically stored in prestressed concrete tanks. Except where general corrosion has taken place, the failure mode of prestressed concrete is inherently safe; such tanks fail by cracking and relief of pressure rather than by sudden or catastrophic ripping. They are durable. They can be built to extreme diameters and heights, and to accommodate extremely high pressures. Some of the elevated water tanks of prestressed concrete in Western Europe are among the most attractive and beautiful of all architectural structures. The future of prestressed concrete tanks, therefore, depends primarily on increased reliability and dependability, especially in regards to corrosion, by following the provisions outlined in Chapter 5, Durability, and this chapter.

REFERENCES

1. FIP, "Recommendations for the Design of Prestressed Concrete Oil Storage Tanks," Federation Internationale de la Precontrainte, London, 1978.

2. Preload, Inc., *"Design and Construction of Prestressed Concrete Tanks,"*, also *Construction Procedures and Specifications,* Preload Corporation, Garden City, New York, undated.

3. *FIP State of Art Report,* "Cryogenic Behavior of Materials for Prestressed Concrete," Federation International De la Precontrainte, 1982.

Prestressed Concrete Pressure and Containment Vessels for Nuclear Power and High-Pressure Gases

17.1 INTRODUCTION

The construction of prestressed concrete pressure vessels is both a challenge to the constructor and a major consideration in the design of the vessel. A pressure vessel is a subsystem within the overall system, and as a subsystem, may involve a complex integration of concrete, prestressing tendons, anchorages, ducts, mild reinforcing steel, penetrations, liner, and cooling system. The successful integration of all of these into a completed vessel requires that the best and most sophisticated techniques of construction engineering, planning, scheduling, coordination, prefabrication, materials handling, and installation all be brought into full utilization.

Pressure vessels present three aspects not normally found in other types of engineering construction:

1. The quality control, accuracy, and tolerance limitations are extremely severe. Regulatory agencies require very demanding Quality Assurance/Quality Control (QA/QC) programs and complete documentation.
2. The work site is very compact and, therefore, highly congested.
3. The time available is very limited and very critical.

The solution, of course, is adequate construction preplanning, using the best planning tools available. Thus, methods can be specified in detail that will assure adequate quality control; forms, supports, and concreting methods can be chosen to achieve accuracy; prefabrication can move much of the labor away from the immediate focal point; and proper scheduling can assure completion on time.

However, it must not be overlooked that we are dealing with real, tangible materials and equipment, and with groups of individual workers and site supervisors. We are subject to all the multitude of practical limitations and variations that occur on all construction; the difference is that with pressure vessels, we cannot tolerate anywhere near the range of variances and mistakes that we accept elsewhere.

It must be again emphasized that all of the above activities, as well as others relating to the main plant itself, are going on simultaneously at one focal point—a cube perhaps 50 meters (160 feet) in each dimension. After discussion of individual activities and the techniques required, further examination will be made of the scheduling, coordination, and management of construction.

Prestressed concrete pressure vessels have been primarily connected with nuclear reactor power plants. This development took place as a result of the demand for larger vessels, capable of withstanding higher pressures. The present generation of boiling water and light-water rectors has utilized a heavy steel primary reactor vessel and a prestressed concrete secondary containment structure. This containment structure serves to contain fission products in case of accident, and thus must also serve as a radiation barrier. Prestressed concrete is thus excellently suited to serve as a containment structure and is being so employed in a large number of nuclear power plants in the United States, Canada, Europe, and Japan. A containment structure is a rather unique structure in that the only time it functions as a pressure vessel is during acceptance and annual tests, and for a short period following an accident.

For the CO_2 gas reactors, a single pressure vessel of prestressed concrete is utilized. Prestressed concrete is ideal for this use because of its behavior during the maximum credible accident in which it goes through progressive stages of loading and failure. The several stages may be as follows:

Stage 1: Elastic behavior.

Stage 2: Cracks form and widen, liner keeps vessel tight. Upon reduction of pressure, cracks close.

Stage 3: Cracks widen, liner ruptures, gas escapes locally, reducing pressure.

Stage 4: Wide cracks, spalling, extensive leaks, but no catastrophic ripping or bursting.

Reactor pressure vessels must withstand large thermal gradients as well as pressure. There are numerous transient stages which have to be thoroughly investigated in design.

Future generations of nuclear power plants will probably involve fast breeders, cooled by sodium, gas, or steam. In the meantime, there is an existing development of advanced reactors, such as the high-temperature gas reactor utilizing helium and the light-water breeder reactors. Far in the future are the deuterium reactors, with their almost unlimited available energy from the oceans and with no radioactive waste. Dual purpose plants, combining desalination and power, and "energy centers" are presently undergoing feasibility engineering studies.

While attention today is concentrated on pressure vessels for nuclear reactors, prestressed concrete offers similar significant advantages to high-pressure vessels for chemical plants and refineries, and pressurized steam vessels for energy storage and for coal gassification vessels. The noncatastrophic mode of failure, the ability to assemble and construct on site, and the lack of limitations on size and pressure make prestressed concrete an ideal material for such uses. The problems of penetrations are no more severe than for nuclear reactor pressure vessels and may even be simpler of solution.

While "prestressed concrete pressure vessels (PCPV)" has become almost a generic term, actually a pressure vessel may embody a large amount of conventional reinforcement, especially high-yield steel, and may act as a partially prestressed concrete structure as it passes through its several "limit states" or behavior stages.

The prestressed concrete pressure vessel requires a "systems engineering" approach to construction. The typical system includes:

1. Concrete, including forms, batching, mixing, placing, curing.
2. Reinforcing steel.
3. Liners.
4. Cooling tubes.
5. Penetrations.
6. Ducts (sheathing, splices, and protective caps).
7. Sheathing couplers, trumpets, transition cones.
8. Air vents and drains.
9. Tendons.
10. Stressing record forms.
11. Bearing plates.
12. Anchorage assemblies.
13. Corrosion protection.
14. Duct filler retaining caps.
15. Equipment selection.
16. Quality-control procedures.

17.2 CONCRETE

17.2.1 Production and Delivery of Concrete to Site

While the design engineer will undoubtedly have selected the aggregate sources, the type of cement, mix proportion, and W/C ratio, it is the province of the contractor to establish methods to produce concrete of consistent, uniform quality. Many properties of the concrete have an important effect on its performance for pressure vessels. These properties include thermal expansion, conductivity, and stability, creep and shrinkage at normal and elevated temperatures, compressive and tensile strengths, modulus of elasticity, Poisson's ratio, and heat of hydration.

The aggregate is normally crushed, washed, and screened at the crushing plant and delivered by cars or trucks to the site, dumped in piles or hoppers at the site, and stored pending elevating to the batching plant. It is obvious that the chances are manifold for contamination, chipping, and dusting, etc.

First concern should be uniformity of production at the quarry to be sure weak or unsatisfactory material from seams cannot be blended in. Depending on the source, a careful geological inspection should be made and a particular block of the quarry should be selected for this production and continuously inspected to prevent accidental blending from other blocks or sources.

At the site, aggregate for the vessel should be stored on concrete slabs or in bins or metal hoppers, not on the ground. For a reactor vessel, the aggregates should be kept separate from all other concrete supplies at the plant.

Aggregates should be protected from excessive heat (by covering or the water soaking-evaporation method) and from ice and snow. Rescreening just above the batch bins will remove chips. The batch bins should be large enough to assure uniformity of moisture content.

For the cement, it is important that it be properly aged and that its temperature be within established limits.

It is important to limit the contamination in the mixing water and in the concrete to prevent the possibility of corrosion of the steel components. Recommended limits (by weight) are:

MIXING WATER

Chlorides as Cl^-	250 ppm
Sulfates as SO_4^-	250 ppm

IN WET CONCRETE

Chlorides as Cl^-	1000 ppm
Sulfates as SO_4^-	1000 ppm

The ambient temperature of freshly mixed concrete should be between 5°C and 15°C (40°F and 60°F). This can be accomplished in summer by using crushed ice in the mixing water and by cooling by evaporation in the aggregate storage. Liquid nitrogen may be used to cool the aggregate or even injected into fresh concrete. Vacuum cooling of the aggregates may be utilized effectively in extremely hot climates. In winter, the mixing water can be heated, and the aggregates kept in warm storage or heated by steam.

For the low-slump concrete which has typically been involved, turbine mixers are preferable to ready-mix trucks. Mixing time should be carefully regulated; variations will produce wide divergence in quality and workability.

Transportation to the vessel site should then be in hoppers, buckets, truck-mounted hopers, etc., which are specially designed for low-slump concrete. These usually have vibrators or screws for discharging. Agitating vibrators or paddles may be employed to prevent segregation and false set.

To prevent segregation, and to increase workability while maintaining the required W/C ratio, admixtures are practically essential. These should contain no calcium chloride and should, of course, be subject to the approval of the engineer.

Strong doses of conventional water-reducing admixtures may be preferable to high-rate water-reducing admixtures because of the potential for slump loss associated with the latter. If HRWR are used, they should also contain a retarder. The use of such "super-plasticizers" enables a more workable mix which can be more easily placed amidst the dense steel. Internal vibration is still required.

This matter of workability is of prime importance to the contractor. The vessel is full of highly congested spots and it is often just at these spots that the best consolidation is needed. It is recommended that, prior to any concreting of the vessel itself, a full-scale mock-up of one area of typical congestion be made and actually poured with the design mix. Then it can be stripped, cut into, and the thoroughness of consolidation verified. This insures that there will not have to be last minute changes in the mix, placement methods, or vibrators, as concreting is underway. In addition, it is the best possible means of indoctrinating the workers as to the why and how of the concreting procedure. This writer has used this mock-up approach very successfully on special and congested prestressed construction.

For the Fort St. Vrain, Colorado, High Temperature Gas Moderated Reactor (HTGR) vessel, the following mock-up procedures were followed:

1. A penetration liner mock-up simulating an actual penetration liner, to demonstrate the adequacy of the concreting procedure in producing sound concrete in the area of penetrations, cooling tubes, and shear anchors.
2. A full-scale mock-up of a 60° sector of the bottom region, extending 5 feet below the bottom head liner and involving 60 cubic meters (80 cubic yards) of concrete.
3. Two mock-up demonstrations for evaluating the technique for applying concrete by special means to the bottom head liner, to demonstrate production of void-free concrete of specified strength in the bottom head.
4. A full-scale mock-up of a 120° sector of the PCPV head to establish the sequence for installing the reinforcing steel and other embedments.

17.2.2 Placement

The concrete pours will have been carefully delineated by the contractor and engineer. Generally, lifts of (1.5 to 2 meters) (5 or 6 feet) are scheduled, and pours in a size range of 100 to 300 cubic meters (130 to 400 cubic yards) per day. Form pressures and deflections, heat and shrinkage are all related to the size and height of lift.

For containment structures, slip forms have often been used, requiring continuous pours over a period of up to 10 days.

Because of protruding ducts and reinforcement, it will probably not be possible to get the bucket closer than about 2 meters (6 feet) vertically to the surface. A rubber elephant trunk extending down from the bucket bottom can be used to prevent segregation during fall.

Thorough internal vibration is essential. In addition, form vibration on the liner plate can be used to insure intimate contact between liner and concrete. The frequency of vibration required depends on the mix; for some very dense harsh mixes, higher frequency and more powerful vibrators may be required. The effect of powerful vibration on form pressures must not be overlooked.

Care must be taken to prevent dislodging or mis-location of ducts or reinforcement during vibration. This may be helped if positions of ducts and critical areas are clearly marked on the side forms.

The curing procedure should be selected so as to insure adequate maintenance of moisture for concrete maturing and may be used to absorb some of the heat of hydration. Circulation of water through the cooling tubes has been employed for heat reduction. Whether water curing or membrane curing is employed, the important thing is to apply it early enough to prevent surface drying. In very hot, dry areas, especially with a wind, this surface drying may present a major problem. Enclosures and scheduling of concrete pours for night may be required. Night pouring also aids in reducing the ambient temperature. Some reactor vessel construction has been completely enclosed and relative humidity and temperature control maintained.

In winter, and especially when cold winds blow, the forms should be insulated. After stripping, the concrete surfaces should be blanketed, in order to prevent thermal cracking.

Construction joints present a major concern. Horizontal surfaces may be jetted so as to expose the aggregate, while the concrete is green (say, 12 hours old), with air and water. Side forms may be stripped at age 12 to 24 hours, and jetted. High-pressure hydraulic jets can clean off laitance and mortar even after a few days of age.

In starting a new pour, the old concrete should have been kept thoroughly wet for 24 hours. The new concrete lift should start with 150 to 200 mm (6 to 8 inches) of the basic concrete mix from which the coarse aggregate has been left out, then proceed with the regular mix. Vibrators should penetrate all the way to the old concrete. Here again, a mock-up pour, including a horizontal construction joint, with all the congestion of ducts and bars across the joint, will prove the proper technique of placement and vibration.

Wet sandblasting can also be used to prepare construction joints, but produces a sand rebound which is difficult to keep out of ducts, etc. Therefore, its use is not recommended.

One of the main problems in all of these operations is keeping mortar and debris from contaminating other surfaces. Ducts must be tightly covered, as must cooling-pipe openings and anchorages. Mortar splash onto reinforcement must be kept to a minimum. Before starting a new pour, the cleanliness of adjacent areas must be restored. Curing water must be kept out of ducts, or well drained.

To ensure complete consolidation of concrete around anchorages and penetrations and the underside of the liner base, grouting has been employed. Holes near the surface are formed by rubber tubes; after concreting, the tubes are withdrawn, and grout is injected under pressure.

For the base of the liner it seems best to set the liner on supports a few inches above the screeded concrete surface, and then inject grout. A thixotropic admixture can be employed to promote flow and eliminate bleed.

Systematic testing procedures must be carried out on all concrete pours. A fully equipped concrete laboratory should be maintained at the site.

Forms should be designed to limit deflection and to be rugged against accidental impact from buckets and vibrators. It is particularly important that anchorages and penetrations be rigidly and ruggedly held. The use of form ties must be approved by the designer, as they represent a heat transfer path. Thus, it may be necessary to use the type that is subsequently removed and, in this case, the holes must be later filled with grout. For this reason, ties within a pour must be kept to an absolute minimum; ties across and above the top of the lift and anchors in a previous lift may eliminate the need for internal ties in a pour.

If, in spite of careful planning and supervision, rock pockets or honeycomb do occur in the concrete, repairs must be instituted. For example, on one containment structure, the first pour was too large, slump too low, and consolidation by vibration proved inadequate. For this containment vessel, a specially formulated epoxy, designed to be stable at 80°C (180°F) was injected. In other pressure vessels, grout injection may be needed, or specific spots may have to be cut out, with keys and a well-prepared surface, and new concrete placed. In such cases, shrinkage of the repair grout or concrete must be reduced by careful control and curing, and by judicious use of non-corrosion-producing anti-bleed admixtures.

The concreting of the typical dome top closure of pressure vessels is very demanding, because of its thickness, the congestion posed by the multiple prestressing ducts crossing at 60° angles, and especially the interference of the standpipes. Because of the thickness of the slab, heat of hydration and the subsequent cooling, which tends to result in cracking due to the many restraints, must be controlled.

For these reasons, concreting has frequently been carried out in lifts of 2 feet. The surface of each lift is jetted clean of laitance prior to placement of the next lift. Care must be taken to ensure thorough vibration of each lift, especially around the standpipes.

In order to overcome the limitation of access, in some European vessels the dome has been concreted in small sectors, each full height. In this latter case, only those standpipes are installed which are in the sector to be concreted. Vertical construction joints are prepared for the next adjoining castings by water jetting of the surface. With this method, however, heat of hydration is an even greater problem than with the multiple lifts.

In all cases, pre-cooling of the concrete mix is beneficial.

17.3 CONVENTIONAL STEEL REINFORCEMENT

There is a strong tendency in design to use large amounts of such reinforcement, particularly with unbonded tendons. This presents a special problem in congestion and support. Reinforcement must be thoroughly tied. Where feasible, prefabricated cages may be installed, as for example, around penetrations. Usually tack welding of reinforcement is prohibited, so crossing bars must be rigidly clamped to keep the cage from distribution.

Splicing of small bars is generally by lapping. For bars of 30 mm (1¼ inch) or larger, splicing should be by welding or mechanical connectors. If butt welding is required, care must be taken not to accidentally burn adjoining ducts, and any tendons must be positively protected. Mechanical connectors, suitably qualified for pressure vessels, are being increasingly used and are recommended since they ensure concentric transfer of force, reduce splitting stresses in the concrete, and reduce congestion.

This reinforcement may be used effectively to support the ducts; in fact, a careful design may produce the necessary framework for duct support. Where bars rest on ducts, a small saddle may be necessary to prevent denting. Once again, the designer must consider the process of construction.

17.4 LINERS AND COOLING TUBES

General practice has been to prefabricate steel liners to the side of reactor vessels, then skid and jack them into position on the concrete base. In other cases, prefabricated sectors are made and lifted into position, requiring jointing at the site. It is important to note that the erection sequence chosen may influence the behavior of the liner.

In most reactor construction there is a very large gantry crane installed to lift the reactor core; this can often be used to lift in the liner assembly.

Liners must be tight; therefore, a high-quality welding technique must be employed with radiographic and other testing. The steel and the welds must have ductile (nonbrittle) behavior at the lowest temperatures to which the liner will be subjected, both while in service and during construction.

It seems most desirable that the cooling tubes be installed on the liner plate in the prefabrication phase. Care must be taken in welding not to distort or wrap the liner excessively. Similarly, stud anchors for the liner should also be installed during prefabrication. Induction welded studs may prove most practicable. Care must be exercised not to create any notch effects in either liner or anchors.

Once the liner is set in general position, it must be jacked and aligned into exact position and held there during concreting (Fig. 17.1). It may be possible to use the cooling tubes as structural reinforcement, or to use the liner rib reinforcement as temporary support. Local deformation due to concreting pressures must be severely limited by adequate bracing, supports, or ties.

When welding on the liner after concrete has been placed (as for example, sealing holes in the base through which grout was placed), extreme care must be taken to minimize heat which may warp the liner away from the concrete.

This also applies to welding of liner plate joints where concreting has been carried to or past the joint. This joint must be detailed properly to prevent warping or pulling away from the concrete.

Figure 17.1. Concrete lifts rise against steel liner plate. Note cooling tubes attached to steel liner.

In the case of containment structures, such seams may be purposely designed for flexibility, with anchorages at the center panels only.

Cooling water tubes must be protected at all stages from internal corrosion, accidental filling with debris, grout, etc. They are generally filled with nitrogen during welding. They are tested with nitrogen for leaks after welding and after concreting. Silica gel may be injected after welding.

Corrosion protection must be provided to the liner assembly (liner, cooling tubes, anchors) during fabrication, assembly, and installation, as well as during service. When organic compounds are specified, surface preparation must be carefully carried out in accordance with specifications, and coating thickness maintained with no "holidays." Additional corrosion protection may be required at local points of concentration, such as low spots, etc.

During concreting, sufficient supports, both ties and internal struts, should be provided to insure that the roundness of the liner is maintained, with no local flattening or bulging.

By applying shotcrete to the liner after attachment of anchors and cooling tubes, the liner may be stiffened and protected from corrosion. The "wet" process is usually preferable to the "dry" process of shotcrete. Shotcrete requires the specification and enforcement of procedures that ensure against entrapment of rebound and nonuniformity. Rebound must be removed from the site. Only skilled and experienced operators should be employed. Careful visual and sonic inspection should be carried out on the completed shotcrete layer to verify uniformity and nonporosity.

Shotcrete should not be used to fill pockets or corners. Shotcrete may incorporate lightweight aggregate when acting as insulation.

Plastic liners are being proposed, including those applied by spray and those put on in the form of sheets with adhesives or anchorages. Extreme care must be taken in protecting plastic sheets during storage and installation, including protection from sun's radiation, to insure no change in color or shade, cracking, blistering, or swelling, since these defects are usually progressive.

The adequacy of plastics is often dependent on the substrata, that is, the concrete surface and the adhesive. Surface preparation must be carried out with agents and techniques that are compatible with the plastic; for example, use of hydrochloric acid to clean the concrete may cause corrosion of stainless steel sheets or disintegration of adhesives, etc. This care applies to the form oils and form parting agents used for the concrete, to the materials and techniques for correction of defects in the concrete surface, and to the chemical and mechanical means for preparing the surface.

During application of the plastic liner, strict adherence must be maintained to the specifications and limitations on the mixing time, temperatures during application, time intervals between successive applications, thickness of coats (or adhesives), qualification testing of applicators, and local tests on the completed liner to check final dry-film thickness.

Procedures must be set up for repairs of defects.

Laps of plastic to steel and of plastic to adjacent sheets or previously applied layers must be given particular care. This also applies to the juncture at penetra tions.

Insulation, in the form of pumice blocks, may be attached to the back side of steel liners.

It has been very difficult to achieve full concrete contact under the base plate liner. Therefore the base liner is usually set up on channel supports, 100 to 150 mm (4 to 6 inches) above the concrete base slab. These also anchor the liner to the base slab to resist deformation during grouting. The liner is then welded and tested. Then grout is injected under the base liner. To offset the problems of bleed and entrapped air, in the past it was usual to perform this grouting in two stages. During the first stage, inflatable tubes or rubber void forms were inserted under the liner plate. After first stage grouting, these tubes were withdrawn and the holes grouted.

In other cases, holes were cut in the liner plate to allow the escape of air and water. Then they were filled level and closed with a welded plate.

The development of new admixtures which promote fluidity yet are thixotropic, so as to prevent bleed, may make it feasible in future construction to carry out the grouting in a single stage.

17.5 PENETRATIONS

These are almost always prefabricated. They may be attached to the liner during prefabrication, but usually are installed after the liner is in position. They require especially careful support to prevent dislocation due to deadweight and concreting. The welds are then made to the liner. Most penetrations are surrounded by mild reinforcement and the prestressing ducts are deflected past them. This results in extreme congestion, yet it is precisely here that the best concrete is required. Thus, the prefabrication of a complete assembly, including encasement in a precast block, seems to be a good solution for large penetrations.

Another alternative has been to install small rubber tubes around the penetration. These are then pulled out after the concrete has hardened and the resulting voids are then pressure grouted.

17.6 DUCTS FOR PRESTRESSING TENDONS

The selection of the proper duct material is of prime importance in determining the quality of final construction, friction losses in prestressing, and even concrete quality. Duct tubes must be able to withstand the external wet concrete pressure without collapse.

Sheathing, trumpets, and transition cones should preferably be of rigid or semi-rigid steel in order to prevent electrolytic corrosion with the tendons. Corrugated plastic ducts may be used for straight runs. However, on curved profiles, the stressing of the tendons often cuts into the walls of the duct. Special duct profiles are under development in an attempt to overcome this deficiency. When organic materials, such as plastics, are used for ducts, or if organic coatings are applied, they must not introduce sulfides, nitrates, or chlorides due to decomposition during the life of vessel.

In some vessels, thin flexible steel tubing has been used. Such tubing is not absolutely watertight, and is liable to local deformation and "wobble." When used, an internal mandrel should be inserted to give support. Inflatable ductubes and light-gauge metal conduit have been used as mandrels. These will follow most curvatures required.

Thin-gauge rigid metal ducts appear to offer the most satisfactory solution. Rigid ducts must be preformed, either in the factory or at the site. Factory preforming is practicable only for reasonable lengths; the longer lengths must be prefabricated at the site.

Joints and seams must be sufficiently tight that they do not leak grout or laitance. All joints and connections should be mechanically jointed, as by sleeves, and taped. Heat-shrink tape appears to perform more reliably than waterproofing tape.

Standard water pipe, preformed to the correct curvature, with screwed or rubber sleeve connectors, has been utilized for tendon ducts. It offers the advantages of positive leak-tight splices, and sufficient strength to prevent denting, crimping, etc., even when handled roughly.

Duct installation is exceedingly tedious and detailed and requires the utmost care to insure proper positioning and support. The usual tolerance on positioning is ± 12 mm ($\frac{1}{2}$ inch) at any point.

Ducts should be protected from corrosion in storage, during installation, and

Figure 17.2. *Thin steel watertight ducts have been capped and taped to prevent debris and water entry.*

after placement. In particular, water must not be allowed to collect in low spots, and the ends must be sealed against moisture entry (Fig. 17.2). Vapor-phase-inhibitor (VPI) crystals may be dusted in prior to sealing.

Ducts must also be protected from dents and deformations during handling, installation, and concreting. No dent should exceed 9 mm ($\frac{3}{8}$ inch).

Extreme care must be taken to prevent debris and mortar from accidentally entering the ducts. The only sure way to do this is to provide and install caps prior to installation of the ducts. Red plastic caps provide an easy means for checking.

Air vents and vent piping must be installed at specified points to permit air to escape during injection of duct filler (grease or grout).

17.7 PRESTRESSING TENDONS

17.7.1 Selection of Tendon System

The selection of the tendon system will generally have been made by the designer in collaboration with the tendon and anchorage manufacturers. Construction considerations include: size of tendons, ducts, anchorages, and bearing plates; stressing, seating, and anchorage procedures; and protection of the tendon system during storage, assembly, and installation.

There is a definite trend toward ever-larger concentrations of force in a single tendon. Several manufacturers offer very large tendons assembled from multiple seven-wire strands; others offer large cables formed from individual wires. Anchorage systems are wedged, swagged, or buttonhead. A new European development uses high-tensile steel strip.

Tendons are subjected to static tests for strength and elongation at rupture, and to dynamic tests of the entire system, i.e., tendon plus anchorages.

Whatever system is selected, its degree of success will be, to a large extent, determined by the accuracy of manufacture, and the attention given to the details of the installation.

The availability of individually greased-and-sheathed strands, which can then be assembled into multiple-strand tendons, offers the potential for lower friction and enhanced corrosion protection.

Conventional wires and strands for tendons should be delivered to the job site in waterproof packs, for which corrosion inhibitor has been provided. (See Fig. 1.6.)

17.7.2 Fabrication

Tendon material is usually fabricated into large cables on the job site. During fabrication and storage, it is essential that the material be protected from corrosion. The storage warehouse, etc., should be heated and dehumidified. Usually this can be accomplished by heaters alone. Temporary coatings for corrosion protection must be compatible with the permanent protection system. For example, a wax permanent system is compatible with an oil-base temporary system, but a Portland-cement grout system is not, unless the oil is first flushed out. Temporary coatings must be designed for a specific time and exposure, and this must be monitored.

During fabrication and again in handling and threading, the tendon cable must be protected from nicks and abrasion. During shipment, the tendon should be protected from a permanent set or kink, notches, and abrasion, and from contamination from the outside (e.g., dirt, smoke, industrial plant effluents).

17.7.3 Installation

A fabricated tendon is usually wound on a drum. This is hoisted onto a temporary platform so that the cable can be fed into the duct. Usually tendons are fed down the walls from the top. The duct is blown clean with compressed air. A messenger wire is sent through the duct, a nose piece (grip) attached, and the tendon pulled into place. As the tendon is drawn in, water-soluble oil or grease should be swabbed on as a lubricant. As soon as the tendons are in place, the ends should be sealed by polyethylene bags.

In some instances, tendons have been installed by pushing in one wire at a time.

Friction can be reduced by pulling the tendon back and forth or even more effectively by pushing the tendon in, by means of a powered roller, concurrently with pulling.

When tendons are fed into the duct, a roller device is often used to prevent abrasion of the tendon. (See Fig. 1.8.) Tendons fed in from the top, leading vertically down the walls, may have to be restrained due to their increasing self-weight.

Anchors are installed, jacks lowered to position, and stressing performed. Stressing of vertical tendons in the walls is typically carried out from a stressing gallery under the base (Fig. 17.3). A field telephone should be provided for instant commu-

Figure 17.3. *Tendons are fed down through the ducts in the walls into the stressing gallery where they will be jacked.*

nication from one end to the other. Usually stressing from both ends is required for the circumferential tendons. Tendons may be overstressed up to 5%, with the engineer's approval, to compensate for seating and friction losses. Complete records must be maintained.

Both elongation and pressure on the calibrated gauge must be recorded; they should check to an accuracy of about 5% (Fig. 17.4). Greater variation than this indicates excessive friction or duct blockage. There are remedies available for emergencies; additional water-soluble oil or grease may be injected to reduce friction and the tendon may be cycled back and forth a short distance.

Generally, at some specified time after initial stressing, a selected number of tendons are re-stressed to compensate for elastic shortening and early creep.

The handling and support of the large hydraulic jacks requires well-planned and engineered temporary devices, since they are often high in the air or overhead in a stressing gallery.

Tendon stressing and re-stressing is performed in accordance with a specified sequence. Usually this requires jacks to be so located as to permit stressing in sequence at multiple points.

Steel wires, strands or bars for prestressing should not be subjected to excessive

Figure 17.4. *Verifying the elongation of the tendons after stressing.*

temperatures (over 450°F), welding sparks, grounding currents, etc. Tendons must be protected when any welding or burning is to be performed in the vicinity.

A great deal of discussion has taken place concerning the problem of individual strands or wires twisting over one another when installed in a curved duct. As far as this author can determine, this is not a serious problem in actual practice and can be ignored except for very short tendons of high curvature. Preformed flattened ducts may be used to spread out the strands or wires around these short curves. (See Fig. 1.13.) With parallel wire tendons, the cable may be bound during fabrication, e.g., by wrapping continuously with soft-iron wire along the length where the curvature is extreme.

Some tendons are especially difficult. These are those which are installed across the dome with three sets crossing at 60° intervals. Thus each duct will have a different curvature. Selection of the duct material and wall thickness is a compromise between flexibility to accommodate the bend and rigidity to prevent excessive wobble and local denting. Duct supports are usually of mild reinforcement. Saddles must be used to prevent denting during concreting.

The dome tendons anchor in a circumferential top ring. Anchorages must be carefully fitted to ensure accurate alignment with the duct. Precast anchorage blocks may be used.

This top ring itself is usually prestressed circumferentially, with the tendons anchoring in buttresses.

In other pressure vessels, the circumferential prestress has been applied by wrapping the vessel with strands. Since the required forces are extremely large as compared with water tanks, the wires are wrapped in multiple layers in curved steel channels. The tension is applied by a caterpillar-tread type of clamp, with the pulling force being resisted by a chain wrapped around the vessel. The applied tension is continuously monitored by load cells and recorded.

A corrosion-inhibiting grease is applied to the wires as they feed into the steel channel.

In order to maximize the force that can be applied in the limited space, high-tensile steel strip has been used in lieu of wires or strands. It is placed in layers in a steel channel, which is subsequently covered and filled with grease.

17.8 ANCHORAGES

The anchorages are, of course, an integral part of the tendon system. (See Fig. 1.9.) Experience shows that troubles and breakage, if they occur, are most often associated with the anchorages. This requires extremely close and detailed inspection at the shop, identification of shipments, protection during storage in a dehumidified atmosphere, care in placing to avoid nicks, etc., or damage from welding, and corrosion protection after.

Bearing plates should have clean surfaces, free from rust and grease before installation, and from concrete grout afterwards. They must be set to the tolerances specified, usually ±12 mm ($\frac{1}{2}$ inch) in length and ± 2° in angle.

The spiral reinforcement at the anchorage serves extremely important functions in confining the bursting zone and in restraining spalling. It must, therefore, be set accurately and held rigidly during concreting.

Buttonheads, if used, should be formed so as to prevent any serious indentation in the wires, with no seams, fractures, or visible flaws in the heads (Fig. 1.10).

Wedge-type anchorages are highly dependent upon accurate fabrication of the cone to very close inside tolerances.

Because of their exposed position, anchorages are more liable to accidental damage after installation than other elements of the vessel system; therefore, they should be protected, either by being recessed in the concrete or by temporary guards, etc.

Anchorages and bearing plates must remain ductile, i.e., not subject to brittle fracture, at the lowest temperatures anticipated during installation and the life of the structure.

Anchorages for externally wrapped tendons may be provided by pegs which fit into slots in the steel channels.

17.9 ENCASEMENT AND PROTECTION OF TENDONS

The use of very large tendons consisting of multiple wires or strands presents a high ratio of steel surface area to total steel cross-sectional area. This means that special attention must be paid to corrosion protection. The continuous curvature of the tendons produces radial bearing stresses, forcing the tendons against the sides of the duct, and restricting the penetration of filling materials.

Special corrosion-inhibiting greases are extensively used for filling ducts and protecting the tendons. It is important that the temperature of the grease and duct be relatively uniform and high enough to maintain fluidity. Tests should be run to determine the permissible temperature variations. The duct can be heated if necessary, by blowing hot air through it.

Likewise, the grease may be heated. Care should be taken against accidental fire.

Grease should generally be injected from one end until a small quantity of grease of the same consistency emerges from the other end. The grease injection should be shut off under pressure.

Other systems in Europe use an electrostatically deposited wax.

Permanent wax or grease protection should have the following properties:

1. Remain free from cracks and not become brittle or excessively fluid over the entire range of temperatures, including those during transportation and storage.
2. Chemically and physically stable for the life of the structure. Stable under total radiation anticipated during life of structure.
3. Nonreactive with concrete, tendons, ducts, etc.
4. Noncorrosive, and corrosion-inhibiting.
5. A barrier to moisture.

It is important that the material be continuous and of the thickness specified after installation (pulling in) of the tendons. Pumps, hoses, etc., should be suitable and adequate over the entire range of temperatures and viscosities expected during installation.

Provisions should be made for removal of spilled material from concrete surfaces.

All greases or other injection materials must be shipped in tight containers that will prevent contamination.

In some European reactor vessels and in at least one United States containment vessel, the stressed tendons have been grouted. In such a case, grouting practice should follow the recommendations of Section 5.10, Grouting. Grout should have a low W/C ratio, be thoroughly mixed by machine, injected under low pressure until grout of the same consistency emerges from the other end, and shut off under pressure. The following limits on contaminants in the grout (by weight) are recommended:

Chlorides as Cl	650 ppm
Sulfates as O_4	800 ppm

The use of a thixotropic anti-bleed admixture is recommended to promote complete filling of the duct.

Extensive tests have shown that, with multiple-strand tendons, a path exists for grout flow regardless of radial force and bearing. With multiple parallel wires, grout flow is impeded to the inner wires and where wires are forced tight against the sides of the ducts. In these cases, special means, such as the use of soft-iron wire spacers, are necessary.

Until recently, for vertical and semi-vertical tendons, bleed and sedimentation could be reduced only by inclusion of a gelling agent plus expansion additive plus a water reducer (but not retarder). Free expansion was permitted through a vertical extension (stand pipe) above the highest point of the tendon. Otherwise, for vertical strand tendons, bleeding of 10% or more often occurred. Today, as indicated previously, admixtures are available which promote fluidity yet gel when the pumping ceases, so as to essentially eliminate sedimentation and bleed. In any event, a field demonstration test, full height, should be required prior to approval of the material and methods.

Grout, properly injected, offers corrosion protection due to alkalinity, heat protection, additional safety in the anchorage zone, and reduction in size of cracking under overload or accident conditions.

The protection of the anchorages is a very important matter. Not only is corrosion more likely here, but the steel is under axial and transverse stresses and hence more susceptible to brittle fracture. Therefore, for pressure vessels, a mechanical seal should be installed around the anchor and the specified material (grout or grease) injected.

17.10 PRECAST CONCRETE SEGMENTAL CONSTRUCTION

There are many construction problems in reactor vessel construction which can best be met by judicious use of precast segments, joined by cast-in-place concrete.

Penetrations, especially large penetrations, are critical areas. The tendons must be deflected around them, thus being spaced more closely. Concentrated mild reinforcement is usually required to resist local secondary cracking. Concrete quality and consolidation must meet the highest standards.

Therefore, it seems logical to precast these penetrations in blocks of a size that can be handled by the available crane capacity. Segments would be concreted in their most favorable position, cured, and set in place. The joints would be treated like other construction joints.

In many cases, it will prove desirable to preinstall the anchorages in precast beams or slabs. These can be concreted in their most favorable position, with anchorages held in exact alignment. The segments are then set in the forms (Fig. 17.5). Thus, they provide the rigidity and resistance to displacement, as well as solving the difficulty of concrete consolidation in congested areas.

Where tendons are anchored to buttresses, these can be precast in large vertical segments. Jointing techniques borrowed from segmental bridge construction can be applied to incorporate the anchorages into the structures.

The fixation of the thin, flexible liner to the concrete presents many problems. Welding procedures are quite demanding in order to provide full penetration welds

Figure 17.5. *Use of precast panels for external skin and anchorage blocks of nuclear reactor pressure vessel.*

without burning through the thin metal. Generally, cooling pipes and anchors are pre-welded to the liner and then embedded in the concrete during casting. During the construction, the liner must be accurately held in position. Insulation may be installed between the liner and the structural concrete.

Thus, systems in which sections of liner plate are embedded in precast shells, which are then assembled and joined, appear to offer advantages.

Such precast panels should have socketed joints to insure biological shielding. In one European system, external hoop rods are affixed and their initial stresses are equalized. They are then jacked apart by multiple hydraulic jacks. The joints and jacking pockets are then filled with nonshrink grout.

17.11 GROUT-INTRUDED CONCRETE

Preplaced aggregate (grout intruded concrete) was employed on the Fort St. Vrain reactor in Colorado. The specially selected aggregate was placed in lifts containing all the elements (ducts, reinforcement, etc.) and, in addition, grouting tubes and instrumentation tubes. A battery of grout pumps was used to pump grout into the aggregate from the base. This method was selected to minimize displacement of inserts and to substantially eliminate shrinkage. The grout was sufficiently fluid to penetrate highly congested areas.

Extreme care was taken to prevent rock dust, chippings, dirt, or other contamination in the preplaced aggregate. The grout, being of relatively small total volume, was able to be carefully controlled as to consistency, viscosity, temperature, entrapped air, etc. Spare pumps and manifolds were provided to ensure that the pour could continue without interruption due to failure of equipment or power supply.

The use of grout intruded aggregate concrete, with carefully controlled rates of placement and pressures, reduced the magnitude of fluid head acting to distort the liner and forms during concreting. Actual pressures, as well as full development of the techniques, were determined in mock-up tests at the site.

17.12 INSULATION

The insulation on the inside of the liner, if required, may be pumice concrete blocks, stainless steel foil, or fibrous ceramic blocks. Attachment is tedious and complex. The constructor can minimize the cost and time required by providing adequate lighting, ventilation, scaffolding, and material handling.

17.13 SCAFFOLDING, HOISTS, CRANES, ACCESS TOWERS

These are required for many phases of the work and with many activities going on simultaneously. The available space is extremely limited and congested by other activities of plant construction. The selection of equipment, its scheduling and most effective utilization deserve the closest possible attention. Practically every item of material and equipment has to be raised by power, and supported during installation. Tendon coils are particularly cumbersome and awkward. The jacks are heavy and must be held at odd angles, and workers require staging from which to affix anchorages, stress tendons, and inject grout or grease.

Such hoists must also be designed for safety, with controls and limits built in to prevent accidents, and with maximum visibility, including night lighting, for the operators.

17.14 INSPECTION AND RECORDS

Inspection includes maintenance of the quality control and assurance program, for all elements entering into the complete system. Complete and accurate records must be maintained for the engineer, owner, inspection agency, and the governmental regulatory bodies. This requires a carefully thought-out system of inspection, identification, quality standards, permissible tolerances, procedure in case of deviation, and records. Records must be maintained of all shop drawings and details, welding procedures, welder qualifications, weld inspection results, radiographs, ultrasonic magnetic particle reports, liquid penetrant test results, concrete batch records, concrete pour check-off lists, vibration times, temperatures, stressing sequences, forces, and elongations, grouting or greasing pressures, temperatures, etc.

One of the most critical of all inspection functions is the thorough check of each pour prior to concreting, to insure proper placement of all embedments, reinforcement, and prestressing appurtenances.

Sampling and testing of tendons and anchorages must be carried out in accordance with the prescribed procedures and recorded in detail.

17.15 SCHEDULING AND COORDINATION

The prestressed concrete pressure vessel is reportedly on the critical path for two-thirds of the total construction time. The total work force at a nuclear plant may reach 2000 to 3000. While the vessel is being constructed, the construction of the reactor core, boilers, boiler shield wall, etc., must also proceed on an integrated schedule.

The "system design" approach means that construction must also proceed on a "system" basis. All possible activities should be dispersed from the immediate site, and prefabrication used to the highest extent possible.

At the vessel site itself, every effort should be expended to assure optimum working conditions, that is, dry, light, clean working conditions, and adequate hoisting and access equipment (Section 17.13).

Scheduling obviously lends itself to the critical-path approach. There are numerous activities, all of which are interrelated, and all of which are essential to the timely and efficient completion of the vessel. Programs are set up and monitored by a computerized technique. The critical path must include working drawings and their approval, computations, submissions, and tests, as well as procurement and construction activities.

However, the practical limitations of critical-path analyses are well known. One of the largest and most successful nuclear reactor power plant builders in Europe controls the work primarily from a detailed bar chart. A detailed critical-path analysis is made before the start of the job, from which the bar chart (operations vs. dates) is prepared. Thereafter, if any activity falls behind, he augments the work force or lengthens the hours of work as necessary to get back on schedule.

It is believed that the critical path is the better scheduling method and can be effectively utilized, provided it is not made in such minute detail, i.e., too many activities, as to prevent its useful comprehension and application in the field.

After the scheduling is prepared, a detailed layout must be made for each stage. Details of duct and penetration placement are prepared for each pour. Often these can be integrated with the forms insofar as layout is concerned. Equipment and access and material flow must be preplanned. Spare parts and spare equipment must be provided for tensioning, injecting, grouting, and instrumentation, so that the schedule will not be delayed by breakdown. Proper tagging and identification, including shipping instructions and description, will obviate delays and confusion.

Typical Stages of Construction

1. Preliminary work at site: roads, warehouses, utilities.
2. Excavation.
3. Base and foundation slab.
4. Foundation steelwork erection (liner supports).
5. Base slab of vessel.
6. Prefabricated liner, move in and place.

7. Final positioning of liner.

8. Fit and weld cooling water pipes. Grout under liner floor.

9. Walls of pressure vessel.

10. Weld upper penetrations.

11. Complete walls.

12. Top slab supports, lift in top forms.

13. Set and weld top dome liner.

14. Concrete top dome.

15. Cure top dome.

16. Remove support.

17. Reactor construction inside.

18. Thread tendons.

19. Stress tendons.

20. Interior completion. Remove forms and scaffolding.

21. Pressure test.

22. Completion of plant ready for operation.

17.16 PROJECTED DEVELOPMENTS AND APPLICATIONS

Prestressed concrete pressure vessels received a great development effort from their application and utilization as containment and reactor vessels for nuclear power. The successful experience of 25 years in design, construction, quality control, maintenance, and operation gives PCPV's high credibility for expanded use whenever high internal pressures must be resisted.

The ability to concentrate very high tensile force in a small space means that for large diameter tanks, substantially higher internal pressures can be resisted by prestressed concrete than by steel plate.

The development of very high-strength concretes offers great advantage for such uses. Among the proposed uses are high-pressure high-temperature steam containers for the storage of energy and similar coal gassification process vessels.

REFERENCES

1. ASME, Boiler and Pressure Vessel Code, Section III, ACI-ASME Committee 359, "Code for Concrete Reactor Vessels and Containments," American Concrete Institute, Detroit, Michigan, 1986.

2. ACI-361.R-86, Composite Concrete and Steel High-Pressure Vessels for General Industrial Purposes," American Concrete Institute, Detroit, Michigan, 1986.

3. FIP, "The Design and Construction of Prestressed Concrete Reactor Vessels," Institution of Civil Engineers, London, 1978, with Addendum 1990.

4. FIP, "Prestressed Concrete Pressure Vessels for Non-Nuclear Thermal Processes," Institution of Civil Engineers, London, 1982.

CHAPTER 18

Prestressed Concrete Poles

18.1 GENERAL

Prestressed concrete has been widely adopted for poles throughout the world, although its use lags in North America due to the up-to-now abundance of trees. The requirements for poles are rather unique structurally and economically. Properly utilized, prestressed concrete is a useful material for this application and may be the most advantageous for a wide variety of situations.

Historically, prestressed poles were first designed and constructed by French engineers in Algeria. The specific environment, blowing desert sand, had destroyed both wood and steel. Prestressed concrete poles have seen 60 years' successful service in Algeria, despite minimum concrete cover over the tendons.

On the Pacific coast of South America, timber poles are cut down by the locals for use as firewood. Termites attack timber poles in Africa. Corrosion attacks steel poles in the tropics and Middle East. Grass and brush fires are a concern in many areas.

Prestressed concrete poles are extensively employed in Europe, even in the Black Forest of Germany, where suitable trees grow alongside. There, a careful decision has been made of relative values: the trees are to be utilized for the interior of housing, and prestressed concrete is to be used for long-term service poles. Similarly, it is interesting to note that one of the few regions of the United States in which prestressed concrete poles are utilized is Oregon. There, also, the true value of a tree is recognized, as is the true value of a prestressed concrete pole.

Prestressed concrete offers durability from corrosion and erosion in desert areas, and ductility and freeze-thaw durability in cold temperatures and in mountainous regions. In areas such as western South America, where timber is unobtainable, the use of prestressed concrete piles prevents pilferage. Prestressed poles are clean and neat in their appearance, both at time of installation and for the succeeding years;

thus, they are highly suited to urban installations. They can be set directly in drilled holes, in the ground, with or without concrete fill.

Prestressed concrete poles have been utilized for:

Railway power and signal lines.

Lighting.

Antenna masts.

Overhead pipeline supports.

Telephone transmission.

Low-voltage electric power transmission.

Substations.

High-voltage electric power transmission.

As compared with conventionally reinforced concrete, prestressing increases the crack resistance, rigidity, and resistance to dynamic loads and fatigue.

18.2 DESIGN

The design criteria for poles vary widely. For those carrying electric current, grounding must be provided. For those supporting wires under tension, the effect of line breakage, including torsion and impact, must be considered. Wind load is a major factor and, in some regions, seismic behavior must be considered.

A recent and extremely important parameter is the effect of collision and the behavior of the pole afterwards; does it fall on the automobile? Does the pole snap off suddenly when overloaded, or does it fail gradually, with large deformations and energy absorption? The prevention of brittle behavior of both types requires that the concrete be confined by stirrups or spirals, and that either mild steel or unstressed strand be incorporated in the critical lower section of the pole.

Most poles to date have been hollow and tapered. The hollow core has been useful in reducing weight and in providing a raceway for electric wires, etc. The taper is a reflection of the reducing bending moments and the desire to reduce wind (and seismic) loads over the top portion.

Cross-arms and diagonal bracing have been constructed of both prestressed concrete and steel. It would appear that, in most cases, use of galvanized steel or wood for this purpose would be proper because of their light weight and ease of assembly by bolting. The one exception may be double-T or H-frame pole structures, where the main cross-arm may be advantageously made of prestressed concrete.

The greatest moment resistance is required at the base, and here both elastic resistance and a satisfactory plastic "toughness" is required. This can be partially met by the inclusion of high-yield passive steel reinforcement or by additional unstressed strand. Confinement is, however, the primary means, usually provided by closely spaced spiral reinforcement.

A great deal of debate has occurred over the cross section of the pole. Should it be round or square? Wind is not normally the controlling factor, rather, torsional strength and bending strength on line breakage. Thus, square and rectangular sections and H-sections are suitable for poles and are widely employed (Fig. 18.1).

Tapering originally presented major problems to the application of prestressing. With full length tendons, the effect of taper is to increase the effective prestress at the top. By conventional design theory, a reduction in bending strength occurs there because of reduced section depth and because of high initial compression. However, recent research has confirmed that the high precompression is largely dissipated by elastic and plastic strain at ultimate load. Creep also significantly reduces the adverse effects of high prestress.

Manufacturing systems have been developed to reduce the effective prestress in the upper portions of tapered poles, e.g., by preventing bond, or by dead-ending or looping some of the tendons at mid-height. Another approach is to use a low degree of prestress and then to add mild-steel bars in the lower portion. This author feels that in many cases "taper" is a concept that is inherited from timber poles (trees) and that a constant cross section may, in many cases, prove a better solution. Naturally, there will be an increase in seismic and wind stresses but, for torsional resistance,

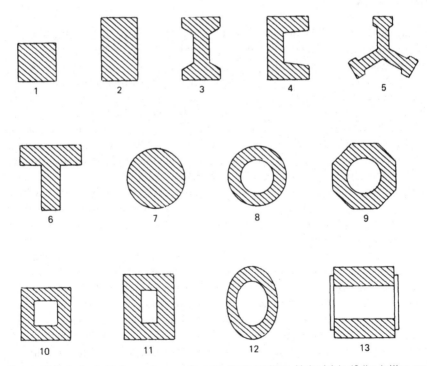

Figure 18.1. *Typical prestressed concrete pole cross sections. Note: (a) in 13 the lattice members are galvanized steel; (b) in 2, 3, and 4, the web may be perforated so that the pole becomes a vertical virendeel truss; (c) most poles are tapered.*

the top must be as strong as the base. Adoption of a constant cross section permits more effective use of prestressing and easier connections.

Aesthetics may, of course, dictate a tapered section.

Wherever the effective section is changed, either in concrete cross section or in embedded steel, the transition should be gradual. Steel reinforcing bars in the lower end, for example, should be terminated at different levels, rather than all at one elevation.

The collision parameter requires toughness and ductile behavior at the base. This can be obtained by a solid concrete section, reinforced so as to have a large plastic deformation after cracking. This reinforcement can consist of the stressed tendons, augmented by additional unstressed tendons, or by mild or high-yield reinforcing steel, or by a structural steel core. With any of these, the spiral confining reinforcement, preferably of cold-drawn wire or other high-yield steel, is essential to prevent brittle compressive failure. With confinement, deflection ductilities up to 6 or more can be obtained.

The weight of poles should be kept as low as possible to ease transportation and erection, and to reduce seismic loads. Lightweight concrete is ideally adapted to this use, provided it is suitably bound by spirals.

For many uses, the poles require a duct-way leading up the pole. This can be formed by a hollow core, an embedded duct, or an external pipe conduit.

If copper wires are used for grounding, they should be insulated from the concrete to prevent electrolytic corrosion of the prestressing tendons since wet concrete can act as an electrolyte.

Since maximum bending moment occurs at the base, the tendons should either be mechanically anchored (as by posttensioning) or have as short a transfer length as possible (e.g., by use of indented strand). Nonstressed steel with mechanical heads can also be used at the base.

Inserts or embedded plates for attaching cross-arms and bracing are usually of galvanized steel.

Lighting poles present special cases, in which the means of supporting the light must be aesthetically blended into the support. For this reason, many are made with curved upper portions of conventionally reinforced concrete or steel or aluminum. However, there is no reason why a central post-tensioning tendon should not run the curve and, in fact, many have been constructed this way.

18.3 MANUFACTURE

Prestressed concrete poles are produced either in fixed forms or by centrifugal spinning. Fixed forms may be tapered on the sides to facilitate stripping. With fixed forms, compaction is accomplished by internal vibration, use of a vibratory table, or vibration of the core.

The deflection of prestressing strands to match the taper is readily accomplished on a long bed by merely casting them tip to tip and then seizing them with wire as they pass through the end gates.

Most poles have employed a tapered core so as to reduce the amount of concrete, hence the weight, and the required prestress. The hole also provides space for an internal conduit.

However, the placing of a conduit in a solid pile is an easy matter and the gain in structural efficiency is increasingly offset by the added labor, complexity, and difficulty of quality control required for the hollow core. Hence except for very large poles, the hollow core may now be an anachronism.

When hollow cores are to be provided, the principal problem is the removal of the core.

Rigid tapered cores are designed for easy removal by such means as:

1. Wrapping with plastic, e.g., polyethylene.
2. Lubricating.
3. Rotating at time of initial set to break bond.
4. Jacking out.
5. Longitudinal vibration during pouring and until set.

Mechanically collapsing internal forms and inflated core forms have also been used.

In one particular manufacturing plant, the exterior and core forms remained fixed, with the completed pole being jacked off longitudinally.

Large poles can be manufactured by using precast segments. The segments can be assembled, jointed, and post-tensioned, giving the ability to use standard elements for a wide variety of pole heights.

18.4 ERECTION

Poles are installed in one of the following ways:

1. Bolted to previously constructed bases.
2. Set in drilled holes in the ground.
3. Concreted in drilled holes in the ground.
4. Concreted in sockets in previously constructed bases.

Under method 2, the hole is usually packed with tamped gravel. Under 3, the concrete may be poured or pumped. Grout may be advantageously used where the annulus is small. This ability to set directly in a drilled hole and be assured of durability is a major advantage of prestressed concrete.

Erection may be by a derrick on a truck, a truck crane, a guy derrick, or a helicopter. The main problems are insure lateral stability, as by use of guys, and to prevent accidental impact of the pole with the crane boom.

Segmental poles have been erected by telescopic action; this appears to be a special case, applicable to very large or unusual poles. Segments have also been joined by welding during erection, or by posttensioning.

Guy wires and stays are often employed, particularly on very high poles and at angles in the alignment of the line. The vertical component of guy stress should be investigated, particularly for high poles, to prevent excessive stresses from leading to buckling. The behavior of the guyed pole after breakage of a line or guy must also be considered.

18.5 SERVICE PERFORMANCE

The prescribed cover differs in the various countries. However, most agree that 20 to 25 mm (0.75 to 1 inch) is adequate.

Longitudinal cracking has plagued highly stressed hollow-core poles having thin walls. The answer seems to be more closely spaced spiral, along with thicker walls.

18.6 ECONOMY

Most objective studies indicate that the first cost of prestressed concrete poles is greater than timber but less than steel. The economic justification, therefore, is usually based on the following:

1. Durability against erosion, insects, corrosion, pilferage, etc. This is of special importance in tropical (jungle conditions), desert, and shore-side environments.
2. Fire resistance. This is of special importance in areas of grass, low brush, and low timber. Ground fires will not normally cause the failure of a prestressed concrete pole.
3. Service life.
4. Aesthetics, especially long-term appearance.

Wider acceptance in the United States must await a reevaluation of true economy, service life, maintenance, installation costs, etc. Nevertheless, the use is growing and, if we can draw valid conclusions from other countries, environments, and economies, prestressed concrete poles will enjoy a major utilization in the future.

Trends of development should include:

1. Reevaluation of taper vs. constant cross section.
2. Reevaluation of the need for a hollow core as opposed to a simple duct.
3. Selection of best cross section: square, circular, rectangular, channel, etc.

4. Development of the most efficient means of insuring against brittle behavior under impact. This appears to require heavy spiral and supplemental mild steel in the lower portion.
5. Extended use of lightweight concrete.
6. Lower costs through mass production.

CHAPTER 19

Pipes, Penstocks, and Aqueducts

Prestressed concrete is widely utilized for the conveyance of water in pipes, penstocks, siphons, and aqueducts in sizes up to 5 meters (16 feet) diameter and more. It has been utilized for large-diameter, high-pressure penstocks for hydroelectric power plants in New Zealand and Scotland. Today's prestressed concrete aqueducts in Italy carry on the tradition of their 2000-year-old Roman ancestry.

Prestressed concrete pipes are excellently adapted to siphons and underwater lines because they combine circumferential strength with longitudinal beam rigidity and strength and durability with weight for stability.

Prestressed concrete pipes lend themselves to factory production. Many proprietary techniques have been utilized for their manufacture. In the following sections a brief description will be presented of a number of manufacturing techniques.

Concrete pipes are cast by the vertical-cast process, the centrifugal spinning process, or horizontally, with a removable mandrel.

The vertical-cast method is used for the larger diameter pipes and penstocks, usually those above 2 meters (80 inches), in lengths of 5 to 8 meters (16 to 26 feet). A machined cast-steel ring is used for the base. Forms must be very rigid, since the concrete is placed very dry with heavy external vibration.

After the central core ring is placed, the welded reinforcing cage is set, containing ducts for circumferential and longitudinal prestressing. The circumferential tendons typically envelope a 240° sector around the periphery, thus overlapping 60° on each other. They are anchored in buttresses (blisters). The outer form is then placed, concrete placed and consolidated, followed by steam curing and stripping. Supplemental water curing improves impermeability.

If the body of the pipe is to be constructed of lightweight concrete, then a small amount of conventional hard-rock concrete may be placed at the top and bottom to provide greater resistance at these critical points to spalling during installation. Internal vibration blends the transition zones effectively.

Joint details may be: flexible, with machined steel rings and rubber O-ring gaskets; or rigid, for mortar joints; or semirigid (lead caulked).

Prestressed concrete cylinder pipe is a variant which is widely employed for low- and medium-pressure pipe.

The manufacturing process can be described in more detail as follows.

Mechanical cast-steel base plates are hydraulically expanded beyond their yield point so as to achieve the required diameter and roundness within narrow tolerances. These rings are then welded to the cylinder. Hydrostatic testing is carried out to ensure watertightness.

The assembly is then set in a casting mold, having machined top plates which control the concrete wall thickness and the position of the cylinder. The concrete core is cast vertically, contained by heavy steel forms. The outside core form may be a steel membrane, permanently incorporated in the pipe. Concrete quality is carefully controlled to prevent segregation as it drops through the pipe length, and to ensure strength and low permeability. External vibration is used to consolidate the concrete, to eliminate rock pockets and honeycomb, and to achieve a fine finished surface free of visible surface defects such as air voids (due to entrapped air or bleed) and sand lenses.

Casting height generally varies from 5 to 7.5 meters (16 to 24 feet) so special care has to be used to prevent discontinuities in the concrete. Careful blending of the aggregates, control of the water-cement ratio, accurate batching, adequate mixing time, and control of the vibration are required. Admixtures which deaerate the concrete mix and which control bleed are needed to achieve the highest quality pipe.

Air voids in the outer surface of the pipe pose a threat of corrosion to the prestressing wire and the embedded steel cylinder. Hence, the control of the concrete core manufacturing process is very critical.

Curing is usually by steam, e.g., a 48-hour steam cure. The core is then placed on a turntable and draws the wire under a carefully monitored tension onto itself as it rotates. More than one layer of wrapping may be required. Each wrap is protected by either sprayed-on mortar or mortar applied automatically by a brush.

The finished pipe is again steam cured. Although it is usually left uncoated, an external coat of bitumastic or epoxy may be applied at this time.

Prestressed concrete pipes, both cylinder and noncylinder, are generally selected because of durability, strength, rigidity, and economy. They have been designed for up to 3 MPa (400 psi) internal hydrostatic pressure. They have been used worldwide, for both freshwater and wastewater transport, and for saltwater cooling systems. Among their most outstanding uses have been the water supply from Saudia Arabia's Eastern Province to Riyadh, and the Great Man-Made River in Libya.

Although prestressed concrete pipes are selected for durability, there have been problems of corrosion, emphasizing the need for thorough engineering and construction control.

The earliest widespread corrosion problems appeared in Eastern Canada, where calcium chloride had been added to the shotcrete to accelerate its strength gain in cold weather. An electrolyte, water, was available by permeation from inside, so all the conditions for corrosion existed and widespread failures ensued.

A large diameter penstock in New Zealand was laid in corrosive soils. The water-permeated concrete absorbed the chloride and sulphate ions. The sulphate led to cracking, and the chlorides to corrosion of the tendons, resulting in a catastrophic failure.

Occasional problems have occurred due to delamination of the exterior shotcrete coating at the ends. Details of the ends should be such as to prevent this. Epoxy coating may be used on the ends.

Internal corrosion of prestressed concrete pipes has occurred at joints and with reinforcing steel in the cores. Usually this has occurred with empty or partially full pipes of seawater, e.g., seawater cooling pipes in Saudia Arabia, and sewage discharge lines in Dublin into which tidal seawater backed up during low flows.

Pipes full of seawater are generally immune from attack on the inner surface due to lack of oxygen. When initial corrosion has been observed at an early stage, further corrosion has often been effectively inhibited by keeping the pipes full.

Special care has to be taken with joint rings, since these are usually connected to the steel. In partially full or empty pipes, the joint rings become anodes, powered by the protected reinforcing steel as cathode. The joint rings are usually protected by cement mortar. Care must be taken in applying this properly. Epoxy mortar may be best. Zinc anodes have been used effectively.

Sewage and wastewaters containing oil can liberate anaerobic bacteria which convert to sulfates in seawater and hydrogen sulfide or weak sulfuric acid when exposed to air. This can attack concrete pipe, especially at the crown, at joints, where the air is entrapped in the joint crevice. This type of attack initially dissolves the concrete, exposing the steel to conventional corrosion.

If it is contemplated that the pipe will be operated less than full of fluid, then an internal coating or PVC liner would appear prudent. Epoxy coating of reinforcing steel of the core and epoxy coating of the joint rings should be considered. Inclusion of PFA in the concrete mix will give greater resistance to attack.

The details of reinforcement of bells for prestressed concrete pipe joints are extremely important. After manufacture of some 80 km (50 miles) of prestressed concrete pressure pipe for a new water supply system for Madras, India, hydrostatic tests showed the bells were not capable of sustaining the full hydrostatic head. Of course, such a defect in design should have been caught at an earlier date, but this example serves to emphasize that a pipe is subject to internal hydrostatic pressure throughout its entire length, including the bells. The bells are further subject to longitudinal shears and local moments. Therefore, this region of discontinuities requires thorough detailing and confirmation by testing.

Alkali aggregate reactions have occurred with prestressed concrete pipes, as with other concrete structures. Pipes however are relatively more susceptible because of the permeation of water, which keeps them partially saturated. Hence alkali reactivity of aggregates is accelerated. Careful attention should be paid to limiting the alkalis in the cement, to evaluation of the reactive constituents of both the coarse and fine aggregate, and to the possibility of moving the resultant mix well beyond the pessimum concentration of reactive silica (about 4%) by addition of up to 30% PFA or 6 to 10% of condensed silica fume. Note that high percentages of condensed silica fume may make the mix too sticky to place, whereas more moderate percentages still provide the necessary durability.

Some desert environments and some effluents may be excessively aggressive to the interiors of the pipes. In that case, interior coatings of epoxy or fiberglass (FRG) or PVC liners may be employed. These may only need to protect a 60° sector at the crown or may be full circle. For extremely aggressive soils, external coatings of bitumastic or epoxy are also indicated.

Centrifugally spun prestressed pipes utilize centrifugal acceleration to consolidate the concrete and to drain water from the inside. One process utilizes, in addition, steel balls which run around the inside and provide additional compaction of the concrete during the later stages.

Longitudinal duct forms must be stressed during spinning, so as to prevent dislocation under the centrifugal force, which may otherwise leave voids behind the ducts. Use of removable formers, which consist of stressed bars, has been effective in minimizing this problem.

Horizontal casting employs a moving mandrel, in a manner similar to that described for hollow-core concrete piles (see Chapter 11). It is usually employed for small diameter pipes.

Prestressed concrete steel-concrete cylinder pipe uses a steel liner, with a mortar lining placed centrifugally on the inside. Then the composite section is wrapped with wire under constant tension, the pipe being turned while a wire or rod is fed under tension. The wire-wound pipe is then given a coat of cement slurry and a mortar coat is placed by spinning brushes as the pipe slowly rotates. This type of pipe is capable of resisting quite high internal pressures without leakage.

This same process may be used with an all-concrete core, wrapped under tension as described above, and coated. In China, the concrete core is constructed of ferrocement, using three layers of heavy wire mesh, and centrifugally spun mortar.

As with all prestressed concrete, calcium chloride must never be incorporated in the mortar coating. One of the most disastrous cases of corrosion occurred with concrete pipe where calcium chloride had been used to accelerate set of the mortar to prevent freezing.

Helical prestress has been applied in Czechoslovakia, the wire feed-off device being moved longitudinally as the pipe is rotated. This gives a triaxial stress condition.

Self-stressing cement techniques have been successfully applied to the manufacture of small diameter pipe in the former USSR. A preformed spiral of high-strength steel is placed in the forms, the concrete made with expanding cement is placed and consolidated, and the completed product is taken through a carefully controlled curing process to ensure the correct degree of expansion after set, thus imparting prestress to the concrete.

Longitudinal prestress improves crack resistance and gives greater beam strength. Posttensioning may be used to join several sections or even a complete pipeline together. In the latter case, tendon couplers are used to join convenient lengths of prestressed pipe. Joints may be concreted, in which case they are usually

75 mm (3 inches) wide, or they may be narrow joints, filled or injected with epoxy. A thin coat of cement mortar is usually not satisfactory. Joints may also be flexible, with the tendons passing through the joints. In these cases, the prestressing is considered to be of temporary benefit only, for aid in setting and adjustment of bearing, backfill, etc., unless positive means are taken to provide corrosion resistance. Simple coating with bitumastic has proven to be insufficient.

An outfall sewer was installed in the ocean off the coast of New Zealand, in which the pipe segments were joined by prestressed tendons to give longitudinal strength during installation.

Precast concrete pipes are heavily dependent on the circumferential steel details. Whether the pipes are prestressed longitudinally or circumferentially, the critical sections are almost always at the bell joints. Here the details are difficult to execute, so as to ensure adequate circumferential confinement. The bell cannot depend on any circumferential or radial support from the adjoining section. The bell is also larger in diameter; hence the circumferential stress may be locally greater. Thus the bell should be given additional circumferential stress, or reinforcement. Longitudinal prestress, if applied, or mild-steel reinforcement should be supplemented in the bell by additional mild steel, anchored at the bell end by bends or mechanical heads.

Aqueduct sections are made by horizontal casting in standard forms. Such units are usually semicircular or trapezoidal. They are conventionally reinforced transversely and pre- or posttensioned longitudinally. They are usually set as simple spans on flexible bearings, with a flexible seal. Box sections, manufactured in a manner similar to box-beam bridge girders, have been used for over-chutes and for underwater sewage mains. They are prestressed longitudinally, so as to give high beam strength. They are particularly useful where the supports must be widely spaced, as with pile-supported piers. Prestressed concrete cylinder piles, with greatly increased spiral reinforcement, have been used in a similar manner for these over-chutes.

By proper balancing of weight and buoyancy, river crossings and siphons may be readily sunk into position, provided adequate means of control are instituted.

Large-diameter concrete penstocks have utilized cast-in-place concrete or precast segments. The circumferential prestressing has been applied by high-capacity posttensioning tendons, curved over 240° to 270° sectors. While frictional losses due to curvature are high, they may be kept within reasonable limits by the use of smoothly curved rigid ducts and by stressing from both ends of the tendon.

Penstocks are especially subject to corrosion unless special precautions are taken. Since they rest on or close to soils, they may have chlorides or sulfates in immediate contact. As penstocks, they are fully saturated; thus the ions can migrate inward even though any permeation is outward. The saturation reduces resistivity, so corrosion of the tendons can and has occurred with disastrous results.

The prevention lies in recognition of the phenomena involved. In temperate and tropical climates, epoxy or even bitumastic coating of the exterior of the pipe may be an adequate answer.

In environments subject to freezing, the phenomena of freeze-thaw attack and delamination, discussed in Chapter 16, Prestressed Concrete Tanks, may require special consideration. Internal membranes appear to be a positive solution. Of course, air entrainment should be incorporated in all concrete mixes.

Penstocks may also be prestressed longitudinally by internal tendons to provide beam action and longitudinal resistance to thrust and gravity forces.

Prestressed tendons are frequently used to anchor penstocks, both concrete and steel, by inserting tendons in holes drilled into rock, anchoring them with grout, and stressing them against an anchor block around the penstock.

Canal linings have been built in Italy by using thin, pretensioned concrete planks. These are usually 2.5 cm (1 inch) or so thick and are pretensioned. They are so flexible that they readily adapt to a semicircular canal cross section, with high compressive stress (and thus impermeability) on the inner face. They must, however, be set on an accurately prepared bed, and the joints between planks must be made watertight by sealants.

While prestressed concrete has not so far been used for the transport of petroleum products, as far as is known, it would seem to have a potential use in swamps, stream crossings, and offshore lines where weight is required to stabilize the lines, and where beam strength may be desirable. The use of a steel liner or diaphragm is indicated in the same manner as for prestressed concrete fuel tanks.

Joint details would, of course, have to be developed to suit the pressures involved and flexibility desired. The ratchet-type jack developed for tensioning bars at a coupler provides one means of constructing a longitudinally prestressed joint.

Prestressed concrete is also suited to the conveyance of some aggressive chemicals, and to the conveyance of salt water at desalination plants. The very highest quality concrete must be used, with a hard-finished surface. A high degree of prestress is desirable. Internal coatings, such as epoxy, or a steel liner will undoubtedly be necessary. Extreme care must be taken in detailing and executing joints and anchorages; thus the longest possible prestressed sections will usually be found to provide the best solution, so as to minimize the number of joints.

The use of prestressed concrete for pipes continues on a major and expanding scale, particularly as growing utilization is made of its structural capabilities.

CHAPTER 20

Railroad Ties (Sleepers)

20.1 GENERAL

Prestressed concrete railroad ties (sleepers) have been extensively utilized in Europe for many years. In the United States, their main line application has been relatively limited to a few heavily trafficked extensions and industrial railroads. A notable exception has been the Amtrack lines from Boston to Richmond. Rapid transit and light-rail have opened a new opportunity: prestressed concrete ties have been utilized for rapid transit in San Francisco, Chicago, Boston, Los Angeles–Long Beach, and Atlanta, among others.

Prestressed concrete offers the advantages of gravity anchorage of the welded rail against longitudinal and lateral displacement, an estimated life approximately twice that of timber, elimination of under-rail cutting under heavy wheel loads, and greatly reduced ballast and track maintenance. The trend to heavier axle loadings and higher speeds is an important impetus for the more extensive use of prestressed concrete ties.

The retarding factor to their more widespread use in the United States has been one of first cost, compounded by the fact that many railroads own their own forests, mills, and treating plants; thus, the comparative costs are sometimes based on unrealistically low timber tie prices. The exclusion of freight charges may artificially benefit either timber or concrete in particular cases. However, when a comparison is based on market prices, and life-cycle costs, the prestressed concrete tie will generally be found to be highly competitive.

Another limiting factor for the wider utilization of prestressed concrete ties has been the failure to date to find a means of satisfactorily replacing individual ties or a few ties in existing timber tie tracks. The greater stiffness of the concrete ties leads to their carrying of an excessive portion of the track load, and hence the individual concrete ties are punished and subject to cracking.

Similarly, where prestressed concrete ties have been installed in a limited length of track, some difficulties have arisen at the transition at the ends to timber ties. A more resilient under-rail pad on the end ties appears appropriate.

Ties for turn-outs and switches present a similar problem: it is costly to manufacture special concrete ties of varying length.

Having acknowledged the problems facing the greater utilization of prestressed concrete ties, the projections on a structural performance and economic basis clearly indicate an imminent widespread change to prestressed concrete.

The heavier axle loads, higher speeds, high labor cost for track maintenance crews, demand for greater safety, need to anchor welded rail, and the growing scarcity, diminishing quality, and higher costs of timber ties make the transition to prestressed concrete only a matter of time.

Prestressed concrete ties are one of the smallest individual items manufactured. They are, however, used in great quantities under extremely severe conditions and are subject to high performance demands. Thus, they deserve the closest attention in development, design, and manufacture.

The exposure conditions include poorly drained subgrades, freeze-thaw, excessive drying, and salt drippings from refrigerator cars. They are subjected to dynamic loadings up to 10^8 to 10^{10} cycles; thus fatigue is a consideration. They must also perform in cases of derailment to insure no loss of gauge, so that the following cars will not also derail.

Fastenings must take accelerations up to 60 g and more. Insulation must be provided between rails to permit electric signal operation.

Ties must perform their function as part of the total load-carrying system which comprises wheel, rail, sleeper, ballast, and road bed. The consequences of failure range from serious to catastrophic.

Initial developments of prestressed concrete ties were based on analogy to timber ties, and attempts were made to duplicate timber tie properties, fasteners, spacing, etc. This proved to be an inadequate approach. More recent studies have been properly based on a systems analysis of the actual materials and performance required.

20.2 DESIGN

20.2.1 Spacing

Timber ties are usually spaced at 50 to 60 cm (20 to 24 inches). With the heavier rails now in use, and a wider base for the prestressed concrete tie [30 cm (12 inches) as against 20 cm (8 inches)], it has been found satisfactory to use 70-cm, 75-cm, and in some cases even 90-cm (28-, 30-, and even 36-inch) spacings. At 75 cm (30 inches), tests show the rail stress is increased only 10% and the ballast bearing pressure is the same (Fig. 20.1).

In China, double-width sleepers are made, and spaced at twice the distance. In Japan and the former USSR, pretensioned slabs have been used, set on accurately graded bitumastic.

Figure 20.1. Pretensioned concrete ties in heavily loaded track for coal railroad in New Mexico.

20.2.2 Shape

Since the most demanding structural loading for ties is under center-bound conditions, attempts have been made to relieve this force in prestressed ties by several means, including:

Wedge-shaped center.
Narrower width at center.
Under-cut center.
Ditching of the ballast at the center.
Creation of a partial hinge at center of tie.

While all these methods have worked in test sections, in actual field experience the results have been somewhat unsatisfactory, due, primarily, to inadequate ballast maintenance, and to longitudinal creep of the whole track system. The alternative has been to make the tie stronger, usually with more prestress.

20.2.3 Length

Under the rail, a high positive moment condition exists. Since this is near the end of the tie, means must be taken to ensure that full prestress is available even after 2×10^{10} or more cycles of loading. This means that the tendons must be anchored well outside the rail. Anchoring may be by mechanical anchors, as in post-tensioning, or by bond, as in pretensioning. In the latter case, the tie must extend beyond the rail by the transfer length.

In pretensioning, a number of steps can be taken to ensure a short transfer length. (This is the transfer length under repetitive loading.)

1. Deformed strand, in which an indented wire is wrapped as one of the six exterior wires of the strand, gives significantly better performance. This type of strand has been manufactured in Japan and is available worldwide.
2. Gradual release reduces the transfer length by 40 to 50% as compared with shock release.
3. Slightly rusted steel has a shorter transfer length than clean bright steel. The difficulty is how to obtain a uniformly "slightly rusted" condition.
4. The surface of the steel, is affected by the wire drawing process and the lubricants employed. This has a great deal to do with the effective adhesion. Lubricants should be removed by cleaning or washing.
5. Strand has a much shorter transfer length than wire.
6. "Dimpled" and deformed wire gives better performance. Ribbed bars have been used in Germany.
7. Strand of slight irregularity in lay appears to perform better than absolutely perfect lay strand.
8. Greater steel tendon perimeter (e.g., more strands or wires of lesser diameter), with the same total effective prestress on the concrete section, reduce transfer length.

In actual manufacturing practice and tests with pretensioning to date, use of an adequate area of deformed strands, gradual release, and a slight increase in tie length have provided fully satisfactory results.

Posttensioned ties are used in Europe: their positive end anchorage is obviously beneficial but the costs of production are significantly more.

In Sweden, a unique two-block tie has been developed and widely used. It is now being used in other countries, including the United States. The two concrete blocks are made by machine, and connected with a galvanized steel tube. The assembly is then post-tensioned with a prestressing bar, and the tube filled with grout or bitumastic.

Duo-block ties have the advantage of great flexibility and eliminate center-binding as a problem. However, the slant of the rails may be distorted in service and this may raise problems for high-speed, heavily loaded traffic.

In Czechoslovakia, longitudinal slabs have been manufactured, prestressed transversely, and set as a unit on prepared and graded ballast. These, of course, give very low bearing pressures, but present serious problems of drainage and ballast maintenance.

In the former USSR, rectangular frames and slabs have been produced, using continuous wire-winding machines, such as the turntable or the DN-7. These are pretensioned and the wires are essentially self-anchoring, so that transfer length is not a problem. Their use does, however, reportedly present some track and ballast maintenance problems, and this solution appears to be more expensive than the standard monoblock ties.

A number of longitudinal track support beams have been proposed and some have been installed on short sections of rocket-launching tracks. Gauge must be maintained by struts and ties. Japan National Railroads has conducted extensive tests on this system in an effort to further reduce labor for ballast maintenance. There is a problem of concentrated bending in the rail and excessive ballast pressures at the ends of segments under high-speed traffic.

20.2.4   Prestress Tendon Profile

By varying the concrete cross section, a tendon profile is achieved in which the center of prestress is lowest under the rail, with approximately centric prestress at the center. Thus, the tendon is effectively "deflected" in relation to the center of gravity of the concrete, although the tendons are straight (Fig. 20.2).

20.2.5   Rail Seat

Current practice is to use direct fastening of rail to concrete on an insulating and bearing pad; thus, the 1:40 inward cant must be cast in the concrete. The seat is recessed so that the lips will provide lateral gauge support against forces in and out.

20.2.6   Fastenings

These must resist several million cycles of up and down loads with acceleration up to 60 g. Various types of wedge and spring clips have been developed to hold the

Figure 20.2. Prestressed concrete ties are shaped for optimum stress distribution.

track to the tie under positive stress. It has been found that a fastening which is neither excessively rigid nor excessively flexible is most satisfactory (Fig. 20.3).

20.2.7 Stirrups

Stirrups have not normally been used in prestressed concrete ties. However, in some designs, they have been employed under the rail-seat to resist shear and at the ends, to confine the tendons, so as to develop the stress in a shorter length.

20.2.8 Insulation

For electrical signal operation, the rails must be insulated. Attempts have been made to develop concrete that will provide adequate electrical insulation even while wet. The inclusion of microsilica in the concrete mix has significantly increased the resistivity.

Insulation (isolation) pads of various materials, such as nylon, rubber, or plastic, are placed under the rail. Also various means, such as pads, are adopted to insulate the rail-fastening assembly. In one case, an insulating pad is placed between the clip and the rail. In another, the bolt and nut are insulated from the concrete. In a third, the insert is insulated from the concrete by coating it with an epoxy.

Figure 20.3. *Installing rail fastening bolts in forms for pretensioned tie manufacture.*

20.3 MANUFACTURE

Many systems for the mass manufacture of prestressed concrete sleepers (ties) have been developed throughout the world. Many of them have proven to be inefficient and lacking in precision control. Yet prestressed concrete ties are extremely demanding on such matters as accurate strand positioning and tensioning, concrete flexural, compressive and bond strength, accuracy of cross section, and tolerances of rail seat and fastenings.

Moderately high- to high-strength concrete is required.

Thorough consolidation by vibration is essential.

To maintain a daily turnover of manufacture, accelerated curing is needed.

Both post-tensioning and pretensioning systems have been employed.

Many European ties embody a heavy but optimally shaped profile, using post-tensioning. The ties are cast in individual molds, generally assembled in groups. They are cast upside down, which ensures accuracy of the rail seat and permits easy removal of the ties from the molds.

Bars are used for posttensioning, with threaded ends and nuts. The ties are demolded and stacked before prestressing. Thus a positive end anchorage is obtained.

A pretensioned tie developed by the author's firm has enjoyed wide acceptance (Fig. 20.4). The 1,500,000 prestressed ties for the Amtrack train system between Boston and Richmond were so produced, as were the ties for the San Francisco Bay Area Metro Transit, a heavy-duty coal railroad in New Mexico, and several light-rail systems. These ties utilize the deformed strand and a slightly longer length than their post-tensioned equivalent, in order to develop full prestress under the rail. They are profiled so as to optimize the prestress at each critical section.

For the Amtrack project, they were manufactured in battery molds. Mechanized equipment was used at every stage so as to achieve high production rates with minimum labor (Fig. 20.5a, b, c, d, e, f, g).

20.4 PROBLEMS AND EXPERIENCE IN USE

Concrete ties have proven in service to maintain better alignment and grade than timber ties and have performed very well in derailment accidents.

Wedge-shaped ties, so shaped to relieve center-binding, have developed some cracks at the top center due to inadequate ballast maintenance.

Torsional cracks have occasionally occurred on a 45° line at top center. This appears to have been only with the wedge-shaped tie with its reduced center cross section. It points up the need for adequate torsional strength in the design.

Under-rail cracking, extending up from the bottom to the bolt or insert holes, has occurred with a few pretensioned ties and is apparently due to bond slippage, i.e., excessive transfer length.

Rail joints have presented problems due to overloading of the adjacent ties. Better rail-joint connections may be the answer, or use of wider ties with more

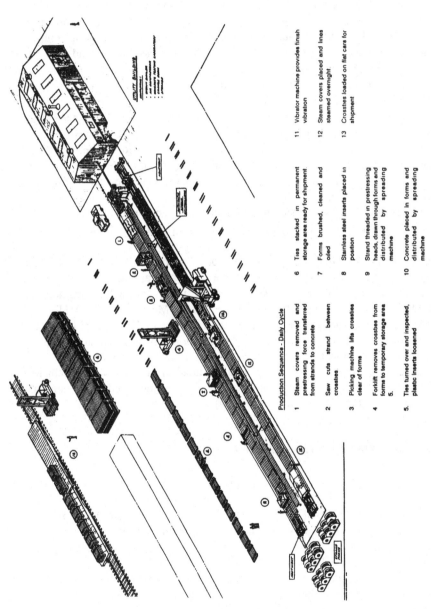

Production Sequence – Daily Cycle

1 Steam covers removed and prestressing force transferred from strands to concrete

2 Saw cuts strand between crossties

3 Picking machine lifts crossties clear of forms

4 Forklift removes crossties from forms to temporary storage area

5. Ties turned over and inspected, plastic inserts loosened

6 Ties stacked in permanent storage area ready for shipment

7 Forms brushed, cleaned and oiled

8 Stainless steel inserts placed in position

9 Strand threaded in prestressing heads, drawn through forms and distributed by spreading machine

10 Concrete placed in forms and distributed by spreading machine

11 Vibrator machine provides finish vibration

12 Steam covers placed and lines steamed overnight

13 Crossties loaded on flat cars for shipment

Figure 20.4. Prestressed concrete crosstie plant: sequence of operations. (Courtesy JF Pomeroy & Co. Inc. and P.J. McQueen).

(a) *Cleaning forms*

Figure 20.5. *Manufacture of Amtrak ties*

resilient pads at either side of the joint. All-welded rail will, of course, largely eliminate this problem.

The transition to timber ties at switches or ends causes overloading of the adjacent concrete ties. Here again, it may be desirable to use more resilient pads or to develop a more flexible tie for this location.

Switch ties can be made of prestressed concrete but require special location of inserts and of course, longer lengths. Prestressed switch ties are now being used on major rapid transit systems.

Heavily loaded ore cars on severe curves have produced flexural cracks. This indicates the need for special design for this use.

Alkali-reactivity has developed in some critical instances where the source of aggregates was changed without adequate testing. This gross oversight has resulted in the rapid failure of a large number of ties which were otherwise satisfactory.

(b) *Installing rail bolts*

Figure 20.5. *(Continued)*

20.5 FUTURE DEVELOPMENTS

Proposed high-speed freight trains will have axial loads up to 360 kN (80,000 pounds) and speeds of 160 km per hour (100 mph). It is probable that wider ties will be used to reduce ballast pressure, and that rail fastenings will be improved to match the greater forces involved.

Fully bonded behavior for pretensioned ties under repeated loadings under-rail may be further assured by such steps as:

a. More heavily dimpled strand or even mechanical clips on the ends of the strand.

b. Increased tie length which, in turn, will also reduce ballast pressure.

c. Use of microsilica in the concrete mix to increase bond and fatigue strength and to increase electrical resistivity.

d. Use of epoxy-coated strand with sand dusted on during the coating process, so as to give enhanced bond.

(c) *Stressing tendons*

(d) *Placing concrete*

Figure 20.5. (Continued)

(e) Covering for steam cure

(f) Removing ties from forms

Figure 20.5. (Continued)

(g) *Shipping ties.*

Figure 20.5. *(Continued)*

More "flexible" ties may be developed for insertion on an individual basis in existing timber tie tracks.

Further development of switch and turn-out ties is in progress.

Wider ties, which may be spaced further apart, are under development.

At the present time, some 50,000,000 prestressed concrete ties have been manufactured and installed, principally in West Germany, the former USSR, and England but more recently in Canada, the USA, China, and Australia.

However, the potential market in the USA has scarcely been touched, with less than 3% of the total ties in service being prestressed concrete. The potential market is enormous.

The demands of rapid transit, higher speed and more heavily loaded trains, as well as rapidly changing economic factors will, undoubtedly, require that a complete nationwide industry be set up to supply high-quality mass-produced prestressed concrete ties.

SELECTED REFERENCES

1. McQueen, J., "Design and Production of Prestressed Concrete Ties for AMTRAK's Northeast Corridor Improvement Program," *Concrete International,* American Concrete Institute, 1984.

2. Freyssinet Internationale, "Prestressed Concrete Ties," Paris, undated.

3. FIP, "Concrete Railway Sleepers," Structural Engineers Trading Organization, London, 1987.

CHAPTER 21

Road and Airfield Pavements

21.1 GENERAL

Prestressing theoretically offers many advantages for pavements. These have attracted many designers and a considerable number of roads, pavements, container storage yards, and airfield runways have been constructed. Yet, so far, only the airfield runways can be said to have justified the cost and efforts. The basic problem stems from the fact that road and pavement construction by conventional means has been highly developed, with mechanized construction processes and established sources of material supply. Prestressed concrete offers a substantial reduction in material cost, but an increase in the cost of skilled labor and supervision and, to date, a lack of mechanization in construction. The potential cost advantages have thus been largely offset or exceeded by the actual inefficiency in construction.

Comparatively short sections of roads have been built in France, Italy, Belgium, Germany, and the Netherlands. Extensive research and development has been conducted in Czechoslovakia, Great Britain, Switzerland, and Japan.

Airfield runways and taxiways, on the other hand, with their much heavier loads, more readily respond to the reduced pavement thicknesses and greater flexibility offered by prestressed concrete. The advantages of a smoother surface, a watertight covering for the subgrade, a longer life, and a substantial reduction in construction joints show major benefits. As a result, a number of major airports have utilized prestressed concrete extensively; these include the airports of Schiphol (Amsterdam), Orly (Paris), Algiers, Kuwait, and La Guardia (New York). Test strips have been installed in Great Britain, Belgium, Biggs Air Force Base and Sharonville (United States), and Germany. Several container yards have been built in Japan.

Pavements are but one part of the system of load-carrying ability. The other part is the subgrade. The interrelationship of subgrade and pavement has been well

520

established by numerous tests and experience for asphaltic concrete pavements, plain, and reinforced concrete pavements. As yet, the data for prestressed pavement and subgrade interaction is relatively scarce.

Because of their thinner section, prestressed concrete pavements are more flexible; thus, they distribute their load more efficiently over the subgrade, resulting in reduced pavement stresses under wheel loads.

21.2 SYSTEMS

Three systems are utilized for prestressed pavements.

21.2.1 Mobile System—Internal Tendons

The pavement slides on the subgrade, expanding and contracting due to temperature and moisture changes. The prestress is imparted by internal tendons, either pretensioned or post-tensioned. The mobility of the slab reduces the amount of prestress required.

21.2.2 Mobile System—External Prestress

In this case, the prestress is furnished by jacks, springs, etc., reacting against abutments and between construction joints, keeping a constant force on the pavement as it contracts and expands. The external forces, hence the prestress, may be adjusted periodically.

21.2.3 Fixed Systems

This may be cast in segments, but the joints are filled so that the entire slab acts as a unit. It is usually prestressed externally between fixed abutments although, theoretically, it could be internally prestressed yet remain fixed due to friction of the pavement on the subgrade. A fixed system is not able to contract or expand.

21.2.4 Design Considerations

The major criteria for which a pavement must be designed are temperature stresses, shrinkage and creep, and wheel loads.

With mobile systems, a major adverse factor is the friction between subgrade and pavement. Techniques to reduce this friction are primary considerations in construction methods and procedures.

In most such installations, a sand layer is placed first, on which the prestressed pavement rides. Contrary to expectations, a friction factor of about 1.0 is usually developed. To reduce this to 0.7 or so, a number of means have been employed between the sand and pavement slab, such as:

1. A single sheet of polyethylene.
2. A bitumastic layer.
3. Double sheets of plastic between sheets of paper.
4. Double sheets of plastic with grease between.

Prestressed concrete overlays have also been considered for strengthening existing runways. In this case, the existing runway is treated as a subbase, and a friction-reducing layer, such as sand, placed between it and the prestressed pavement.

Internal prestressing can be by means of post- or pretensioning. With post-tensioning, the ducts are inserted in the slab, usually with tendons installed in order to minimize "wobble." Both longitudinal and transverse ducts are installed. The transverse ducts are most easily supported by the side forms and "dobe" blocks; the longitudinal tendons can then be tied to them.

A major concern is to keep the duct size as small as possible. Rectangular ducts can be used. Rigid ducts are superior to flexible ducts because of less wobble and less friction. Obviously, in as thin a section as a pavement, the ducts must be accurately positioned and maintained in position during concreting. The benefits of greased-and-sheathed individual strands have yet to be exploited.

With pretensioning, usually the longitudinal tendons are the only ones pretensioned. They are tensioned against abutments as much as 3000 feet apart or more, the abutment being concrete piers or steel H-piles. Transverse ducts are then usually supported by the longitudinal tendons.

After pouring, shrinkage cracks must be prevented. Thus, it may be desirable to pour at night. Curing must be instituted and carefully maintained. It has often been found advisable to use saturated cotton or burlap mats which are, in turn, covered by a polyethylene sheet to prevent drying out.

A little prestress should be released into the slab as soon as possible to offset shrinkage stresses. This can be done at ages of 1 or 3 days, with final prestress released at, say, 7 days.

With overlay pavements, two-way pretensioning can sometimes be employed, with the tendons anchored against the original slab.

In the Netherlands, an 1100-meter long roadway pavement was pretensioned around a curve, with the strands being deflected around sheaves affixed to an anchored horizontal frame.

An effective and economical means of providing transverse prestress, which has been developed and used in Germany, is the use of "tensile element" bars, small bars which are given a high prestress and then placed in the slab. Through the mechanism of creep, they induce a degree of prestress into the pavement slab (Section 3.6). These bars act primarily as passive reinforcement but, unlike conventional reinforcing bars, actively participate in resisting shrinkage.

Another interesting solution for roadway pavements has been a network of diagonal post-tensioning, with the angle and spacing varied to fit the longitudinal and transverse prestressing required.

Internally stressed pavements cannot buckle, and have usually shown greater elastic and ultimate strength than the design required. They do require fairly sizable quantities of prestressing steel. Repairs to small areas are readily made; major repairs may be extremely difficult.

Externally stressed pavements offer the possibility of substantial savings in steel. The external prestress permits variable application, so that a little stress can be induced shortly after initial set and, during service, the stress can be varied to meet the change from summer to winter. This external prestress can be induced by permanent hydraulic jacks, by coiled steel springs, by flat jacks, or by pneumatic tubes. These tubes can be incorporated in the construction joints and are very efficient. The only difficulty is a tendency of the tubes to be pinched; in fact, all forms of applying external prestress except, perhaps, flat jacks, are subject to possible damage in service and must be protected.

Externally stressed pavements can buckle or "blow-up" under a combination of high stress and high temperature. This can be prevented by proper design, accurate construction, and careful adjustment of the external prestress forces as required. The incorporation of some steel in the pavement will enhance both its ultimate strength and its buckling resistance.

These pavements are easily repaired; the external prestress is temporarily released and the repair accomplished by normal methods.

The major problem lies in the joints. Research is currently under way to develop reliable, effective joints.

One means to prevent buckling is to deepen the slab as, for example, by making a hollow-core or box slab, with the hollow cores being utilized for utilities, etc. On the Mont Blanc tunnel approach (Italy to France), the box sections were used for ventilation ducts, as well as roadway slabs.

Prestressed concrete pavements tend to curb or wrap upward at the edges. Therefore the pavement should be thickened at the perimeter.

21.3 ARTIFICIAL AGGREGATES

Since a major portion of the stresses in prestressed slabs is due to temperature, it is obvious that a concrete with lower thermal response would be desirable. Expanded shale, slate, and clay aggregates (lightweight aggregates) have a reduced thermal response and, also, provide better insulation, so that the lower surface of the slab, in contact with the subgrade, is not subjected to as great a variance, particularly, the short-term variances which are most troublesome. However, most lightweight aggregate, when used in pavements, is subject to "plucking" erosion under traffic. An epoxy overlay may prevent this. The inclusion of microsilica greatly enhances the bond and may overcome "plucking."

Strains due to temperature are transformed to stress in direct proportion to the modulus of elasticity. A concrete, such as lightweight concrete with its low E, can reduce temperature stresses by up to 30%.

Prestressed lightweight-aggregate concrete offers the following advantages for pavements:

1. Lower modulus of elasticity.
2. Better insulating qualities.
3. Reduced thermal response.
4. Better skid resistance.
5. Improved durability under deicing salts.

An experimental roadway employing lightweight aggregate has been built in Belgium.

21.4 PROBLEM AREAS

The two major problem areas for prestressed pavements are the construction joints and edge warping.

Construction joints must permit movement, must transfer shear, must be water-resistant and, most difficult of all, must respond to wheel loads in a similar manner to the adjoining pavement, i.e., they must not be too rigid. Combinations of steel, rubber (neoprene), and concrete are utilized. The joints must be designed so that dirt, spalled concrete fragments, etc., cannot become lodged and prevent movement.

As mentioned earlier, edge warping is a problem due to a combination of shrinkage, temperature, and prestress. Thickening of the edge helps by increasing the deadweight. Properly proportioned, it may effectively balance the warping effect.

21.5 PRECASTING

Precast slabs have been used for both roadway and airfield pavements. The largest such installation is at Kuwait, where triangular precast slabs were employed. Post-tensioning was in the spaces between adjoining slabs, thus permitting three-dimensional prestressing. Use of precast slabs minimizes shrinkage and creep, but requires extreme care in alignment and bedding.

The over-water runways at LaGuardia Airport, New York, are a marine structure as well as a runway. This project is discussed in Chapter 13. A combination of precast beams with a composite cast-in-place slab was employed. This, in turn, was post-tensioned in two directions. The use of the cast-in-place permitted an accurate and smooth surface to be achieved while enjoying the structural and economical advantages of precasting. The project has proven eminently successful over its 30 years' life and has been structurally reinforced to carry much heavier loads than originally contemplated in design.

21.6 MECHANIZATION IN CONSTRUCTION

There is a definite need for constructors to develop mechanized construction means which will make prestressed pavements competitive in cost with conventional pavements. This will enable the many advantages of prestressed pavements to be realized and thus increase the volume of use, justifying further cost-saving mechanization. The technique of tensile element bars, mentioned earlier, is one such step. Longitudinal pretensioning appears to offer significant cost advantages. Lightweight aggregate concrete appears particularly applicable, especially with the addition of microsilica. Individual strands, pre-greased and sheathed, would solve space problems.

Whichever system and design is adopted, construction planning, equipment, and techniques have an unusually great influence on the economic and structural success of the pavements.

REFERENCE

1. AC1 325, "Prestressed Concrete Pavements," American Concrete Institute, Detroit, Michigan, 1989.

CHAPTER 22

Machinery Structures

22.1 GENERAL

Prestressing and prestressed concrete techniques offer a potential major break-through in the design and construction of machinery structures, including foundations, beds, frames, and, in some cases, even the working parts. This is a new concept and one generally not familiar to machinery designers and constructors. At the same time, the quality control and tolerances required are substantially more rigid than those usually required for civil engineering structures.

Prestressing enables the achievement of results which are otherwise unobtainable with concrete. Foremost among these is reduced deformation in service, a property of increasing importance for automated machinery. Enhanced ductility and fatigue endurance through three-dimensional prestress are other highly important properties. Thermal insulation and reduced response to thermal changes can be attained through the use of specially selected and manufactured aggregates.

A machinery structure is a dynamic system that includes soil, foundation, bed, frame, machine, and product. By means of prestressing, the foundation, bed, and frame can be made to act as one.

Fatigue is presented in a new dimension in machinery structures. Civil engineering structures may be designed for perhaps 2×10^6 cycles; a machine may require design for 30×10^8 cycles. Prestressing can be used to keep the cyclic stress range entirely in compression, thus greatly increasing the fatigue resistance.

Dynamic response is determined by the relation between the frequency of the machine and the natural frequency of the system. This latter can be modified by changing the mass and/or the rigidity. Prestressing enables blocks (masses) to be added in such a way as to work integrally with the structure.

Dynamic response is also determined by the modulus of elasticity. Selection of the proper aggregates (e.g., granite, expanded shale, etc.) gives a means of controlling this modulus through a range of 100% or so.

Prestressed concrete concepts enable the designer to control the behavior of the structure so that it will have fully elastic behavior up to cracking; thence a plastic-elastic behavior (with microcracks) giving damping and ductility with reduced stiffness, from which the concrete will substantially recover upon reduction of load; and, finally, a plastic (ductile) failure mode. The achievement of such a spectrum requires the careful proportioning of stressed and unstressed reinforcement. Combinations of stressed tendons with wire mesh have proven useful in this regard.

Prestressed concrete machinery structures are usually designed to be crack-free under normal operation, as cracking reduces stiffness and fatigue strength. A structure which is in resonant frequency with the exciting force up to cracking will undergo a substantial frequency change on cracking; thus, it possesses a built-in-safeguard against progressive dynamic failure.

Prestressing makes it possible to stress together segments to form a foundation structure of almost any size or shape which will still act monolithically.

Prestressed concrete machinery structures can be designed with elastic deflections as low as 1/7000 of the span by mobilizing the participation of the entire foundation block.

Thus, in summary, prestressed concrete offers these advantages:

1. High energy absorption before rupture.
2. High internal damping.
3. High fatigue strength.
4. Safety through the automatic pretesting of materials, and in the inherent ductile mode of failure (noncatastrophic).
5. Stiffness.
6. Precision.
7. Ability to design and modify dynamic response.
8. Unlimited shape and size.
9. Low deformability under repeated short-term loadings.
10. Chemical resistance.
11. Noise reduction.
12. No condensation.
13. Not notch-brittle at low temperatures.
14. Economy.

Prestressed concrete does possess some disadvantages as compared with steel. These are mainly associated with strains. Strains include those due to volume change, such as shrinkage, temperature, and creep. They also include elastic changes under prestress and under dead and live loads. These all can be controlled as noted later.

Another disadvantage of prestressed concrete may be its greater size as compared with cast steel. This may mean that the machine itself exerts a greater bending force, or that the foundation must be larger. Proper rational design makes beneficial use of this larger size and mass to reduce foundation pressures and vibration, and overcomes the bending moments by prestressing.

On occasion, the machinery structures may be tied to rock or very firm soils by prestressing, so as to make the mass of the rock an integral part of the structure.

Prestressing is an excellent means of securing inserts and of bolting the machine to the frame or bed or foundation in such a way as to prevent differential movement and fatigue. By securing high tensile bolts under tension, dynamic "chattering" is prevented.

22.2 CONSTRUCTION

Prestressed machinery structures are inherently more important and higher-priced structures per unit of volume than most civil engineering structures. Thus, much more sophisticated manufacturing techniques are appropriate and justified.

First, aggregates, cement, and mix can be selected to meet the specific properties required (compressive and tensile strength, modulus of elasticity, minimum volume change, low thermal response, self-insulating qualities, etc.).

Because of the generally small size of structure or segments, mixes can be carefully vibrated and compacted, with a very low W/C ratio.

Proper curing can readily be provided and controlled.

Forms (molds) can be made of machined cast steel, adjusted by machine screws to exact position, within a tolerance of 0.2 mm (0.008 inch).

By means of such procedures, concrete segments can be cast to an accuracy of about 0.25 mm (0.01 inch).

An alternate means of concreting is the use of prepacked aggregate concrete. In this case all reinforcement, ducts, anchorages, and inserts are accurately set. Crushed rock is carefully screened to remove all fines, then is placed in the forms, and the inserts rechecked. Grout is then intruded through preplaced tubes. This is a special grout, possessing great fluidity, either through the addition of admixtures to reduce surface tension, or by mixing to a colloidal state.

With prepacked concrete, the form pressures may be quite high, especially in confined and localized areas, so forms must be very rigid and unyielding. Concrete-backed steel or FRG sheets make excellent forms.

To reduce heat of hydration in large blocks, the aggregates may be pre-chilled. Ice may be used to lower the ambient temperature of the mix to 10 to 15°C. Cement (50 to 60°F) may be low-heat or large-grain size (low Blaine fineness number). Liquid nitrogen may be introduced into the fresh mix to achieve a fresh concrete temperature of 5°C (40°F).

Water may be circulated through the duct tubes or through special cooling tubes to remove the heat of hydration. If the duct tubes are to be used for this purpose, they should be of rigid tubing with screwed joints.

Shrinkage may be minimized by a very low W/C ratio, such as 0.33, obtained by use of HRWR admixtures. Microsilica plus PFA may be effectively used to promote high strength. Shrinkage may be minimized by sealing the surface to prevent moisture loss. The proportion of microsilica can be raised as high as 20% to give ultra high strength and abrasion resistance.

22.3 TENDONS

Prestressing tendons and their anchorages must be selected for proper dynamic behavior. This requires that the system be proven under dynamic testing and that there be adequate elongation at rupture.

Both bonded and unbonded tendons are used. The use of unbonded tendons permits the re-stressing and/or replacement of tendons during the lifetime of the structure. Techniques for grouting of bonded tendons are discussed in Section 5.10 and techniques for protecting unbonded tendons in Section 5.11. In any event, the anchorages must also be protected from corrosion, fire, and impact. For these reasons, recessed anchorages are preferable. See Section 5.5.3.6.

22.4 TRIAXIAL PRESTRESS

The behavior of concrete under multi-axial loading is radically transformed. Local bearing stresses can be increased up to perhaps three times the compressive strength of the concrete by use of spiral reinforcement. Proper shaping of the concrete can permit greatly increased stress and strain, as is shown by the plastic hinges of concrete, which develop apparent stress levels of 140MPa (20,000 psi) and rotational strains up to 4%. These are, in effect, notches in the external surface which are confined by the cover concrete so as to create triaxial stress conditions.

Triaxially stressed concrete has a high elastic limit, high stiffness, and superior fatigue endurance.

One means of triaxial prestressing is by using tendons in three or more planes. Stressing in two of the planes may be obtained by curving one set of tendons with overlapping anchorages, or else by using continuous wire-winding. With heavy spiral reinforcement of high-strength wire, longitudinal prestressing to a high degree produces radial prestressing also, through Poisson's effect.

22.5 FERRO-CEMENT

For certain structures, frames, etc., requiring great ductility, the combination of ferro-cement techniques with prestressing offers interesting possibilities. Layers of mesh are placed in the concrete so as to give a steel proportion of approximately (470 Kg/m³ (800 lb/yd³). To properly place the concrete, it is usually applied in layers or coats, by hand, alternating with the mesh.

Approximations of this phenomenon can be achieved in more conventionally placed concrete by placing one or two layers of mesh close to the tensile surface, or by the use of finely divided wire fibers in the mix.

22.6 COLD WORKING

One of the most interesting aspects of prestressed concrete for machinery structures is the cold-working phenomenon. Repeated loading and unloading cycles show a gradual stabilization against creep. This may be achieved more readily by a combination of alternating heating and stressing treatments. Obviously, any such process has to be applied in the direction or axis of final stress and is most practicably applied to segments.

22.7 SEGMENTAL CONSTRUCTION

The joining of segments can best be achieved by the dry-joint or epoxy-joint techniques described in Section 4.3. Dry (exact fit) joints may be obtained either by casting in sequence against the previous segment and match-marking, or by casting in machined cast-steel forms as described earlier in this present chapter. A 0.4-mm ($\frac{1}{64}$-inch) tolerance is obtainable by this latter method and has generally proven adequate for assembly of segments with dry or glued joints. This is greater accuracy than can generally be obtained by grinding.

If grinding is employed, controls must be set up to compensate for wear of the grinding wheels, and means established for determining a plane of reference. The attendant difficulties usually limit tolerances obtainable to about 1.0 mm ($\frac{1}{25}$ inch).

22.8 UTILIZATION

Prestressed concrete has been utilized for forging hammer blocks. One of these has had over 300 million cycles in service, with accelerations up to 100 g, and is still uncracked.

Structures subject to shock loading, such as rocket test stands and explosion test chambers, are particularly suited to prestressed concrete because of microcrack resistance, chemical inertness, high thermal resistance, safety in explosion, and noise reduction.

Large foundations can be held to very rigid tolerances of accuracy, which is especially important for extrusion beds, and automated machinery.

Large-diameter hydraulic presses have been made, with internal pressures up to 40 MPa (6000 psi). The prestressed concrete cylinder head is manufactured, then a machined steel liner placed and grouted. The piston is likewise of triaxial prestressed concrete, with a machined liner placed over it and joined by grouting. The steel serves as wearing surface and seal, but the prestressed concrete resists the pressures and shock loading.

Turbine blocks may require adjustment in order to change from a resonant frequency. Additional masses (blocks) of concrete may be added to the block and prestressed to it, thus "tuning" it. This is especially valuable when employing a turbine block of lower frequency than the turbine itself, since the lower frequencies are harder to accurately predict and may require field adjustment.

To readily permit such adjustment, it may be desirable to install additional, unfilled ducts in the original block, into which tendons may later be inserted.

Frames for presses appear to be an excellent application for prestressed concrete because of rigidity and freedom from distortion.

Boring machines have had their pillars prestressed to the table or bedplate to prevent distortion.

A 10,000-ton testing frame, a steel strip mill, a torsion-stretcher bed, and a 600-ton hydraulic press are among the many successful applications of prestressed concrete. A prestressed concrete hyperbaric chamber has been constructed in England.

There is an obvious relation between the use and techniques of prestressed concrete for machinery structures and for prestressed concrete pressure vessels (Chapter 17).

22.9 SUMMARY

Prestressing enables low deformability and controllable dynamic response to be achieved in machinery structures. As such, it has moved into a field of high sophistication requiring the greatest of care and control in design and manufacture. It is probable that the increased demands for automated machinery and ever larger machines will force the greater utilization of prestressed concrete for machinery structures. It is important, therefore, that the design and construction be undertaken by those thoroughly versed in the properties, techniques, and art of prestressing.

CHAPTER 23

Towers and Special Structures

23.1 TOWERS

Towers serve the joint purposes of theme monuments, television and microwave transmission, and quite frequently, restaurants as well. A substantial number of these spectacular structures 400 to 500 meters (1300 to 1600 feet) high have been constructed in Western Europe, the former USSR, and China. Prestressing has been essential to overcome the bending moments introduced by wind and, in some locations, earthquake.

Similar structural aspects are encountered with intake towers in reservoirs, where the dynamic loadings must also consider the added mass effect of the water. In some reservoirs, the potential for landslides into the reservoir, generating a solitary wave, must also be considered (Fig. 23.1).

Stacks have historically not been prestressed, but seismic considerations are now presenting a need for post-tensioning.

The construction demands for prestressing are similar to those of shafts of tall offshore structures.

Ducts should be rigid steel, preferably of 2.0-mm (0.08 inches) thickness, water-tight, with threaded or sleeved couplings. Each length of duct should be delivered with a plastic cap to prevent the entry of foreign material.

The strands are usually threaded in from the top, using a funnel to prevent kinking or abrasion of the strand or wire. On one recent tower in Beijing, a messenger line was run in from the top and the strands pulled up from the bottom, eliminating the need for a large hoist on top.

Posttensioning is often done from a gallery at the base, built into the foundation slab, in order to facilitate access.

Grouting of the tendons is subject to the same problems of bleed and sedimenta-

Figure 23.1. *Tall intake tower for reservoir in Venezuela is post-tensioned to resist earthquake and waves generated by landslide.*

tion noted for offshore structures and nuclear reactor containment structures, only intensified by the greater height.

To avoid excessive bursting pressures, one or more intermediate grouting stations may be employed.

The dangers of the high pressure required for tall vertical tendons is that, if grout leakage occurs, for example, at a faulty coupling, the pressure may be enough to produce a delamination in the concrete. This in turn exposes a greater area to the action of the pressure.

This phenomenon, which has proven very expensive to correct, explains the rationale behind the use of rigid steel watertight tendons, and screwed or sleeved couplings.

Plastic ducts are not recommended, since they tend to reflect any crack in the concrete such as that induced by thermal strains or prestressing restraint.

Towers, typically having relatively thin concrete walls, are subject to very high compression. Under service conditions, the prestressing will add to the compression resulting from bending, while under ultimate conditions, about two-thirds of the prestress will be relieved by shortening. Additional confining steel can be used in zones subject to high compression.

Since large bars, under high compression, tend to produce splitting at their blunt terminal ends where they are spliced, mechanical joints or milled ends should be employed. Where smaller bars are lap-spliced, ties should bind the bars tightly at both ends of the lap.

Through-wall confinement will augment the concrete core's ultimate capacity and provide ductility, as well as counter laminar splitting.

In those cases where the tower is founded on rock, vertical tendons (anchors) may be drilled in and spliced or lapped with the concrete tower's vertical tendons.

The base of high towers is often designed as a shell and behaves as a tension ring. Circumferential prestressing is usually employed, stressed around 180° overlapping sectors between buttresses located on 60° arcs.

23.2 SPECIAL STRUCTURES

There are, of course, a great many special applications of prestressing and pre-stressed concrete. New uses will constantly continue to arise and be exploited by this technique and concept.

23.2.1 Monumental Statues and Ancient Monuments

Monumental statues present difficult and unique structural problems, exposed as they are to wind loading and seismic forces while at the same time not necessarily being designed for the most effective structural behavior and position. Fortunately, as Freyssinet pointed out many years ago, nature utilizes bones and tendons in living forms; in a statue the concrete or bronze, etc., may replace the bones, and the tendons may be of high-strength steel. Nevertheless, nature provides dynamic response to maintain balance, while a statue must remain fixed. Hence, the design of a prestressing system for such a statue may present a challenging and intriguing exercise.

Prestressing techniques are similarly used to restore and preserve ancient stone monuments and structures, many of which were toppled by earthquakes such as the great one that shook the Middle East in the 6th century. Techniques include drilling through the existing segments, and post-tensioning, with particular emphasis on protection of the tendons from corrosion. Stainless steel and even titanium have been used. The interstices of the existing structure should first be filled with grout or injected with epoxy under carefully controlled minimum pressures. Deflections, etc., should be carefully monitored during all stages of grouting and stressing.

23.2.2 Girders and Beams

Prestressed concrete girders and beams have recently been designed as structural skid beams to support oil production equipment in the arctic, because of favorable behavior at low temperature and rigidity. Prestressed concrete offers substantial economies over steel girders, fabricated from low-temperature steel, even after consideration of freight costs.

23.2.3 Underground Concrete Walls

Underground concrete walls are widely employed, being constructed by use of a slurry to hold the soil while the excavation, reinforcing, and concreting is carried out. Post-tensioning has been ingeniously employed to strengthen the wall in vertical bending. The duct, with tendon inserted and an anchor on the lower end, is tied to the reinforcing cage. After the wall is concreted, the tendon may be stressed from the top.

The profile of the tendon may be safely exaggerated as compared to typical horizontal beams, and, since this stressing is usually for temporary purposes only, the cover may be minimal. The reason is that the horizontal component of the prestress is resisted by the passive pressure of the soil.

23.2.4 Stone

Prestressing techniques have also been applied to stone. The problem has been the nonuniformity of stone and planes of weakness and lamination. The actual prestressing consists of drilled holes, with post-tensioned tendons and grouted joints.

23.2.5 Mass-Production of Small Items

Many small mass-produced items lend themselves to prestressing. Fence posts and grape stakes are typical of small linear items suitable for pretensioning in mass production. Multiple-form techniques are used. Manhole covers are an illustration of the application of circumferential stressing, with wires applied under tension in a groove in a precast plate. The groove is later filled with mortar. Such items become economical only when produced in large numbers by mechanized processes.

An interesting application of prestressing is to horticultural (greenhouse) planks. Such planks must be resistant to rot and constant humidity; they must be nontoxic and noncorrosive. Narrow planks of prestressed lightweight concrete can be lifted by two workers, and answer the needs of the environment.

23.2.6 Supports for Use in Mines

Prestressed concrete beams are being used in the mining industry to replace the traditional timber supports. Not only is adequate elastic strength required, but also ultimate strength. This large ductility range can best be obtained by incorporation of unstressed strand or mild steel in substantial quantities, in addition to the stressed tendons, and by providing confinement of the concrete core. Lightweight concrete is well adapted to this application because of its greater deflection under load, fire resistance, and lower handling weight. Such mine support timbers and tunnel liners are durable, fire-resistant, and economical.

23.3 PRESTRESSED FORMS AND BEDS

Prestressed and ferro-cement techniques have been applied to the construction of both stressing beds and the forms or molds for production of prestressed concrete. Consideration must be given in form design to the degree of flexibility or rigidity required for operations. Surface effects must also be considered, and concrete forms may require a surfacing of epoxy or plastic to permit easy stripping and removal of the product.

When precast pretensioned forms are incorporated in a permanent cast-in-place structure, additional benefits result. A very durable skin is provided. If highly stressed, some of the tensile resistance is transferred to the adjoining cast-in-place concrete, permitting greater strains before cracking.

23.4 SPATIAL STRUCTURES

In cable-supported structures, tents, pavilions, etc., tendons are prestressed and anchored so as to work against one another, and to support the membrane covering. Particular care must be taken to prevent corrosion, as by the use of galvanized or plastic-encased tendons, and to prevent abrasion where tendons cross. Such structures have been of great interest at recent World Fairs.

Mast-supported structures with three-dimensional cable stays to support concrete shell roofs are moving from the exhibition stage to practicable and economical utilization for pavilions, markets, and parking garages.

The "Pyramid" at the entrance to the Louvre in Paris is a spectacular illustration of the use of tendons stressed in three dimensions to secure truly plane surfaces for the sides of the multi-paned glass structure. (See also Section 10.12.)

23.5 UNDERPINNING

Significant reductions in the cost and time required for underpinning existing structures have been attained by the use of post-tensioning. Instead of excavating underneath the existing footings, and attempting to regain by jacking any ensuing deformations, columns or shafts may be constructed alongside, e.g., by drilling, insertion of a reinforcing cage or structural steel member, and concreting. These insertions are then connected to the original column by short post-tensioned tendons.

If the tendons are installed on a saddle profile, the original column may be raised a small amount by the stressing, thus restoring the original state of compressive stress and countering any settlement. As in all underground uses of prestressing, corrosion must be considered if tendons will be left exposed for any length of time. Bars are less susceptible than strand but are not as flexible around saddles. Tendons may be protected with grease, epoxy coating, bitumastic, or galvanizing. Permanent underpinning tendons should be encased in grout.

23.6 FUTURE APPLICATIONS

These will continue to be developed as engineers, architects, and contractors become more familiar with high-performance concrete and prestressing techniques. As each new application is made, the composite experience of the profession is enhanced and the horizon ever widened.

It must be emphasized in all these new applications that basic principles of stress and strain, Poisson's effect, and creep still hold and that the details must be thoroughly developed to ensure against local defects which could lead to unsatisfactory performance, catastrophic or progressive collapse.

CHAPTER 24

Cracking and Corrosion of Prestressed Concrete Structures

24.1 GENERAL

Prestressed concrete structures are subjected to many of the same phenomena causing damage to conventional reinforced concrete structures. In addition there are a number of problems which are peculiar to prestressing and which need consideration. It is these which will be dealt with in this chapter. Some of these have been previously mentioned in earlier chapters but are repeated here for completeness. Those which are general, e.g., corrosion of reinforcing steel, freeze-thaw attack, alkali-aggregate reaction, etc., will not be presented here except where they interact with the prestressing or require special treatment.

24.2 CRACKING DUE TO PRESTRESSING

Although one purpose of prestressing is to prevent cracks, especially flexural bending and shear cracks, improperly detailed and applied prestress can cause cracks.

24.2.1 Circumferential Cracks and Spalling Around Anchorage

These are due to the well-known anchorage bursting stresses. Spiral coils are typically supplied as part of the anchorage hardware to confine the concrete and enable it to take the high bursting strains.

24.2.2 Splitting Between Tendons

Between large tendons or large groups of smaller tendons, tensile strains may develop, especially near the anchors. To resist these, transverse mild-steel bars, often made up as an orthogonal grid, are needed.

24.2.3 Cracking Behind Anchorages

Posttensioned tendons are often anchored within a concrete structure or element. These may be the positive moment continuous tendons in a long-span continuous girder, anchored near the quarter points, or anchorages tying the shaft of an offshore shaft to the roof of the base caisson, or similar intermediate anchors. When the prestress is applied, the concrete around the tendon and in front of the anchorage is compressed. It shortens, and in doing so, tries to shear along the sides of the anchorage and pull away from the concrete behind the anchorage.

This tension must be resisted by adequate passive (mild steel) reinforcing steel. A common rule, in the absence of detailed strain compatibility analyses, is that the capacity of the added mild steel should be equal, at working stress, to one-half the initial tendon force. These bars should be run far enough to develop their ultimate tension, and their terminations should preferably be staggered.

If the tendons are not properly anchored as described above, these tension cracks behind the anchorages can develop into shear cracks, endangering the girder as a whole.

Lateral shear is critical because these intermediate anchorages usually require a bolster. When groups of tendons are anchored at the same bolster, they may not only rupture the concrete behind in tension but shear the sides.

Bolsters must therefore be well-anchored down by stirrups and be tied laterally by shear steel in the slab. They should preferably be staggered.

24.2.4 Cracking Due to Abrupt Curvature of the Tendon Out of the Plane of the Slab

This detail occurs frequently at bolsters for intermediate anchorages. The curved tendon shortens and, in so doing, creates a high radial force trying to burst the tendon out.

These zones require adequate stirrups or other transverse ties to prevent cracking and spalling out at the inside of the bend.

24.2.5 Delamination

In thick-walled structures, especially at corners, more than one layer of tendons may be used, curved in either the horizontal or vertical plane. When these are not fully parallel, and they generally are not, they create transverse forces tending to split the wall in the middle, i.e., to cause delamination.

Similar delamination may occur in the deck of long-span continuous bridges, especially over the piers where, due to negative moment, the tendons are densely clustered. Any eccentricity, even minor, may produce through-thickness tension forces. Due to the space occupied by the ducts, the net concrete area may be unable to provide the necessary tensile capacity, resulting in delamination in the middle of the slab.

In cast-in-place segmental bridge construction, the longitudinal ducts will typically be supported at the end bulkhead but may sag in between. This creates a sharp although small break in direction at the joint. When the tendon is stressed, the normal force may lead to laminar crack in the deck slab.

Delamination may also occur in thick-walled structures, such as offshore structures, which are subjected to very high compression, e.g., under hydrostatic loads.

Delamination is difficult to detect visually, as it may create only minor surface spalling or none at all. The hand-held hammer, or its mechanical equivalent in the case of very thick walls, will reveal the internal delamination by the hollow sound. A chain dragged on the deck has a similar response.

Delamination is also the most difficult damage to repair, since the area is large and the lack of through-thickness reinforcement may make injection of epoxy difficult. Any pressure applied serves to widen the crack. Preinstallation of stitch bolts is the only positive method.

Delamination can be prevented by the provision of through-thickness steel, in the form of stirrups, T-headed bars, or anchored dowels.

24.2.6 Cracks Between Looped Tendons

In deep walls and deep webs of girders, it is common practice to loop the tendon in an U-curve at the bottom, running both legs vertically to the top. Similar loops in the horizontal plane may be used in slabs.

Where adjacent loops overlap, ꟺ, then the lateral forces offset each other. However, when, as is often the case, the loops do not overlap but are separate, U U, then there are large transverse forces created between the loops at the bottom, often resulting in a crack between the vertical legs.

Transverse reinforcement should be provided to resist this splitting force.

24.2.7 Cracks Due to Shear Lag

Where the prestress is concentrated near the center of the structural element, for example, in the spine beam or bottom slab of a trapezoidal box girder bridge element, the pre-compression may not extend all the way to the edge of the overhanging deck flange. Any tensile forces due to deadweight bending or thermal strains may result in edge cracking.

Thus good practice is to ensure a minor amount of longitudinal prestress at the outer edge of the flanges, or at least mild-steel reinforcement. In its absence in the design, the contractor may find its voluntary inclusion a wise precaution.

24.2.8 Splitting Cracks at Reentrant Angles

Where there are V or Y configurations, as for example, in two intersecting walls, prestress tendons running through the legs will tend to pull the angles into a straight line, causing spalling and pull-out. This is, in effect, an exaggerated case of the phenomenon discussed in Section 24.2.4. This same problem can also occur during the stressing of curved tendons, when any eccentricity, even slight, or local bearing may cause flexural cracking. Heavy stirrups are required, tying the entire wall together through its thickness, concentrated especially at the throat of the Y or V. If practicable, the change in direction can better be resisted by making the angle change in small increments.

24.2.9 Cracks in Bearing Area Due to Restraint of Shortening

When beams or girders are manufactured, they initially sit flat upon their soffit. As the prestress is applied, the member cambers upward, rotating around the end corners and placing all the load on these small areas. The result is often shear cracking and crushing at the corners.

Mild-steel reinforcement must be provided, crossing the potential sloping crack at close spacing. In extreme cases, such as very large girders, a softener block, such as wood, may be needed to distribute the concentrated load.

24.2.10 Deflected Strand Pretensioning, Release of Prestress, and Deflecting Force

Pretensioned girders and heavy beams are usually stressed with deflected strands, the deflection being at one, two, or more points. In between the tendons are chords.

When the concrete has cured, the longitudinal prestress must be released. Yet the deflection devices will anchor the girder so that it cannot shorten, hence cannot be pre-compressed. Conversely, if the deflection devices were to be released first, the stressed strands would break the member upwards, with large flexural cracks at each deflection point.

The prevention is to arrange the deflection devices so that they can travel the short horizontal distance needed to pre-compress the member. This can be done with rollers, or inclined wire rope ties, details varying with the member's size and the deflection force required.

The longitudinal prestress can then be released first.

24.2.11 Cracks Due to Local Restraint of Forms or Embedments, or From Steam Curing

When prestress is applied to the member, it shortens. This shortening may be restricted locally by joints in the forms or by embedments, resulting in cracks which typically radiate out from the corners of the restraint.

The same phenomenon can occur during manufacture, if steam curing is applied. The steel form or embedment expands more rapidly than the concrete.

The prevention is to use sponge rubber gaskets or wood fillet strips, so as to accommodate the local deformations.

24.2.12 Flexural Cracking Due to Incorrect Support

When prefabricated prestressed concrete members are transported, it may not be practicable to support them at the exact location for which they were designed. Support may be required in from the ends. When this exceeds the capacity of the member in cantilever, since the prestress is not relieving the stress but augmenting it, a negative moment flexural crack may occur.

In practice, this problem may require supplemental mild steel.

24.2.13 Cracks at Picking Points of Prefabricated Members

When prefabricated members such as piles are removed from their forms, a high negative moment may be introduced at the picking insert. This is not only due to dead load, but also to impact, acceleration, and suction.

Mild steel or short lengths of unstressed strand can be placed at the top of the member at the picking points, as supplemental steel, to prevent these cracks.

24.2.14 Cracks Due to Overpressurizing When Grouting

The typical small size of ducts as related to the member being posttensioned means that the radial stresses when grouting can easily be sustained by the concrete or the membrane strength of the duct. However, if there are cracks into which the grout can intrude or splices in ducts which allow the grout to escape into joints, the pressure acting over an ever-increasing area can propagate the crack. Overpressurization has proven especially critical for tall vertical ducts. This is the reason for recommending completely watertight sheaths of substantial thickness for vertical ducts of offshore structures and towers.

Plastic ducts unfortunately tend to "reflect" cracks in the concrete.

24.2.15 Cracks in Piling

These are cracks peculiar to the special conditions of piling installation, and include rebound tension cracks and vertical splitting cracks. They are discussed in detail in Section 11.8.

24.2.16 Delamination Due to High Compression Normal to Posttensioning Ducts (Excessive Multi-Axial Stress)

When concrete members are loaded to or near their design compressive capacity in one direction and then posttensioned to a high compression in the orthogonal direction, delamination may occur, due to induced tension on the third axis as a result of the Poisson effect.

This same effect may be created when large tendons on one axis are posttensioned and grouted and the member is then loaded to very high values of compression on the orthogonal axis. In this case, the high compression acting on the grouted duct tends to cause it to oval, creating a splitting tension in the third axis.

In both cases, use of closely spaced stirrups will arrest and control the laminar splitting.

24.2.17 Cracks Due to Fatigue

This is a problem which formerly was encountered only in reinforced concrete structures subject to large numbers of cycles of repetitive load, such as the girders of railroad bridges, reinforced concrete piles under sustained pile driving, and floating concrete structures in a wave environment. Fatigue was seldom recognized as the cause.

Prestressing is especially effective in such installations because, through it, a state of precompression is induced so that the cycles are primarily "low compression–high compression." Concrete has a much lower endurance for cycles which enter the tension regime, i.e., from tension to compression and back into tension.

Recently the aggravated behavior of submerged concrete has been recognized, where the pore pressure generated has led to early cracking. The cracks then suck in water on each cycle, and when they close, cause hydraulic fracturing of the cement.

The degree of prestress should therefore be selected such that reversal of stresses does not occur. When such is inevitable, as it is in the case of driven piling in water, then sufficient steel area needs to be provided to prevent exceeding the yield stress in the steel. Thus any cracks will close.

24.2.18 Crushing and Shear of Posttensioned Concrete During Stressing

In girders designed to be heavily stressed, the temporary compressive stresses should be computed on the net section, allowing for the area taken out by ducts. When this rule has been disregarded, the remaining concrete has sometimes been overloaded in compression, resulting in failure in compressive shear.

24.2.19 Cracking Due to Restraint from Shortening

Structural systems with othogonal or intersecting elements may be so rigidly held in position that the shortening due to prestress is prevented. Sometimes this restraint is due to the weight of the member and its friction, but most often it is due to the intersecting members and structural elements at the ends.

Thermal strains can similarly result in cracks when the cooling after expansion is similarly restrained.

24.2.20 Cracking Due to Restraint of Reinforcing Bars

Excessively large bars immediately under posttensioning anchorages cannot deform plastically as does the concrete. This may produce local cracking.

Smaller bars, well-distributed, and mesh are more able to conform to the plastic deformation of the concrete.

24.2.21 Cracking Due to Overstress by Prestressing

Where the dead load of the member will not act to resist the effect of prestressing or where the live load is large in relation to the dead load, then it may be necessary to apply the prestress in stages so as to avoid overstressing, resulting in cracking in negative moment at a location designed for positive moment or vice versa.

In the case of bridge span construction by the cantilever-suspended span method, the cantilevered arms do not have their full dead load resistance until the suspended span is erected. During this interim stage, the arms cannot be fully prestressed without cracking, unless adequate opposing reinforcement is provided to offset the overstress. Alternative methods are to tie the arm down to the base or shaft by an inclined rigid leg or to only partially stress the member prior to erecting the suspended span.

24.2.22 Cracking Due to Freezing of Water in Ducts

It is very difficult to keep rainwater and curing water out of ducts, especially those which terminate in a bridge deck, unless they are capped by tight-fitting plastic caps. These should be fitted prior to delivery to the point of installation. With red or yellow caps, it is relatively easy to observe a missing cap.

If water freezes in the ducts, its expansion may cause cracking. Cracks extending through the wall, for example, the web, are relatively easy to repair, whereas those in a slab may be laminar and require stitch bolting.

24.2.23 Cracks Following Posttensioning Ducts

These may be due to freezing of water or fresh grout in the ducts (see Section 24.2.22), overpressurizations during grouting (see Section 24.2.14), or settlement of the fresh concrete below a large duct. This last may in turn be due to a high water-cement ratio, or lack of vibration, especially with very thin webs.

Such cracks can also occur when the prestressing force is too concentrated, for example, when several large tendons are close together, or when posttensioning tendons supplement pretensioned tendons, without any enclosing or confining mild steel. Cracks can be prevented by enclosing the highly stressed zone with stirrups.

24.2.24 Transverse Cracks Due to Steam Curing

In the manufacture of pretensioned concrete elements, especially long, heavy ones, longitudinal deformation may be prevented by friction of the member on the soffit or in the forms. When the concrete is still fresh, and is heated by steam curing, it expands. If, after the concrete has hardened, but not yet cooled, the tarpaulins or covers are removed, then the thermal contraction may cause transverse cracks. These can be prevented by not exposing the element or its upper surfaces to the weather until some or all prestress has been released into the member.

This phenomenon is aggravated in the winter, when cold dry winds blow across the moist heated surface, and drying shrinkage combines with thermal contraction.

24.3 CORROSION

24.3.1 General

This section will address the special problems of corrosion of prestressing tendons and anchorages. Many aspects of this matter have been addressed previously in Chapter 5 on Durability, where provisions are recommended for preventing corrosion. Others have been addressed in sections on specific applications.

The experience of 40 to 50 years of prestressing has shown a remarkably low incidence of corrosion of tendons. Nevertheless, the consequences of failure are so severe and critical that every effort must be made to prevent corrosion. The individual wires are small, and they are under high stress. Fortunately, there is some evidence that cold-drawn wire and strands have surface characteristics that minimize the onset of corrosion.

24.3.2 Pretensioning Tendons

Pretensioning tendons, encased in dense concrete, which has had sufficient cement paste to have penetrated the interstices between the wires of each strand under intense vibration, are well-protected against corrosion. Ends of the strands which become exposed will corrode, but only to a depth of 3 to 5 mm ($\frac{1}{8}$ to $\frac{1}{4}$ inches) if the grout has filled the interstices. Where concreting has been less effective, corrosion may eventually work back along the center wire, especially in a saline splash zone environment subject to freezing and thawing.

Daubing the exposed ends of the strand with epoxy is often practiced and will prevent surface rust staining. However in a freeze-thaw environment, the epoxy will be wedged off. This can be prevented by burning the strands back about 10 to 12 mm and filling with epoxy mortar.

24.3.3 Posttensioned Tendons

Posttensioned tendons are typically protected by the duct sheath, plus the injected grout, all encased within a concrete member of low permeability. Even when bleed water or trapped water in the duct has reabsorbed into the concrete, leaving an air-filled void, the high pH and the limited supply of oxygen leads to a slow rate of corrosion.

However, the few problems which have occurred in practice have been associated with such voids. That is why it has been recommended not to flush the ducts with water, to cap them to exclude water entry, to flush well with grout, and to use a nonbleed or low-bleed grout.

Problems of corrosion of tendons have been reported where the grout was mixed with seawater or where the tendons had been saturated with seawater after installation but prior to grouting. This latter is one exception to the rule against flushing with fresh water, but in this case, the subsequent grout injection should waste at least 100% or more in order to ensure that water pockets are "dragged" out of the duct.

Tendons which have been stressed and left ungrouted for long periods have suffered corrosion and failure. In the reported cases, the atmosphere was one of frequent fogs which had entrapped the corrosive effluent gasses of nearby refineries. Every effort should be made to grout as soon as possible after installation.

Use of water-soluble oil, as recommended under "Posttensioning," (see Section 2.3.3) will minimize the risk prior to grouting.

Problems have been reported in Europe, especially with hot-rolled tendons, where they had been improperly stored, subject to mud, water, and atmospheric corroding substances, including sea and air. Storage should always be above ground, protected from rain and the atmosphere, and in adverse environments, dehumidified.

Wire-wrapped tanks, penstocks, and pipes have had perhaps the least satisfactory record of durability. The fact that they are more or less continuously saturated with water, an efficient electrolyte, makes them especially susceptible.

In one case, calcium chloride was used to accelerate the set of the shotcrete cover in the cold climate. In another, the ground in which the penstocks rested was heavily contaminated with chlorides.

In a number of early cases, sewage leaked through the inner tank wall due to thermal and shrinkage cracks during construction. In other cases, there were laminar cracks between the wire wrapping and the shotcrete cover, allowing water contaminated with salt and sewage gasses to seep down to the steel.

24.3.4 Posttensioning Anchorages and Fittings

These should be stored and protected as the high-quality machinery which they are. After installation, protection is normally by epoxy coating, usually further enhanced by concrete encasement. The encasement should be tight, and configured to shed water, so as to prevent moisture from running down cracks along the encasement to the anchorage. The exterior should be sealed.

Anchorages are very prone to corrosion because the large concentration of steel acts to generate a high cathodic potential.

24.3.5 Unbonded Tendons

Single strand tendons, greased and encased in paper, have not had a fully satisfactory record. Chloride ions have penetrated at the anchorages and have migrated under the paper sheath, leading to corrosion and eventual failure. In some cases the individual strands have failed explosively, although fortunately the energy in one strand is relatively low.

These have now been superseded by single strand tendons which are greased and sheathed by polyethylene, extruded over the tendon. These have given excellent performance throughout the sheathed length.

The danger zone with all unbonded tendons is at the juncture with the anchorage, where the strands or wires are bared in order for the wedges to grip them. If the anchorage pocket leaks, water can gain access to the short length of unprotected strand, leading to corrosion there. Unfortunately, at the anchorage, there is available sufficient steel to generate the cathodic-anodic reaction.

For the single-strand sheathed tendons, details are now commercially available to enable a watertight connection to be made. For multi-strand tendons, a similar watertight connection should be made and indeed, the necessary fittings are now available from the manufacturers of posttensioning systems.

Plastic or steel caps can be fitted over the anchor plates and filled with grout or grease to protect the ends of the strands.

By the use of plastic fittings, electric isolation can be obtained.

24.3.6 External Tendons

Special care must be taken at the anchorages of external tendons, since potentials can be set up where the tendons enter the concrete and are bared to a new environment. Another critical location is at the juncture of the polyethylene and the steel tubes, where it is essential that integrity be maintained. This is why the conservative practice of running the polyethylene through the steel tube and belling the entrances of the steel tube is recommended, since carelessness or accident can separate even the best-made splice.

Unprotected external tendons, used in an early nuclear reactor containment structure, did suffer corrosion from a humid saline atmosphere acting on highly stressed tendons. The solution in this case was to dehumidify the atmosphere surrounding the tendons. Similarly, in the Hood Canal Bridge, where the external tendons were protected only by bitumastic, corrosion did occur just inside the concrete encasement at the anchorage.

REFERENCES

1. ACI 224, "Cracking of Prestressed Concrete," American Concrete Institute, Detroit, Michigan, 1989.
2. Gerwick, B. C., Jr., "Causes and Prevention of Problems in Large-Scale Prestressed Concrete Construction," *PCI Journal*, May–June 1982.
3. Shupack, M., "Corrosion Protection of Unbonded Tendons," *Concrete International*, American Concrete Institute, Detroit, Michigan, February 1991.

CHAPTER 25

Maintenance, Repair, and Strengthening of Existing Structures

25.1 GENERAL

Prestressed concrete structures must be maintained and, when damaged by any cause, repaired so as to continue to perform their functions.

As noted in the previous chapter, damage may arise as a result of numerous causes, both external actions and environmental degradation. It is important to ascertain the cause and the extent of the problem so that repairs may be effective.

25.2 MONITORING

Maintenance procedures for important structures should include periodic monitoring with emphasis on potential problem areas, such as bearings, anchorages, and expansion joints. Penetrations may produce local cracking or corrosion.

A well-planned inspection procedure is very valuable. This should be initially prepared by the designer and updated for the particular structure as experience indicates. It should indicate areas of importance and vulnerability, describe potential types of cracking or deterioration, indicate their significance, and recommend methods of repair. Periodic visual inspection on a detailed basis will reveal early signs of corrosion through rust staining or cracking.

Where corrosion has occurred or is suspected, monitoring can employ the electrolytic half cell method, to determine the voltage being generated and to determine whether chloride penetration has penetrated sufficiently to set up a potential for active corrosion. Other electrical techniques can be used to determine the actual rate of corrosion.

Cores may be extracted to determine the chloride penetration or depth of carbonation.

Various techniques are under development to determine internal delamination, which is often (but not necessarily) due to corrosion. These include ultrasonic pulse echo, ground radar, infrared scanning, and impact echo. However still the most practicable methods are the use of the hand-held hammer and the chain drag, as interpreted by the human ear.

25.3 MAINTENANCE

This includes cleaning of expansion joints, removal of any foreign material, and, where appropriate, washing of the structure to remove salts or other deleterious materials which may have accumulated.

25.4 REPAIRS

As indicated earlier, repairs must be directed not only to restoring the structural integrity, but where feasible, to preventing continuing damage.

In the case of cracking through the thickness of walls or slabs, repairs may take the form of epoxy injection (Figs. 25.1, 25.2).

Where cracks in prestressed concrete marine piling were determined to be due to a combination of thermal strains plus freeze-thaw cycling, FRG jackets were installed and the annulus filled with an insulating grout containing tiny glass balls.

Figure 25.1. Epoxy-injection of through-thickness cracks in an offshore structure.

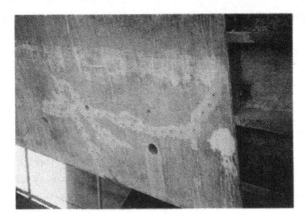

Figure 25.2. *A bridge girder, cracked by freezing of water in the ducts, is repaired by epoxy injection.*

Where delamination has occurred due to overstressing or internal freezing or other actions, it may be necessary to first stitch the slab or wall together at close intervals before injection of epoxy or grout. The pressures used should be strictly limited; otherwise the crack will just be progressively widened and extended by the hydraulic head acting on the large crack surface. The basic rule is that there must be steel crossing the crack before injection and that the pressure acting on the tributary area must not stretch the steel beyond yield. Multiple injection points should be used, starting at the perimeter of the delaminated zone (Fig. 25.3).

Where it is impracticable to follow this rule, gravity flow alone may be used to flow in a low viscosity, high capillarity epoxy or grout.

Corrosion of reinforcing and prestressing steel may be arrested by cathodic protection, provided there is full electrical continuity. It is difficult, however, to design a cathodic protection system that will adequately penetrate the metallic sheaths used in post-tensioning.

Replacement tendons, when required, are usually applied inside the box girder or alongside a girder stem. Deviator blocks and anchorages, if not previously provided, must be anchored by drilled-in bolts.

In some cases, it may be necessary to construct additional concrete beams alongside the existing member, tied to it by drilled and grouted studs. These new beams can then be post-tensioned so as to stress both the new and the old concrete.

Where the existing reinforcing bars are corroded, the concrete cover should be removed by chipping with hand-held tools or water jet, so as to avoid propagating cracks into the concrete core. The concrete should be removed to give 25 mm (one inch) clear behind the bar. The bar should then be cleaned to bright steel by wire brushing. Zinc bracelets (strips 20 mm or ($\frac{3}{4}$ inch or so wide) should be affixed at each end and the bar coated with epoxy. Some authorities recommend two coats.

The cavity should then be filled with fine concrete (10 mm ($\frac{1}{2}$ inch) coarse

Figure 25.3. *Epoxy injection of a laminar crack is carried out with extreme care to prevent propagating the crack, since there is no through-thickness reinforcement.*

aggregate or just sand cement mortar). Special proprietary mixtures are available which have high bond, low shrinkage, and low permeability (Fig. 25.4).

The repair itself should be externally coated by epoxy paint. In areas subject to freezing, use latex or silane instead of epoxy, so as to allow the water vapor to escape.

An all too frequent problem is that ducts have not been completely filled with grout. In this case, two holes should be drilled and a fluid grout containing an anti-washout admixture should be injected. A hydrophobic epoxy can also be used. The vacuum grouting process allows the regrouting to be carried out through one hole. The vacuum is also useful in removing any trapped water.

Drilling into a duct must be done with care so as not to damage the tendons. A water jet can be used to drill down to the duct, as can a slow speed drill.

Many cases of corrosion have occurred to posttensioned floor slabs, especially those constructed prior to 1980 when paper sheathing was used. The new plastic sheathing appears to prevent most such occurrences. Parking garage slabs have incurred early corrosion as a result of the tires of cars bringing in remnants of salt from deicing application on the streets and bridges.

The repair of such corroded tendons requires detailed engineering analysis and planning. In general, the slabs should be shored all the way to ground level prior to any repair operations.

Then the prestressing tendons should be detensioned. Where the entire slab is to be removed and replaced, this detensioning can be done by a transverse saw cut or cuts through the slab. One cut may not suffice, due to the friction of the paper sheath and the curvature. No one should stand above or below the slab being cut as it is possible, although unlikely, that a tendon will snap out of the slab. Timber shields should be preinstalled at each end in case the cut tendon flies out.

If the slabs are supported by a prestressed girder, consideration must be given to

Figure 25.4. *After replacement of damaged reinforcing steel bars, repairs are completed by progressive filling with special concrete, in this case, grout-intruded aggregate.*

the possibility of an explosive upward break in the girder due to removal of the dead load. It may be necessary, for example, to tie the girder down at midpoint. Also, the transverse lateral stability must be assured, with bracing installed as necessary.

When only a few tendons are corroded, it may be feasible to replace one tendon at a time. The existing tendon is detensioned, a new plastic sheathed tendon spliced on, and the new tendon pulled through as the old one is pulled out. However, the paper sheathing may jam. It has therefore been proposed to rout or broach the existing duct by a drill so as to remove the old paper and slightly enlarge the duct.

If only a local zone is corroded, as for example, at an anchorage, the tendons may be detensioned, a short gap cut back in the concrete, the wires cleaned and sheathed, a new anchor block cast, a new anchor re-attached, and the tendon stressed.

New tendons can be installed in a slab one at a time by trenching, down from the top over the support, up from the bottom in the center. Plastic sheathed strands can be installed and the cuts patched by cement mortar.

Prestressed wire-wound tanks have occasionally developed corrosion problems. The cover concrete should be removed by water jet and the wires cleaned, epoxy coated, and the zone patched by shotcrete. If the corrosion is serious or if several wires are broken, it may be necessary to install new external post-tensioning, e.g., plastic sheathed strands.

Reference is made to Sections 13.4 and 14.9 where further information is presented concerning repairs.

25.5 STRENGTHENING OF EXISTING STRUCTURES

Prestressing techniques have been employed with great success to correct excessive and undesirable deflections in existing structures. They have also been used to strengthen existing concrete and steel structures to carry additional loads or to permit the removal of intermediate supports. See also Section 13.7.7.

Tendons are usually applied externally, and may or may not be subsequently bonded to the structural members. Such tendons must be protected against corrosion by means such as galvanizing, greasing (with provision for maintenance), or concrete encasement.

Where the excessive deflection was due to creep, care must be taken not to over-correct, that is, not to produce excessive camber and creep in the other direction. Fortunately, this is usually offset by the deadweight of the deck fill or topping, which have been poured on the sagging girders.

External prestressing may be used on the inside of box girders or the outside of I-girders to increase the capacity of existing bridges and to provide improved resistance to fatigue and cracking. In some bridges, such as the Tagus River Bridge in Portugal, this future strengthening was foreseen and provision was made for anchorages and deviator blocks. In most cases, however, the external post-tensioning requires the addition of anchorages and deviator blocks. These may be tied to the existing girder by mild steel or short posttensioning bars.

On occasion, it will be found practicable to drill through an existing diaphragm and to utilize it as an anchorage or deviator block.

Great care has to be taken with corrosion protection at the discontinuity where the external tendon enters the concrete. Here the tendon may not only be subject to a change in the chemistry of the environment, but may also be subjected to combined stresses from bending or bearing and possibly to fretting.

The original Hood Canal Floating Bridge in Washington was strengthened and stiffened to improve its dynamic behavior by external tendons run at mid-height through the pontoon. When one-half failed in a storm exceeding the design storm, the 400-meter long tendons let go at one anchorage and their stored energy destroyed the intermediate bulkheads. The new Lacey V. Murrow Bridge, currently under construction across Lake Washington, near Seattle, has been designed to prevent progressive failure of this nature by incorporation of intermediate anchorages.

The San Francisco–Oakland Bay Bridge was reconstructed to remove the middle columns from the lower deck and to permit truck loading on the upper deck. This project was carried out under full traffic.

A number of techniques were employed, including external stressing by bar and strand tendons and reinforcing existing steel transverse beams by stressing a high-yield strength plate and locking it by high-strength bolts to the original beam. In one section, external beams were added by segmental construction, to augment the existing transverse girders. In the tunnel portion of this Trans-Bay crossing, precast slabs were placed transversely, the two halves making up the full width except for a gap of about 1 meter (3 feet) in the middle. Each slab contained post-tensioning bars, unbonded in ducts, but anchored at the outside ends. The bars were spliced to each other at mid-section by a right hand–left hand sleeve. Special jacks then forced the slabs apart so as to stress the bars.

The secret of execution was assuring that each spliced bar, before jacking, had the correct initial tension. This was done by a torque wrench, which tightened the sleeves to a uniform initial tension.

The gap was then filled with concrete and its curing and strength gain accelerated by encasing it in a steam jacket for 18 hours.

These transverse slabs contained diaphragms which abutted similar diaphragms in the subsequent slabs. Each pair of slabs was then posttensioned to the adjoining pair so as to give longitudinal continuity.

All concrete in the reconstruction program was high-strength structural light-weight concrete. This reconstruction was completed prior to the 1989 Loma Prieta Earthquake and was not involved in the localized failure in the East Approach span, which received such wide publicity. No prestressed elements were damaged.

When new concrete is added alongside existing concrete, bond between the two must be established. Consideration must be given to the different moduli and creep characteristics of the old and new concrete.

In other strengthening projects, similar techniques have been used to improve the capacity of existing steel girders. Some have employed the stretched alloy plate system described above, augmented by composite behavior with the existing concrete deck slab.

Existing steel girders may be strengthened by external tendons. Buckling of the compression flange can be prevented by composite action of the concrete slab. Web buckling can be prevented by placing reinforced concrete between the flanges, tied on by steel bolts or posttensioning bars.

It has been found necessary in some bridges to enhance the shear connection between the concrete slab and the flange of the steel girder by drilling holes in the slab and either welding studs or installing high-strength bolts.

External prestressing of the exterior girders only of a multigirder bridge has been found effective in reducing the stresses in the interior girders as well.

25.6 RETROFIT OF STRUCTURES TO RESIST SEISMIC FORCES

Posttensioned tendons may be effectively used to provide increased torsional shear resistance to existing cap beams of bridges. They may be threaded through drilled holes in the webs of the adjacent girders and, after stressing, encased in concrete.

External tendons are often used to upgrade the seismic resistance of framed buildings, particularly reinforced concrete framed structures, by placing them as diagonal X-braces in selected bays of the perimeter walls. It may be necessary to tension both diagonals simultaneously to avoid imparting excessive deflections and stresses in the columns.

Seismic restrainers typically utilize prestressing strand to control the movement of steel and precast concrete girders and to prevent their falling off their supports. Similar restrainers are used at expansion joints in bridge structures.

In retrofitting the footings of bridges where overturning or rocking is a potential failure mode, pre-compressed pile tie-downs may be installed. After drilling through the existing footings, conventional steel pipe piles are installed deeply enough to anchor the footing against uplift, then partially filled with concrete. Prestressing rods are now anchored into the top portion of the pile by grout, and then stressed against the footing. Thus a force is applied to the footing sufficient to prevent uplift, but the length of bar subject to elongation is limited. Similarly the compression applied to the existing piles is limited, thus avoiding their overloading.

REFERENCES

1. Tracy, R., Lozen, K., Zeort, K. R., "Restoration of a Deteriorated Post-Tensioned One-Way Slab," *Concrete International,* American Concrete Institute, Detroit, Michigan, March 1992.

2. Nehil, T., "Rehabilitating Parking Structures with Corrosion-Damaged Button-Headed Post-Tensioning Tendons," *Concrete International,* American Concrete Institute, Detroit, Michigan, March 1992.

3. Poston, R., McCarthy, O. J., and Schupack, M., "Repair of Wire-Wound Prestressed Concrete Tanks," *Concrete International,* American Concrete Institute, Detroit, Michigan, March 1992.

4. *FIP Guides to Good Practice,* "Repair and Strengthening of Concrete Structures," Thomas Telford, London, 1991.

5. Barchas, K., "Repair and Retrofit Using External Post-Tensioning," *Concrete Construction Magazine,* July 1991, pp. 536–539.

6. ACI 546, "Repair of Concrete Bridge Superstructures," American Concrete Institute, Detroit, Michigan, 1989.

CHAPTER 26

Demolition of Prestressed Concrete Structures

Prestressing stores up an enormous amount of energy in the structure, which must be largely or wholly released at the first stages of demolition. There is a fundamental distinction as to how rapidly the energy will be released between bonded and unbonded tendons. In the case of the former, if the bonding remains intact, as for example with pretensioned strands or with well-bonded post-tensioning tendons over an adequate length, the release will take place incrementally and not explosively.

Conversely, an unbonded tendon, when suddenly released, will fly out as a deadly projectile. A 100-meter long unbonded posttensioning tendon has 2.5 MN-m of energy (over 2,000,000 ft lb).

It is necessary that the demolition process be carefully evaluated and "designed" to insure against personal injury and unwanted damage to adjacent structures. Adequate blocking and matting must be placed behind the anchors and procedures must be adopted to cause any sudden release of energy to fly inward.

It usually is necessary to shore under the span to support the dead load once the tendons are released. Then unbonded tendons may be released one by one by jacking, provided the anchorage can be accessed and a grip engaged. In the majority of cases, unfortunately, this will be impractical.

Unbonded tendons may be exposed by chipping a trough into the concrete just ahead of the anchorage and heating it gradually with a yellow flame (about 500°C) (900°F) so as to elongate the strands and thus release the stored energy.

Once the prestress is lost, the girder or structure will sag onto the falsework. This may not occur evenly, possibly overloading individual shores.

One means of overcoming this is to wedge up from the falsework so that the girder is tightly supported along its full length. However, if the falsework or shoring is not continuous but only supports the girder at discrete points, an analysis must be made as to the maximum load that could progressively be transferred to each support.

Pretensioned tendons usually develop bond over distances of about one meter (three feet). Bonded posttensioning tendons may require several meters. Even unbonded tendons may not be capable of full release due to curvature friction and jamming of the sheath.

Thus release points must be selected that will ensure adequate detensioning.

While detensioning by heating is the most effective way, it is slow and tedious, especially with a multi-strand tendon. Alternatively, individual wires may be cut by wire cutters one at a time. Individual strands of a multi-strand tendon may be cut by burning, one at a time. While jamming and friction will probably prevent premature release, to ensure safety, timbers or mats should be placed over both anchorages to prevent accidental missile effects.

Once detensioning has been achieved, the concrete may be removed by conventional means of concrete demolition.

When bridges or buildings are to be completely demolished, explosives are the preferred means because of the ability to demolish the entire structure in one operation. This in turn can, in many cases, improve safety for the workers.

What must be done, in the case of post-tensioned structures, is to insure against missile or projectile effects.

In these cases, shoring will not be employed. Charges must be planted and fired in a sequence, by means of delays, that will ensure the detensioning prior to general demolition, even if only a few milliseconds earlier. In such cases, charges will be planted adjacent to the anchorages at both ends and these will fire first, so as to effectively cut and detension the tendons at each end.

When anchorages have been released, there may still be a great deal of energy stored up in the middle of the member due to bonding, intended or unintended, but this will usually not involve any projectile effects. It usually will be released progressively during subsequent demolition. Workers must be protected from local bursting effects, up and down.

In the case of external tendons, these can best be removed by slow heating, to avoid explosive release. If the project is to remove and replace existing tendons, then this should be done one at a time, i.e., one tendon is removed and a new one installed and tensioned, before a second is removed.

When demolition is confined to limited areas or zones, a saw cut should be made along the perimeter of the zone to prevent spalling and disruption beyond the cut. This saw cut can be made with diamond or Carborundum blades to a depth of 40 mm ($1\frac{1}{2}$ inch) (just shy of encountering the steel). Then hand-held pneumatic chisels can break away the concrete, exposing the tendons.

See also Section 4.9 concerning cutting of prestressed concrete members.

REFERENCES

1. Corlett, M. S., "Demolition of a Major Prestress Bridge," *FIP Notes*, Federation Internationale de la Precontrainte, London, 1987/1.
2. *FIP Guides to Good Practice*, "Demolition of Reinforced and Prestressed Concrete Structures," 1982, Thomas Telford and Sons, London.

CHAPTER 27

Prestressed Concrete in Remote Areas

Prestressed concrete is obviously the same material and involves the same techniques and construction methods, whether constructed in urban centers or the most remote corners of emerging countries. However, the advantages of prestressed concrete in such remote areas are often even more dramatic than in highly developed centers. These advantages include:

Maximum use of local materials.

Maximum use of indigenous labor.

Minimum import of special materials (tendons) and equipment (jacks, etc.)

Ability to set up in remote and difficult places.

Development of a local industry that can be continued and expanded as a basis for housing, local industry, and public works.

Low cost.

Durability with minimum maintenance.

Many of these advantages are typical of all concrete construction, whether prestressed or conventionally reinforced. Prestressing reduces the steel consumption in weight to about one-sixth and in cost to about one-half. Prestressed concrete members themselves are lighter, have longer span potential, and are more durable.

The problems in prestressing in remote areas are the lack of local industry and material support, the lack of skilled and semiskilled labor, the lack of technical skills, the lack of transport, the difficulty of quality control and, most of all, the lack of construction management. Many remote areas are exposed to extremes of temperature and humidity, increasing the problems of durability, or are desert areas, with problems of obtaining sound aggregates and fresh water, or arctic environments, with problems of winter construction (Fig. 27.1).

Figure 27.1. *Pretensioned concrete manufacturing facility in Middle East produces precast elements for marine structures and bridges.*

The greatest single problem has been the production of high-quality concrete. Coarse and fine aggregates must be tested for soundness and for alkali-aggregate reactivity. Fine aggregates must, in addition, be tested for chlorides, fluorides, and excessive silt or organic material. In desert areas, despite the high cost of fresh water, aggregates may require washing with fresh water to remove salt.

In extremely hot areas, such as those of the Middle East, aggregates and concrete must be cooled. Water soaking (with fresh water), vacuum cooling, shielding of stockpiles from the sun with aluminum or galvanized sheathing, and mixing with ice are techniques used to bring down the temperature of the fresh mix. Steel forms may become excessively hot; these can be pre-cooled by a water spray. After pouring, a fog spray should be applied almost immediately, followed by moist steam curing and/or water cure. Water cure should be a minimum of 7 days (3 days if supplemental to steam cure). Use of a retarding admixture may be found beneficial in preventing flash set. Many of these problems may be minimized if the concrete pour takes place in the early hours of the morning (about daybreak).

Prestressing steel may suffer serious corrosion in transport to remote sites, particularly in the tropics. It should be wrapped in heavy export wrapping, protected by water-soluble oil or with VPI crystals sealed inside. Packaging and handling must be such as to prevent rupture of the package.

In the tropics it may be difficult to find water that is free from excessive organic materials and dissolved chemicals. Frequently, however, water which is discolored is found satisfactory upon tests and in actual experience. The emphasis thus is on proper testing of all materials.

Cement must be export-packaged and preferably palletized and wrapped so as to insure against moisture. The age should always be determined and be within manufacturer's limitations. The type of cement and its fineness should be carefully selected. In particular, ASTM Type V cement with its near zero tricalcium aluminate (C3A) is excellent for sulfate resistance in fresh water but develops high permeability for chlorides and should not be used where saline waters or exposure are present.

The fineness of cement affects its rate of hydration, and hence the heat developed. A coarsely ground ASTM Type II cement will generally perform best in hot environments, both tropic and arid.

The replacement of 15 to 30% of the cement by pulverized fly ash (PFA) will reduce the heat generated, increase the impermeability, and increase the resistance to sulfate attack and alkali-aggregate reaction. Its use is generally recommended for the tropical, hot arid, and temperate environments.

Reduction of the ratio of water to total cementitious materials is of even more value in the hot arid environments than in the temperate zones. Water/Cement (W/C) ratios of 0.37 are practicable for cast-in-place concrete through the use of high-range water-reducing (HRWR) admixtures. Ratios of 0.32 to 0.33 are practicable for precast products. However, in hot climates, the HRWR admixture (superplasticizer) must contain a retarder so as to prevent premature slump loss.

In hot arid coastal zones, such as those along the Red Sea and Arabian/Persian Gulf, chloride attack on the reinforcing steel dominates the concerns over durability. Aggregates must be washed with brackish water, mixing water should be distilled, and the several steps outlined above must be rigorously followed. In addition, in the splash zone, that zone washed by seawater intermittently, and for a few meters above, coating of the concrete with epoxy appears to be fully warranted as a means of preventing chloride penetration.

Decks of over-water and near-water bridges and platforms are susceptible to accumulation of chlorides through successive deposition of salt from the low summer fogs. Periodic washing, with seawater, will remove the accumulations. A sealant such as silane may be very beneficial.

Epoxy coatings deteriorate over time due to ultraviolet radiation and silanes lose their effectiveness due to their solubility in water, so both have to be replaced at periodic intervals.

It follows that protection of prestressing steels in such environments is very critical. They must be stored so that chlorides cannot be deposited on them through fog, spray, or the atmosphere. The author visited one job where the tendons were being stretched out and made up on the beach (the only flat place), and allowed to be covered by the high tide! Especial care must be taken with the anchorages. Recessed anchorages, sealed by epoxy, are essential.

Epoxy coated wire and strand have been introduced by several manufacturers. One brand uses zinc-enriched epoxy, applied by the electrostatic fusion process. Another dusts fine sand onto the hot epoxy: it embeds itself and develops much enhanced bond. The use of epoxy coated wire and strand is still in its development stage and expert opinions are divided as to its utility and performance. The use which appears most appropriate is the transverse stressing of concrete decks of bridges.

In arctic environments, aggregates must be selected for freeze-thaw durability, a property for which tests may not be conclusive. Experience in previous use for concrete structures, whether prestressed or not, is the best indicator. Air entrainment should always be employed.

Historically air entrainment has been measured in the fresh concrete by an air meter. However, this does not distinguish between entrapped and entrained air. Only the latter is beneficial.

Recent experience, especially in Canada, has led to the requirement for petrographic examination of cores from the concrete structure (or test specimens), to ensure that the size of the bubbles and their spacing are at optimal levels to prevent freeze-thaw damage.

A special risk is posed by marine structures, such as prestressed concrete piles, in Northern environments. First, if the piles are hollow, during freezing weather the inside may be warm and humid while the outside is cold and dry. The result may produce thermal cracks. The prevention is to provide adequate amounts of spiral confinement so that the circumferential steel stays below yield. Thus, any tiny cracks which open will be pulled closed when the weather warms.

The second risk is freeze-thaw attack. While the air temperature may remain below freezing, the pile may thaw twice a day as the tide rises and covers it. Thus, in a sustained period of freezing weather, say 60 days, the pile may see 120 cycles of freezing and thawing. Add to this the fact that saturated concrete is much more susceptible to freeze-thaw attack and one can see why unprotected piles and sea-walls deteriorate within a very few years.

Air-entrainment plus the use of highly impermeable concrete through low water/cement (W/C) ratios and the addition of microsilica is the best protection. Hollow-core piles should be avoided. Epoxy coatings may be counterproductive since the water vapor migrates to the cold face, freezes, and pops off the epoxy with a sliver of concrete attached. Silanes are appropriate since the water vapor can escape. However the amount of absorption of the silane will be minimal with impermeable concrete. Silanes are water-soluble and have to be renewed every few years.

While the recommended measures of carefully monitored air entrainment plus microsilica should prove adequate in most instances, less well controlled piles have been adequately protected by encasement in a FRG with the annular space filled with epoxy containing tiny glass spheres: these serve as an insulator.

Production in arctic environments is typical of all cool- and cold-weather concreting activities. Steam curing is definitely indicated. Care must be taken to cool gradually to prevent excessive thermal strains upon removal from the steam chambers. Particular care should be taken to prevent drying-shrinkage cracks from cold dry winds immediately after removal from curing. Since drying-shrinkage and thermal strains are additive, continuation of water curing, as by soaking blankets or sealed thermal blankets, may be appropriate.

Management, training, and direction of unskilled indigenous labor is generally easier than anticipated if properly planned and organized. At least one skilled prestressing ironworker foreman and one skilled concrete foreman (supervisors) are desirable. A skilled crane operator and a mechanic are essential. These workers can direct the actual crews, train them, and supervise their work. Invariably, the indigenous personnel are eager to learn. They learn best from highly skilled expatriate working supervisors, who can actually show them how to do each of the required tasks (Fig. 27.2).

There is one fundamental rule to apply: train each unskilled indigenous worker for one task only. Tasks should be subdivided into detailed categories such as:

PRESTRESSING IRONWORKER

Spiral making.   Mild-steel fabrication.
Spiral tying.   Mild-steel placing.
Strand laying.   Placing of lifting loops.
Strand anchoring.   Burning of strands.
Stressing and releasing.

Figure 27.2. *Manufacture of prestressed concrete piles in Middle East plant is carried out by workers from nine countries.*

In a skilled labor market in a highly developed country, one prestressing iron-worker may do all of these tasks in a given day; in an emerging country, one person should do only one of these tasks. It will be found that within a reasonable period, the indigenous worker will become very efficient in the one task. Because of lack of flexibility in assignment, manpower requirements may run about three to five times that of a plant in a developed country, and may in addition have special labor requirements such as watchmen, clerks, security, etc. As time goes on and men become more experienced, particularly if they have some education, their flexibility will improve.

Care must be taken to respect local craft divisions. Traditionally, certain tasks are performed in a certain way, by a specific group. As long as the end product is of adequate quality, there is no reason to upset these patterns. When this is recognized, new tasks or higher quality production can be fully mechanized, with the most modern equipment, and this will usually be accepted. On one job in Southeast Asia, for example, rock was hand-"pitched" from a quarry, trucked to the river in $\frac{1}{2}$-ton lorries, loaded into canoes by basket, taken across the river, and unloaded by basket to a modern crushing-screening washing plant. Remarkably, the finished aggregate cost about the same as in developed countries. This result is quite typical; despite much lower labor costs, costs of finished products are often in the same general range in widely varied economies.

At the engineering-management level, most developing countries have trained engineers who can be a most valuable asset and who will quickly absorb specialized skills. One highly qualified prestressing engineer and one manager are essential, and therefore are usually from the contractor's permanent organization. Accounting and purchasing roles may be handled by expatriates with indigenous assistants. If all expatriate personnel take the time and patience to train as well as to supervise, the whole job will become easier.

The engineer must serve as quality-control manager and must be firm and rigid in his acceptance or rejection, but take pains to show why, and wherever practicable, to demonstrate the reason.

Equipment sent to a foreign location should be new, of 10 to 20% larger capacity than required, in order to minimize wear. It should be equipped with properly marked and protected spare parts. If used equipment is to be considered, it must be thoroughly overhauled and placed in first-class condition. Additional spare parts may be indicated with used equipment. Instruction booklets and parts catalogues must accompany the equipment, with a duplicate set in the home office to enable ordering by number. Contact should be established with local distributors to determine the spares and parts carried by them and to be sure they match the particular models.

Small tools and miscellaneous equipment and supplies are troublesome due to loss, theft, and careless misuse. Adequate supplies must be available, properly marked, and warehoused. Particular care has to be taken with such special items as strand vises.

When procuring materials from other countries, such as prestressing steel and cement, a thorough analysis has to be made to ensure the properties are known to the designer and constructor. Engineers become accustomed to trade practice, and tend to identify products by one property only, which may work in their home country but may lead to serious difficulty in other countries. Fortunately, most exporting countries are able to supply to international standards, such as ASTM or RILEM, (the European Specifier of Materials and their properties). Currently, the European Community is establishing "Euronorms" and it is likely that these will be adopted by ISO, the International Standards Association.

Steel, for example, may be customarily classified by yield point. In some countries, however, impact strength, ductility, chemistry, and rolling practices may have to be investigated. Prestressing tendons are usually classified by ultimate strength. However, elongation, ductility, and surface characteristics may vary.

Cements require investigation of their chemistry, fineness, setting characteristics, behavior in steam curing, etc. Admixtures are small in quantity and several recognized ones are distributed internationally. However, the same name may represent somewhat different properties, depending on the country of origin.

Mild-steel reinforcement may be deformed or smooth, and of widely varying yield strength, ultimate strength, and ductility, depending on whether it is rolled locally as mild (soft) steel or re-rolled from rails (hard steel). Yield points of steel in developing countries range from 160 MPa (22,000 psi) to 500 MPa (75,000 psi).

Wire rope varies widely, and only well-known, well-identified brands should be employed in handling and erection. Lifting eyes should preferably be of strand loops. Local steel plates may be susceptible to brittle fracture under impact.

Steel forms may be obtained in many foreign countries. The plates and sheets may vary widely. If they have laminations, they will cause the concrete to stick in the forms. Fabricating practices vary also; tolerances and grinding of joints must be clearly directed. Finally, to ensure compliance with specifications, an international testing laboratory should be engaged to verify compliance.

Many excellent products (materials, forms, equipment, etc.) are ruined in shipping. Export packaging must be specified, with special precautions to minimize damage in handling. The moral is: "If it is possible to mishandle a package, it will be mishandled."

Prestressed concrete is a basic structural material upon which the emerging countries may base the growth of their infrastructure and buildings. It offers wide application, it supports and stimulates the local economy, it trains indigenous workers in many skills, and it provides the highest quality of completed structures. Engineers and constructors from developed countries can take great pride in participating in this challenging extension of technology to the developing countries and in seeing the immediate and dramatic evidence of benefits to the country and its people.

Prestressed Concrete—
Implications and Prospects

Prestressing has been a revolutionary development, first as a material, but more importantly as a concept. As a concept, prestressing is valid and applicable to practically all materials and structures and machines. It is pro-active in that it forces anticipation of the actual behavior of the structure during its lifetime; it requires consideration of several "limit states" in the performance of structures under loads, and compels an evaluation of the mechanism of failure under ultimate load.

Prestressed concrete has important implications in the economic field, causing changes in the constitution and importance of whole industries. It is a means by which more structures, housing, utilities, and facilities can be made available to more people.

Prestressed concrete has important social implications. It has a decentralizing effect in that numerous relatively small plants can compete efficiently with large centralized production. It develops a widespread technological and engineering ability, and encourages innovation and continued new development.

Worldwide, prestressed concrete offers a major opportunity for the developing nations to meet their physical needs and to develop an indigenous skilled labor force.

Internationally, prestressed concrete has jumped the boundaries of nation, race, and political system. Engineers and constructors worldwide are working together on common problems. This is, in the author's opinion, the only true road to peace, where people of all nations are involved in common endeavor. For people, being what they are, rise out of their self-concern and group antagonism only when faced with larger problems and opportunities.

What are the roadblocks and dangers to such a glowing future? First, the inadequacy of management in the prestressed industry. The industry has grown on the enthusiasm and efforts of individuals, challenged by this new material and new concept. But these individuals are not necessarily trained as managers. By their failure to develop the marketing and financial aspects of their business, they risk losing control to less imaginative and enterprising firms. Too many, the author among them, have set their sights more on new techniques and products than on aggressive expansion to fill a need. Both aspects are essential for success of the industry. National organizations, such as the Prestressed Concrete Institute, have recognized the gap and are taking strong steps to overcome it.

A second major roadblock is age. Prestressed concrete is a new industry and the men who started it were young in both age and outlook. Now success has been achieved and the industry leaders have aged. There is a discernible trend to be satisfied with present scope, products, techniques. Failure to maintain the spirit behind prestressing will lead to its assuming a passive role.

The Federation Internationale de la Precontrainte has recognized this problem and is encouraging the national groups to assign younger persons a more active part in industry and technical association matters. Prestressing has been an emotional enterprise; it would never have achieved its present status, nor opened windows on the future, had the pioneers been motivated solely by practical and materialistic motives.

To look at specific areas of needed development, the first is research. There is much we do not really understand about prestressed concrete and even less about other prestressed materials. High-strength concretes, approaching the properties of aluminum, appear practicable with further applied research. Polymers, stabilized by irradiation or thermal treatment, offer increased strength and impermeability. Artificial aggregates can be developed which possess the desired properties of insulation, thermal expansion, etc. Very high degrees of prestress offer practical applications for increased durability and for composite action with cast-in-place concrete. Use of precast segments of high-strength concrete, perhaps coated with epoxy for added protection in view of the thin cover, offer exciting possibilities for long-span bridges, trusses, and space structures.

The application to machinery frames and supports, and eventually to machinery itself, offers a whole new field of utilization as well as improved stability and tolerance control.

The application of prestressed concrete to ship hulls and to offshore floating structures is again a new concept and a new industry. The traditional weight limitations no longer apply to very large structures, and the rigidity, durability, low maintenance, and favorable behavior at low temperature make the oceans a great opportunity for the future utilization of prestressed concrete.

In its industrial organization, the prestressing industry must look more and more to the systems approach to construction, to compatibility and integration of the entire facility. Too much emphasis to date has been placed on a one-to-one replacement of structural steel or reinforced concrete. What is needed is a careful study of the overall function and services required. A structure, whether it be a ship, a building, or housing, is an operating unit, not a static collection of beams and walls. We refer to the "skeleton" of a building and some European languages use the word "carcass." The skeleton is the frame upon which a living body is hung and integrated. So it is not enough to develop a skeleton structure; the goal, rather, should be a complete, living structure, which requires that the systems approach be employed.

However, this does not mean that the "system" must be comprised entirely of prestressed concrete, or that it must be a rigid system. As set forth in Section 10.11, "Systems Building (Technical Aspects)," prestressed concrete can develop as a compatible component of many different systems.

The industry must also develop a more efficient marketing approach, which includes performance criteria. Efforts to contain prestressed concrete within code provisions and engineering structural concepts developed for other materials leads not only to lack of economy but also to poor performance, as for example, at connections. Codes have now been modernized and up-dated to include prestressing. However, more effort still has to be expended in engineering education, not only of students but also of practicing engineers. The architectural profession has made striking but not widespread use of prestressing. Familiarity with the potential and practical use of prestressed concrete is as yet not disseminated throughout the profession. There also remain the mechanical engineering, mining and petroleum engineering, naval architectural, etc., professions to which prestressed concrete has made little or no approach. If the industry is to maintain its rate of growth, its marketing efforts must be intensified along the lines indicated and along others as they appear.

Prestressing thus is both a philosophical concept and a highly practical material. Concepts can be transformed into reality only by the properly directed efforts of construction engineers. The most beautiful building or spectacular bridge can be a failure if practical, down-to-earth principles and precautions are ignored or violated by the constructor. Conversely, the constructor has the opportunity to be an essential participant in the creation of these outstanding structures. Thus, construction in prestressed concrete is the translation of ideal and concept into existence and reality, a proper challenge and opportunity for creativity.

In the short interval between the writing of this book and its publication, dramatic new progress has been reported from research on materials. Tendons of carbon fibers and polyaramids such as Kevlar have been developed to the stage where they are being applied in experimental and prototype structures. Corrosion-resistant low carbon steel, treated by a two-phase quenching process (see section 5.3.12) gives great promise for both reinforcing bars and high strength wire for tendons. Microsilica produced from rice husks is now commercially available. Early research shows this material produces a significant increase in impermeability of the concrete as compared with mineral-sourced microsilica, due to higher surface activity. These revolutionary developments in materials technology will inevitably lead to even further advances in prestressed concrete construction.

APPENDIX

Approximate Conversions Between SI Units and Customary U.S. Units

SI to U.S.	Quantity	U.S. to SI
One mm = $\frac{1}{25}$ inch One meter = 40 inches One meter = 3.28 feet	Length	One inch = 25 mm One foot = 300 mm = 0.3 meter
One m² = 10 feet²	Area	One foot² = 0.1 m²
One m³ = 35 feet³	Volume and section modulus	One inch³ = 16,000 mm³
One N = 0.225 lbs. One kN = 225 lbs. = .225 kips One MN = 225 kips	Force and weight	One lb. = 4.5 N One kip = 4.5 kN One ton (short) = 9 kN = .009 MN
One N-m = 0.7 lb.-feet One kN-m = 0.7 kip-feet	Bending moment and energy	One lb.-foot = 1.35 N-m One kip-foot = 1.35 kN-m
1 kPa = 21 psf 1 MPa = 145 psi	Stress and modulus of elasticity	1 psf = 0.048 kPa = $\frac{1}{20}$ kPa 1 psi = .007 MPa
One metric ton = one long ton One metric ton = $\frac{1}{100}$ MN	Displacement	One ton (long) = one metric ton One long ton = $\frac{1}{100}$ MN
$T(C)° = [T(F)° - 32] \times \frac{5}{9}$	Temperature	$T(F)° = \frac{9}{5}T(C)° + 32$

Ben Gerwick, Jr., was born in 1919 in Berkeley, California and received his degree in Civil Engineering at the University of California at Berkeley in 1940.

Following service in the US Navy, he entered Ben C. Gerwick, Inc., a marine and construction firm founded by his father in 1926.

In the early 1950's, he became interested in the potential of prestressed concrete and converted the company's existing precast concrete manufacturing plant into the then new technology of pretensioning. Early work was the development of prestressed concrete piles that the firm also installed. Later the firm developed the deflected-strand process for pretensioned bridge girders, the precast match-casting process, and pretensioned railroad ties. Overseas the company participated in setting up prestressed concrete fabrication plants in Kuwait and Singapore, and domestically, in projects including the overwater extension of La Guardia Airport in New York.

Mr. Gerwick was active in the Prestressed Concrete Institute, serving as its President in 1957, and in the International Federation of Prestressing, serving as its President from 1974 to 1978.

In 1967, the firm became part of Santa Fe International and Mr. Gerwick was given responsibility for international construction. In 1971, he joined the faculty of the University of California at Berkeley as a Professor of Civil Engineering.

Concurrently, he set up a specialized consulting engineering practice, continuing the former construction company's name of Ben C. Gerwick, Inc.

1971 also saw the publication of the first edition of *Construction of Prestressed Concrete Structures*.

As a consultant, Mr. Gerwick participated in the development of offshore concrete oil platforms in the North Sea, of floating concrete structures for cryogenic gas storage, and of the first long-span cantilever segmental bridge in the United States.

569

His work on offshore platforms led to extension of prestressed concrete to offshore structures for the arctic and subarctic to resist sea ice and icebergs.

He has been a consultant on major prestressed concrete bridges in Europe, the Middle East, and Asia, as well as in the United States. His work has included prestressed concrete offshore terminals and floating bridges.

Among his honors and awards are membership in the National Academy of Engineering, and Honorary Membership in the Concrete Societies of Great Britain, Germany, Sweden, Norway, and France; and Honorary Membership in the American Society of Civil Engineers, the American Concrete Institute, and the Prestressed Concrete Institute. He has received the Freyssinet Medal from the International Federation of Prestressing (FIP), and the Medal of Honor from the Prestressed Concrete Institute.

He currently continues both professional and consulting activities in prestressed concrete and marine and foundation practice.

This requires extensive travel, in which Mr. Gerwick is often accompanied by his wife, Martelle. They reside in Oakland, California, and are periodically visited either there or at their vacation home in the Southern Oregon Mountains by their four children and seven grandchildren.

INDEX